Fiber Optical Parametric Amplifiers, Oscillators and Related Devices

Fiber optical parametric amplifiers (OPAs) show great potential for applications in high-speed optical communication systems. This is the first book to provide comprehensive coverage of the theory and practice of OPAs and related devices, including fiber optical parametric oscillators (OPOs).

Following an introduction to the field, the theory and techniques behind all types of fiber OPA are covered, starting from first principles. Topics include scalar and vector OPA theory, the nonlinear Schrödinger equation, OPO theory, and the quantum noise figure of fiber OPAs. The challenges of making fiber OPAs practical for a number of applications are discussed, and a survey of the state of the art in feasibility demonstrations and performance evaluations is provided. The capabilities and limitations of OPAs are presented, as are the potential applications for both OPAs and OPOs and the prospects for future developments in the field. The theoretical tools developed in this text can also be applied to other areas of nonlinear optics.

This book should provide a valuable resource for researchers, advanced practitioners, and graduate students in optoelectronics.

MICHEL E. MARHIC is Chair Professor at the Institute of Advanced Telecommunications at the University of Wales, Swansea. He was awarded a Ph.D. in Electrical Engineering from the University of California, Los Angeles, in 1974. He has held several positions in the academic world in the US over the years. He is the co-founder of Holicon, Holographic Industries, and OPAL Laboratories. A senior member of the IEEE and a member of OSA, he has taken out eight patents and authored or co-authored about 300 publications. His recent research interests include optical communication systems and nonlinear optical ineractions in fibers.

Fiber Optical Parametric Amplifiers, Oscillators and Related Devices

MICHEL E. MARHIC
University of Wales, Swansea

CAMBRIDGE
UNIVERSITY PRESS

University Printing House, Cambridge CB2 8BS, United Kingdom

Cambridge University Press is part of the University of Cambridge.

It furthers the University's mission by disseminating knowledge in the pursuit of
education, learning and research at the highest international levels of excellence.

www.cambridge.org
Information on this title: www.cambridge.org/9780521861021

© Cambridge University Press 2008

This publication is in copyright. Subject to statutory exception
and to the provisions of relevant collective licensing agreements,
no reproduction of any part may take place without the written
permission of Cambridge University Press.

First published 2008

A catalogue record for this publication is available from the British Library

ISBN 978-0-521-86102-1 Hardback

Cambridge University Press has no responsibility for the persistence or accuracy
of URLs for external or third-party internet websites referred to in this publication,
and does not guarantee that any content on such websites is, or will remain,
accurate or appropriate.

Contents

Acknowledgments		*page* ix
1	**Introduction**	**1**
	References	7
2	**Properties of single-mode optical fibers**	**9**
	2.1 Mode profile	9
	2.2 Loss	10
	2.3 Propagation constant and dispersion	10
	2.4 Longitudinal fluctuations of the zero-dispersion wavelength	14
	2.5 Temperature dependence of the zero-dispersion wavelength	14
	2.6 Fiber birefringence	15
	2.7 Nonlinearities	17
	2.8 Types of fiber used for OPA work	21
	References	27
3	**Scalar OPA theory**	**31**
	3.1 Introduction	31
	3.2 Types of fiber OPA	31
	3.3 Derivation of the OPA equations and γ	35
	3.4 Scaling laws	40
	3.5 Solution of the two-pump OPA equations	42
	3.6 Theory for one-pump OPA	60
	3.7 Case of no dispersion with loss	69
	3.8 Solutions for degenerate OPAs	70
	3.9 Conclusion	75
	References	75
4	**Vector OPA theory**	**78**
	4.1 Introduction	78
	4.2 Isotropic fibers	78
	4.3 Fibers with constant birefringence	93

	4.4	Fibers with random birefringence	98
	4.5	Conclusion	107
		References	108

5 The optical gain spectrum — 110

	5.1	Introduction	110
	5.2	The effect of pump power on gain bandwidth	110
	5.3	The effect of fiber dispersion on the gain spectrum	111
	5.4	OPAs with similar gain spectra	120
	5.5	Equivalent gain spectra for OPAs using pumps with different SOPs	123
	5.6	Saturated gain spectra	125
	5.7	Fibers with longitudinal dispersion variations	134
	5.8	Fibers with constant linear birefringence	140
	5.9	Few-mode fibers	143
		References	144

6 The nonlinear Schrödinger equation — 146

	6.1	Introduction	146
	6.2	Derivation of the NLSE for an isotropic fiber	147
	6.3	Derivation of the NLSE for a birefringent fiber	148
	6.4	Analytic solutions of the scalar NLSE	151
	6.5	Including the Raman gain in the NLSE	156
	6.6	Numerical solutions of the NLSE by the split-step Fourier method	157
	6.7	Applications of the SSFM to fibers with longitudinal variations	162
	6.8	Sources of SSFM software	166
	6.9	Conclusion	167
		References	167

7 Pulsed-pump OPAs — 169

	7.1	Introduction	169
	7.2	The quasi-CW regime	169
	7.3	The split-step Fourier method	172
	7.4	Important pulse shapes	173
	7.5	Examples of pulsed-pump OPAs	178
	7.6	Conclusion	185
		References	185

8 OPO theory — 187

	8.1	Introduction	187
	8.2	Fiber OPO theory	188
	8.3	Conclusion	192
		References	193

9 Quantum noise figure of fiber OPAs — 194

9.1 Introduction — 194
9.2 Quantum-mechanical derivation of the OPA equations — 195
9.3 Noise figure of non-degenerate fiber OPAs — 200
9.4 Wavelength exchange — 203
9.5 Noise figure of degenerate fiber OPAs — 205
9.6 Effect of Raman gain on OPA noise figure — 206
9.7 Conclusion — 209
References — 209

10 Pump requirements — 211

10.1 Introduction — 211
10.2 Pump power requirements — 211
10.3 Polarization considerations — 213
10.4 Pump amplitude fluctuations (pump RIN) — 213
10.5 Pump phase or frequency fluctuations — 219
10.6 Conclusion — 223
References — 223

11 Performance results — 226

11.1 Introduction — 226
11.2 Pulsed devices — 226
11.3 CW devices — 233
References — 245

12 Potential applications of fiber OPAs and OPOs — 249

12.1 Introduction — 249
12.2 OPAs in optical communication — 249
12.3 OPAs in high-power wavelength conversion — 273
12.4 OPOs — 275
References — 275

13 Nonlinear crosstalk in fiber OPAs — 281

13.1 Introduction — 281
13.2 Four-wave mixing — 281
13.3 Cross-gain modulation — 292
13.4 Coherent crosstalk — 298
13.5 Cross-phase modulation — 299
13.6 Conclusion — 300
References — 301

14 Distributed parametric amplification — 303

14.1 Introduction — 303
14.2 DPA experiment in 75 km of DSF — 304
14.3 Possible extensions of DPA — 309
14.4 Conclusion — 311
References — 312

15 Prospects for future developments — 315

15.1 Introduction — 315
15.2 Fibers — 315
15.3 Pumps — 319
15.4 SBS suppression — 320
15.5 Integrated optics — 322
15.6 Pump resonators — 324
15.7 Discrete or distributed parametric amplification? — 326
15.8 Conclusion — 329
References — 329

Appendices

A.1 General theorems for solving typical OPA — 333
A.2 The WKB approximation — 340
A.3 Jacobian elliptic function solutions — 344
A.4 Solution of four coupled equations for the six-wave model — 355
A.5 Summary of useful equations — 358

Index — 361

Acknowledgments

Thanks are due to K. K.-Y. Wong, J. Nielsen, M. Jamshidifar and P. Voss for proofreading parts of the manuscript. I would also like to thank G. Kalogerakis, J. M. Chavez Boggio, T. Torounidis, K. K.-Y. Wong, and P. Voss for providing some of the figures.

1 Introduction

Fiber optical parametric amplifiers (OPAs) exploit nonlinear optical properties of optical fibers. Their operation is based on the third-order susceptibility $\chi^{(3)}$ of the glasses making up the fiber core. While this nonlinearity is relatively weak in silica-based glasses, the small cross-sections, low loss, and large lengths available with silica-based fibers can lead to sizeable effects, even with moderate pump powers. While silica-based fibers are currently the most widely used, parametric amplification is also possible in any other type of fibers, some of which have very high nonlinearity coefficients. Fiber OPAs have important characteristics, which make them potentially interesting for a variety of applications, as follows.

Gain bandwidth increasing with pump power
In principle, this provides a means for making amplifiers with a bandwidth of several hundred nanometers, while using just one or two pumps. This could provide a substantial bandwidth increase compared with the currently popular erbium-doped fiber amplifiers (EDFAs) and Raman amplifiers (RAs), which have basic bandwidths of the order of tens of nanometers.

Arbitrary center wavelength
The gain region(s) can be centered about any arbitrary wavelength λ_c. The only constraint is that the fiber must have a zero-dispersion wavelength (ZDW) λ_0 close to λ_c. This feature is also available with RAs but not with doped-fiber amplifiers. It could be used for providing gain in regions where practical amplifiers currently are not available.

Large gain
It is relatively easy to obtain a large gain. Continuous-wave (CW) and pulsed gains of 70 dB have been demonstrated in single devices. The fact that the gain exists only for signals propagating in the same direction as the pump is an advantage, as it minimizes the possibility that oscillations will occur between small reflections at the ends of the OPA.

Wavelength conversion
If a signal is injected into a fiber OPA, an amplified signal emerges from the output together with a new wavelength component, the idler. This feature can be used for generating new wavelengths. If the signal is modulated, its modulation is transferred to

the idler; this can be used for changing the optical frequency of a signal, which could be a useful operation in communication systems. The conversion of unmodulated waves can also be used for generating wavelengths in regions where suitable light sources are lacking, in a variety of other applications.

An important point is that high signal-to-idler conversion efficiency can be obtained, since it is almost as large as the signal gain and can therefore be as high as 70 dB, even for CW operation. This feature is not available with other wavelength-conversion methods, such as those using semiconductor optical amplifiers (SOAs) or $\chi^{(2)}$ in materials such as periodically poled lithium niobate (PPLN).

Spectral inversion
In fiber OPAs, wavelength conversion is accompanied by spectral inversion, which means that the idler spectrum is symmetric to the signal spectrum with respect to the center frequency. This property can be exploited in communication systems to combat potentially detrimental effects such as dispersion.

Phase conjugation
This is another property of the idler: its phase is opposite to that of the original signal. This can be used for counteracting some nonlinear effects that affect the phase of waves, such as cross-phase modulation.

High-speed optical signal processing
Because the fiber nonlinearity is virtually instantaneous, high-speed modulation of the pump will generally result in modulation of both the signal and the idler. For example, if the pump is intensity modulated, so will be the signal and idler. This can be used for (i) shaping or reshaping signal pulses (regeneration), (ii) demultiplexing high-speed TDM signals, and (iii) optical sampling.

Low noise figure
Fiber OPAs can exhibit noise figures (NFs) below 4 dB, i.e. comparable with those of EDFAs. In addition, when they are operated in a phase-sensitive manner their NFs fall below 3 dB and approach 0 dB, which is of course the best that can be achieved with any type of amplifier.

Unidirectional gain and spontaneous emission
The parametric gain of fiber OPAs is available only for signals co-propagating with the pump; this makes it possible to obtain very high single-pass gains, as reflections are not amplified. Likewise, the spontaneous emission associated with the amplification is emitted only in the direction of the pump; this can be exploited for making efficient amplified spontaneous emission (ASE) sources.

Compatibility with all-fiber devices
Silica-based fibers, and some others, can be fusion-spliced to common optical fibers and can thus be incorporated into assemblies consisting of other fiber components or

fiber-pigtailed components. Such all-fiber devices benefit from increased stability compared with systems having discrete bulk-optic components.

High-power capability
Single-mode fiber lasers can now generate average output powers in excess of 1 kW. This indicates that, if needed, fiber OPAs could in principle also operate with these high pump powers. While such powers are excessive for telecommunication applications, they would be well suited for wavelength conversion, which could generate several hundred watts of average power at new wavelengths.

Distributed amplification
Just as in Raman amplification, parametric amplification can also be implemented along transmission fibers instead of between them.

Quantum effects
The fact that a signal photon and an idler photon are emitted simultaneously leads to the possibility of making sources of correlated photons; this could find applications in the emerging area of quantum communication.

These characteristics of fiber OPAs, taken alone or in combination, indicate that they could match or exceed the performance of other existing devices in various applications. While it was recognized early on that parametric amplification in optical fibers is a potentially important mechanism for amplifying optical waves for a variety of applications, its practical development was almost halted for 15 years, while EDFAs and RAs were being developed. The reason for this is that the considerable potential of fiber OPAs is difficult to exploit, for a variety of reasons, in particular: (i) the lack of fibers with a high nonlinearity coefficient γ; (ii) the need for phase matching, difficult to maintain because of longitudinal variations of the fiber ZDW λ_0; (iii) the lack of suitable pump sources, particularly for CW applications.

For these reasons, the early work with fiber OPAs was done with pulsed lasers, with peak powers of up to 1 kW. By using such high powers, it was possible to obtain a high gain in just a few meters of fiber. Also, a high pump power produces large amounts of self-phase modulation (SPM) and cross-phase modulation (XPM), which relax the phase-matching conditions and increase the gain bandwidth. Finally, with short fibers the effect of longitudinal variations in λ_0 is reduced. With these conditions, it is thus possible to obtain nearly ideal performance of fiber OPAs. Parametric amplification in optical fibers was first observed this way by Stolen in 1975 [1]. Lin *et al.* obtained high gain in a 50-m-long communication fiber, sufficient to yield substantial ASE at the wavelengths of peak gain [2]. Similar work was done by Washio *et al.* [3] and by Dianov *et al.* [4].

In the early work, the emphasis was on achieving wavelength conversion between two widely spaced but fairly narrow spectral regions. The location of these regions with respect to the pump was adjusted by one of two means. In communication fibers, by tuning the pump wavelength λ_p in the vicinity of λ_0, it is possible to obtain gain regions located several hundred nanometers away from λ_0; these can be moved by tuning λ_p.

This is the method used in [2] and [3]. Alternatively, by using a polarization-maintaining fiber (PMF) with a large birefringence, it is also possible to obtain large wavelength shifts, owing to the role of birefringence in phase matching when the pump and signal have orthogonal polarizations. In this case tuning is more difficult, because it must be accomplished by changing the birefringence; this can be done by applying a stress perpendicular to the fiber [5] or by heating the fiber [6]. By this method, gain regions around 904 nm and 1292 nm were obtained with a pump at 1060 nm [5]. In spite of these early accomplishments, in the mid-1980s fiber OPAs did not compare well with RAs [7]. In addition, with the advent of the enormously successful EDFAs in the late 1980s, there seemed to be little need for developing an alternate type of optical amplifier and, as a result, research activity on fiber OPAs waned.

Nevertheless, the ability of phase-sensitive fiber OPAs to achieve a zero dB noise figure, as well as to generate squeezed light, sparked activity on this type of OPA. In 1990, Marhic *et al.* and Bergman *et al.* simultaneously reported the first demonstrations of phase-sensitive amplification in nonlinear Sagnac interferometers [8, 9]. This work was followed by some activity on phase-sensitive fiber OPAs, but their delicate nature made application to mainstream communications problematic.

In 1996 Marhic *et al.* reconsidered phase-insensitive fiber parametric amplification and showed that by tuning λ_p near λ_0 it is in principle possible to obtain gain regions that are tens and even hundreds of nanometers wide, even with commonly available communication fibers and reasonable pump powers (of the order of 1 W) [10]. Such bandwidths exceeded those of the EDFAs and RAs that were currently used in optical communication, and this indicated that fiber OPAs could play an important role as amplifiers in future optical communication systems. Spurred in part by these prospects, as well as by steady advances in the required components, the development of fiber OPAs was resumed a few years ago and is now intensifying.

An important factor in this recent activity has been the fabrication of fibers specifically designed to have a high γ (of the order of 20 W^{-1} km^{-1}, i.e. about ten times larger than that for typical communication fibers), as well as λ_0 values around 1.55 μm. The abbreviations HNLF (highly nonlinear fiber) or HNL-DSF (highly nonlinear dispersion-shifted fiber) are often used to refer to such fibers. Having $\lambda_0 \approx 1.55$ μm is very important from a practical standpoint, as it allows the user to obtain good phase matching over wide regions in the vicinity of 1550 nm, which is currently one of the most important wavelengths in optical communication. The first highly nonlinear fibers of this type were made in 1995 by Holmes *et al.* at BT Laboratories in the UK [11]; several other manufacturers now make similar fibers. These fibers have made it possible to bring the performance of fiber OPAs much closer to their theoretical potential, particularly in terms of bandwidth.

Experiments in this wavelength range have also been greatly facilitated by the availability of components developed for optical communication systems, such as tunable lasers and EDFAs, which can be used to generate the narrow-linewidth high-power pumps required for fiber OPAs. In particular, the output power of EDFAs has been climbing steadily in recent years, while their prices have remained stable: today one can purchase a 3 W EDFA for the price of a 100 mW EDFA about five years ago. This trend has allowed experimentalists to upgrade their equipment and move to steadily higher pump powers, which are very important for developing high-performance fiber OPAs.

In summary, with the increasing availability of suitable equipment, significant progress has been made in recent years towards the realization of the potential of fiber OPAs. Important demonstrations have been the following.

- **A pulsed bandwidth of up to 400 nm** Fiber OPAs with pulsed gain bandwidths ranging from 200 nm to 400 nm have been demonstrated in recent years [12, 13]. This work used a pulsed pump, with about 10 W peak power, which is too high for optical communication applications.
- **A CW bandwidth of 100 nm** A 100 nm bandwidth was demonstrated with a single-pump OPA using a 4 W pump [14]. The CW operation means that such devices are compatible with applications in optical communication.
- **A continuous gain of 70 dB** [15] This was accomplished with a 2 W pump and shows that, if necessary, fiber OPAs can exhibit gains as large or larger than can be obtained with other types of fiber amplifiers, such as EDFAs or Raman amplifiers. Similar gains have also been obtained with a pulsed pump [13].
- **A noise figure of 3.7 dB** [16] This result is important because it demonstrates that fiber OPAs can have noise figures similar to those of EDFAs and Raman amplifiers.
- **Polarization-independent operation** [17, 18] As is the case for many other optical communication components and subsystems, it is desirable to have polarization independence for fiber OPAs. This means that the gain should be independent of the state of polarization (SOP) of the incident signal. Fiber OPAs (like Raman amplifiers) in their simplest forms are strongly polarization dependent. However, by using techniques such as polarization diversity [17, 18], it is possible greatly to reduce the polarization dependence.
- **High-speed optical processing applications** Fiber OPAs have been used to demonstrate a large variety of high-speed optical-processing applications, such as wavelength conversion, the demultiplexing of time-division-multiplexed (TDM) signals, optical sampling, signal regeneration, multiband processing, etc.
- **Distributed parametric amplification** Amplification in typical transmission fibers, 75 km long, has recently been achieved [19]. Also, phase-sensitive amplification was obtained in a 60-km-long fiber [20]. These results indicate that potentially parametric amplification in transmission fibers could compete with distributed Raman amplification and could eventually provide transmision links having nearly ideal noise properties.

With the demonstration of these features, all of which are necessary for practical applications, it is now becoming apparent that fiber OPAs may begin to find applications in optical communication, as well as in other fields. Thus it was felt that this was an opportune time for a book on the subject, which would gather much of the knowledge on the subject in one place as well as highlight some of the remaining challenges. Such a book could be of interest to engineers and researchers in related areas, graduate students involved in research on fiber devices and communication systems, etc.

In this book we present the basic types of fiber OPA, review the theory of their operation, present the experimental state of the art, and conclude with speculations about possible future directions for the field. In Chapter 2 we review the main properties of single-mode fibers, as is necessary for an understanding of fiber OPAs. In Chapter 3 we present the main analytic scalar solutions known in the field: these range from simple exponential solutions, when loss and pump depletion are ignored, to various types of special functions when such effects are introduced. The motivation for treating analytic solutions in such detail is based on the belief that they are essential tools for the interpretation of numerical simulations and can provide results quickly where they are directly applicable. In Chapter 4 we extend the formulation to the case where the waves are in arbitrary SOPs and we see how to reuse the results of Chapter 3, with suitable modifications. In Chapter 5 we introduce fiber dispersion and investigate how it governs the shape of the gain spectra. We study in detail the limits of negligible pump depletion and strong pump depletion and see how to optimize spectrum shapes for various applications. In Chapter 6 we discuss the nonlinear Schrödinger equation (NLSE), which is an important tool for studying field propagation when closed-form solutions cannot be used. We present the results of numerical simulations obtained by the split-step Fourier method (SSFM) for realistic fibers with longitudinal variations in λ_0 and in residual birefringence. In Chapter 7 we consider the case where the pump is pulsed. We show how one can use quasi-steady-state solutions when the time variations are relatively slow and the SSFM when they are fast. In Chapter 8 we explain how a fiber optical parametric oscillator (OPO) is made from a fiber OPA by adding optical feedback. We study important aspects such as the oscillation threshold, output power, and conversion efficiency. In Chapter 9 we present in detail some quantum-mechanical aspects of fiber OPAs. Starting from first principles we derive the theoretical 3 dB limit for the NF of fiber OPAs used either as amplifiers or as wavelength converters; we also consider the case of degenerate fiber OPAs, which have a 0 dB NF limit. In Chapter 10 we consider the requirements for some key aspects of the pumps of fiber OPAs, specifically power, linewidth, and amplitude stability. We investigate in detail the impact on OPA performance of some of these aspects, including the conversion of pump frequency or intensity modulation into signal intensity fluctuations and hence NF degradation. In Chapter 11 we present the results of experiments that have been performed to verify some fundamental properties of fiber OPAs, including the maximum gain, the gain bandwidth, and the noise figure. In Chapter 12 we discuss the potential application of fiber OPAs to various areas of optical technology, including optical communication, high-speed optical signal processing, quantum communication, and high-power wavelength conversion. In each area we present recent results obtained by various research groups. In Chapter 13 an aspect of fiber OPAs that is very important for fiber communication, namely crosstalk between signals in systems using wavelength-division multiplexing (WDM), is treated. We investigate the role of key parameters such as dispersion, the number of pumps, etc. and discuss strategies for minimizing this crosstalk. In Chapter 14 we present the topic of distributed parametric amplification (DPA). We treat it separately because this application involves amplification in a long communication fiber rather than amplification in a discrete device placed between such fibers, as in Chapter 12. Finally in Chapter 15 we put forward possible research and development areas in which activity could be helpful for realizing the full potential of fiber OPAs.

References

1. "Phase-matched stimulated four-photon mixing in silica-fiber waveguides," Stolen, R. H. *IEEE J. Quantum Electron.*; 1975; vol. QE-11, pp. 100–3.
2. "Phase matching in the minimum-chromatic-dispersion region of single-mode fibers for stimulated four-photon mixing," Lin, C., Reed, W. A., Pearson, A. D., Shang, H. T. *Opt. Lett.*; 1981; vol. 6, pp. 493–5.
3. "Efficient large-frequency-shifted three-wave mixing in low dispersion wavelength region in single-mode optical fibre," Washio, K., Inoue, K., Kishida, S., *Electron. Lett.*; 1980; vol. 16, pp. 658–60.
4. "Induced four-photon parametric processes in glass fibre waveguides," Dianov, E. M., Zakhidov, E. A., Karasik, A. Ya., Mamyshev, P. V., Prokhorov, A. M. *Pis'ma v Zhurnal Eksperimental'noi i Teoreticheskoi Fiziki*; 1981; vol. 34, pp. 40–4.
5. "Phase-matched light amplification by three-wave mixing process in a birefringent fiber due to externally applied stress," Ohashi, M., Kitayama, K., Ishida, Y., Uchida, N. *Applied Phys. Lett.*; 1982; vol. 41, pp. 1111–3.
6. "Phase matching in birefringent fibers," Stolen, R. H., Bosch, M. A., Lin, C. *Opt. Lett.*; 1981; vol. 6, pp. 213–5.
7. "Raman and four photon mixing amplification in single mode fibers," Pocholle, J. P., Raffy, J., Papuchon, M., Desurvire, E. *Optical Engineering*; 1985; vol. 24, pp. 600–8.
8. "Optical amplification in a nonlinear fiber interferometer," Marhic, M. E., Hsia, C. H., Jeong, J. M. *Electron. Lett.*; 1991; vol. 27, pp. 210–1.
9. "Squeezing in fibers with optical pulses," Bergman, K., Haus, H. A. *Opt. Lett.*; 1991; vol. 16, pp. 663–5.
10. "Broadband fiber optical parametric amplifiers," Marhic, M. E., Kagi, N., Chiang, T. K., Kazovsky, L. G. *Opt. Lett.*; 1996; vol. 21, pp. 573–5.
11. "Highly nonlinear optical fiber for all optical processing applications," Holmes, M. J., Williams, D. L., Manning, R. J. *IEEE Photon. Technol. Lett.*; 1995; vol. 7, pp. 1045–7.
12. "200-nm-bandwidth fiber optical amplifier combining parametric and Raman gain," Ho, M. C., Uesaka, K., Marhic, M., Akasaka, Y., Kazovsky, L. G. *J. Lightwave Technol.*; 2001; vol. 19, pp. 977–81.
13. "Wide-band tuning of the gain spectra of one-pump fiber optical parametric amplifiers," Marhic, M. E., Wong, K. K. Y., Kazovsky, L. G. *IEEE J. Selected Topics in Quantum Electron.*; 2004; vol. 10, pp. 1133–41.
14. "Broadband single-pumped fiber optic parametric amplifiers," Torounidis, T., Andrekson, P. A. *IEEE Photon. Technol. Lett.*; 2007; vol. 19, in press.
15. "Fiber-optical parametric amplifier with 70-dB gain," Torounidis, T., Andrekson, P. A., Olsson, B. A., *IEEE Photon. Technol. Lett.*; 2006; vol. 18, pp. 1194–6.
16. "Measurement of the photon statistics and the noise figure of a fiber-optic parametric amplifier," Voss, P. L., Tang, R. Y., Kumar, P. *Opt. Lett.*; 2003; vol. 28, pp. 549–51.

17. "Polarization-independent one-pump fiber-optical parametric amplifier," Wong, K. K. Y., Marhic, M. E., Uesaka, K., Kazovsky, L. G. *IEEE Photon. Technol. Lett.*; 2002; vol. 14, pp. 1506–8.
18. "Polarization-independent two-pump fiber optical parametric amplifier with polarization diversity technique," Kalogerakis, G, Marhic, M. E., Kazovsky, L. G. In *Proc. Optical Fiber Communication Conference*, March 2006, Anaheim CA; paper OWT4.
19. "Transmission of optical communication signals by distributed parametric amplification," Kalogerakis, G., Marhic, M. E., Wong, K. K. Y., Kazovsky, L. G. *J. Lightwave Technol.*; 2005; vol. 23, pp. 2945–53.
20. "Inline frequency-non-degenerate phase-sensitive fibre parametric amplifier for fibre-optic communication," Tang, R., Devgan, P., Grigoryan, V. S., Kumar, P. *Electron. Lett.*; 2005; vol. 41, pp. 1072–4.

2 Properties of single-mode optical fibers

While parametric interactions between different transverse modes are possible in multi-mode fibers [1], most of the recent work on fiber OPAs has been done with single-mode fibers and there are good reasons for believing that they will continue to be the most important medium for the foreseeable future. For this reason, in the rest of the book we will deal almost exclusively with single-mode propagation and in this chapter will concentrate exclusively on the properties of single-mode fibers that are essential to the understanding of fiber OPAs.

A fiber supports the propagation of a single mode when the light wavelength is in a certain range. On the long-wavelength side, a symmetric fiber can in principle support propagation at any such wavelength. In practice, however, losses due to material absorption or bends will make propagation at long wavelengths increasingly difficult. On the short-wavelength side, the fundamental mode can also propagate at any wavelength; however, below the cutoff wavelength λ_{co} a second mode can also propagate. When this occurs, coupling between the two modes (induced by bends) results in reduced power in the fundamental mode and therefore an increase in loss. In addition, the material may also exhibit increased loss at short wavelengths. Fibers designed to operate over a particular range of wavelengths are optimized by taking into account these two high-loss regions.

2.1 Mode profile

The fundamental mode profile in single-mode fibers is generally well approximated by a circularly symmetric mode with a Gaussian profile. Thus the local irradiance (power per unit area) $I(r)$ is well described by

$$I(r) = I_0 \exp\left(-\frac{2r^2}{w^2}\right), \tag{2.1}$$

where r is the radial distance from the fiber axis and r_0 is the distance at which I drops to $1/e^2$ of its on-axis value, I_0. The quantity $2w$ is known as the mode field diameter (MFD).

The total power P carried by the mode is given by

$$P = \int_0^{2\pi} d\theta \int_0^\infty I(r)r\,dr = \pi w^2 I_0/2, \tag{2.2}$$

where θ is the angle around the z-axis; P is the same as that for a beam of uniform irradiance I_0 and radius w.

2.2 Loss

All fibers attenuate the power of waves passing through them. There are several possible causes for this attenuation: Rayleigh scattering by sub-wavelength density fluctuations; scattering from the roughness of the core–cladding interface; scattering from the domain walls in polycrystalline materials; absorption by dopants or impurities; multiphonon absorption by lattice vibrations at long wavelengths.

At a particular wavelength the power attenuation is characterized by an attenuation constant α. If $P(z)$ denotes the power of the wave at a distance z along the fiber, its evolution is governed by the differential equation

$$\frac{dP}{dz} = -\alpha z, \qquad (2.3)$$

with solution

$$P(z) = P(0) \exp(-\alpha z), \qquad (2.4)$$

indicating an exponential decay. If SI units are used in Eq. (2.4), α is in nepers m^{-1}. In practice it is more common to use decibels (dB) than nepers. Then, if α' denotes the attenuation constant in dB m^{-1}, we have, approximately, $\alpha' = 4.3\alpha$. The power $P(z)$ can be written alternatively as

$$P(z) = P(0) \times 10^{-\alpha' z/10}. \qquad (2.5)$$

Fiber attenuation always depends on wavelength, and the variation can be quite complicated over wide wavelength ranges. For example silica, the most transparent fiber to date, exhibits an attenuation as low as 0.15 dB km^{-1} at 1550 nm but the value is much higher above 2 µm and below 200 nm.

2.3 Propagation constant and dispersion

Let us consider for now an ideal lossless fiber. If a monochromatic wave with frequency ν (angular frequency $\omega = 2\pi\nu$) propagates as a single mode along the fiber, the spatio-temporal dependences of all its field components are harmonic functions of the distance along the fiber z and the time t. For example, the dominant electric field component can be written as

$$E(z, t) = E(0, 0) \exp[i(\beta z - \omega t)], \qquad (2.6)$$

where $E(0, 0)$ is a complex phasor representing the initial amplitude and phase of the wave. The real electric field is $\mathrm{Re}\{E(z, t)\}$.

The quantity β is the propagation constant or wavevector and is a function of frequency. From β one can calculate the effective index $n_{\mathrm{eff}} = \beta c/\omega$, the phase velocity

2.3 Propagation constant and dispersion

$v_{ph} = \omega/\beta = c/n_{\text{eff}}$, and the group velocity $v_g = d\omega/d\beta$. Also important is $\beta^{(m)} = d^m\beta/d\omega^m$, particularly for $m = 1, 2, 3, 4$. We note that $v_g = 1/\beta^{(1)}$.

When n_{eff} is not constant the mode is said to exhibit dispersion. This generally leads to the distortion of a modulated signal as it propagate through the fiber, since its different frequency components travel with different speeds. Distortion of the signal envelope occurs if there is group velocity dispersion (GVD), i.e. if $\beta^{(2)} \neq 0$.

Fibers often exhibit a wavelength where $\beta^{(2)} = 0$ in a region of interest. It is called the *zero-dispersion wavelength*, often denoted by ZDW or λ_0, and is of great importance in optical communication, because signals with carrier wavelengths close to it may experience little distortion as they propagate over even long distances.

The zero-dispersion wavelength λ_0 is also very important for some fiber nonlinearities. A group of several waves traveling near λ_0 can maintain phase synchronism over long distances, a condition that can lead to efficient nonlinear coupling and nonlinear energy exchange between the waves.

As we will see in detail later on, the location of the pump(s) with respect to λ_0 is of critical importance when it comes to optimizing the shape of the OPA gain spectrum. The values of $\beta^{(3)}$ and $\beta^{(4)}$ near λ_0 are also very important in this respect.

While knowledge of $\beta(\omega)$ is in principle sufficient to characterize fiber dispersion completely, a number of related quantities are often used, which are given in terms of wavelength rather than frequency. It is useful to know how to go quickly from one set of quantities to the other. For this purpose, we present here some of the main relationships.

The starting point is that the free-space wavelength λ is related to ω by

$$\omega = \frac{2\pi c}{\lambda} \tag{2.7}$$

where c is the speed of light in vacuum. We then have

$$\frac{d\omega}{d\lambda} = -\frac{2\pi c}{\lambda^2} = -\frac{\omega}{\lambda}. \tag{2.8}$$

An interpretation of $\beta^{(2)}$ is that if two pulses with carrier frequencies ω_1 and ω_2 separated by a small $\Delta\omega = \omega_1 - \omega_2$ are launched simultaneously into a fiber of length L, at the output they will be separated in time by the group delay difference $\Delta\tau$, which is given by

$$\Delta\tau = \frac{L}{v_{g1}} - \frac{L}{v_{g2}} = L\left[\beta^{(1)}(\omega_1) - \beta^{(1)}(\omega_2)\right] \approx L\beta^{(2)}(\omega_1)\Delta\omega. \tag{2.9}$$

The chromatic dispersion coefficient D is defined in such a way that $\Delta\tau = D\Delta\lambda L$, where $\Delta\lambda$ is the wavelength spacing between the two carriers. From these two expressions for $\Delta\tau$, we conclude that

$$D = \frac{\Delta\omega}{\Delta\lambda}\beta^{(2)} = -\frac{2\pi c}{\lambda^2}\beta^{(2)} = -\frac{\omega}{\lambda}\beta^{(2)}. \tag{2.10}$$

Another important quantity is the dispersion slope $D_\lambda = dD/d\lambda$. It is generally useful to know its value at λ_0. For single-mode fiber (SMF) it is of the order of 0.07 ps nm^{-2} km^{-1} and for highly nonlinear fiber (HNLF) it is of the order of 0.02–0.03 ps nm^{-2} km^{-1}.

The relation between D_λ and $\beta^{(3)}$ is more complicated than that between D and $\beta^{(2)}$: it is

$$D_\lambda = \frac{\omega}{\lambda^2}\left(2\beta^{(2)} + \omega\beta^{(3)}\right). \qquad (2.11)$$

Hence D_λ depends on both $\beta^{(3)}$ and $\beta^{(2)}$, except at λ_0. From Eqs. (2.8) and (2.9) we obtain

$$\beta^{(3)} = \left(\frac{\lambda^2}{2\pi c}\right)^2 \left(D_\lambda + \frac{2}{\lambda}D\right). \qquad (2.12)$$

Finally, in OPA work it is often desired to calculate $\beta^{(4)}$ from plots of D and D_λ, which are generally provided by manufacturers. It can be shown that [2]

$$\beta^{(4)} = -\frac{\lambda^4}{(2\pi c)^3}\left(6D + 6\lambda D_\lambda + \lambda^2\frac{dD_\lambda}{d\lambda}\right). \qquad (2.13)$$

This expression contains $dD_\lambda/d\lambda$, which needs to be calculated from the data, as the slope of the plot of D_λ at the desired wavelength. Equation (2.13) is useful when D, D_λ, and $dD_\lambda/d\lambda$, are accurately known at the wavelength of interest.

Often manufacturers provide only data at discrete wavelengths for D or for the time delay τ. In that case it is more straightforward to use a polynomial fit for either of these quantities than to use Eq. (2.13). (Whenever possible, it is better to use τ than D, because τ is obtained directly from the measurement apparatus whereas D is derived from τ by means of some numerical technique that may introduce uncertainty when higher-order derivatives are calculated.)

When D is available, it should first be converted to $\beta^{(2)}$, and $\beta^{(2)}$ should be plotted versus $\Delta\omega = \omega - \omega_r$, where ω_r is the reference frequency at which one wants to calculate the higher-order derivatives. In the units commonly used, the conversion to $\beta^{(2)}$ is given by

$$\beta^{(2)}[\text{m}^2\,\text{s}^{-1}] = -\frac{10^{-32}}{6\pi}D[\text{ps}\cdot\text{nm}^{-1}\,\text{km}^{-1}]\,(\lambda[\text{nm}])^2, \qquad (2.14)$$

where the units for each quantity are written after it in a square bracket. The conversion from λ to $\Delta\omega$ is given by

$$\Delta\omega[\text{rad s}^{-1}] = 6\pi \times 10^{17}\left(\frac{1}{\lambda[\text{nm}]} - \frac{1}{\lambda_r[\text{nm}]}\right). \qquad (2.15)$$

Using a second-order polynomial fit for $\beta^{(2)}(\Delta\omega)$ one obtains a result of the form

$$\beta^{(2)}(\Delta\omega) = a(\Delta\omega)^2 + b\Delta\omega + c. \qquad (2.16)$$

We also have, however,

$$\beta^{(3)}(\Delta\omega) = \frac{d\beta^{(2)}}{d(\Delta\omega)} = 2a\Delta\omega + b$$

and

$$\beta^{(4)}(\Delta\omega) = \frac{d\beta^{(3)}}{d(\Delta\omega)} = 2a.$$

2.3 Propagation constant and dispersion

Hence $\beta^{(4)}$ is equal to twice the coefficient of $(\Delta\omega)^2$ in the second-order polynomial fit of $\beta^{(2)}$.

Similarly, starting from τ gives

$$\tau = \frac{L}{v_g(\omega)} - \frac{L}{v_g(\omega_r)} = L\left[\beta^{(1)}(\omega) - \beta^{(1)}(\omega_r)\right]. \tag{2.17}$$

If we expand τ/L in a power series of $\Delta\omega$ and truncate it after the cubic term, we obtain the approximation

$$\frac{\tau}{L} \approx \beta^{(2)}(\omega_r)\Delta\omega + \frac{\beta^{(3)}(\omega_r)}{2}(\Delta\omega)^2 + \frac{\beta^{(4)}(\omega_r)}{6}(\Delta\omega)^3. \tag{2.18}$$

A fourth-order polynomial fit of τ/L versus $\Delta\omega$ yields $\tau/L = a'(\Delta\omega)^3 + b'(\Delta\omega)^2 + c'\Delta\omega + d'$, however. Hence we see that $a' = \beta^{(4)}/6$, i.e. that $\beta^{(4)}$ is six times the coefficient of $(\Delta\omega)^3$ in the third-order polynomial fit of τ/L versus $\Delta\omega$.

To obtain $\beta^{(4)}$ in $m^4 s^{-1}$, one should express τ/L in $m s^{-1}$ and $\Delta\omega$ in $rad\ s^{-1}$ as in Eq. (2.15).

On occasion it is of interest to calculate higher-order dispersion terms. Then the polynomial fits must be extended to higher orders. In the general case we can approximate τ/L by its Taylor series expansion about ω_r, truncated after N terms, i.e.

$$\frac{\tau}{L} \approx \sum_{j=0}^{N-1} \frac{\beta^{(j+1)}(\omega_r)}{j!}(\Delta\omega)^j. \tag{2.19}$$

We can also use an N-term polynomial fit (of order $N-1$) to approximate τ/L, i.e.

$$\frac{\tau}{L} \approx \sum_{j=0}^{N-1} a_j^N (\Delta\omega)^j. \tag{2.20}$$

Equating the respective powers of $\Delta\omega$ on the right-hand sides of Eqs. (2.19) and (2.20) we obtain the set of relations

$$\beta^{(j+1)}(\omega_r) = j! a_j^{N-1}, \qquad j = 0, 1, \ldots, N-1. \tag{2.21}$$

So in this manner we obtain approximations for $\beta^{(1)}(\omega_r)$ through $\beta^{(N)}(\omega_r)$. As an example, to calculate $\beta^{(6)}(\omega_r)$, we need to do this using $N \geq 6$.

It is important to note that we have provided several ways to calculate a particular $\beta^{(m)}(\omega_r)$ but that they all yield only approximations of the actual value. Deciding which is the best approximation is a difficult task. In general, it is in fact good practice to use N values no larger than m when calculating $\beta^{(m)}(\omega_r)$.

When calculating numerical values of $\beta^{(m)}$, it is common to give them either in SI units, or in picoseconds and kilometers. The conversion between the two is given by

$$\beta^{(m)}[ps^m\ km^{-1}] = 10^{12m+3}\beta^{(m)}[s^m\ m^{-1}]. \tag{2.22}$$

The numerical values for $\beta^{(m)}$ are generally quite small in the SI system but they are closer to 1 with the other choice of units.

2.4 Longitudinal fluctuations of the zero-dispersion wavelength

So far we have assumed that the fiber dispersion properties do not change along the fiber length. In reality it is known that, because of the limitations of the manufacturing process, it is not possible to maintain the fiber core diameter perfectly constant along the fiber's length. Since the core diameter affects the waveguide dispersion, this causes longitudinal dispersion fluctuations. One way to quantify this, particularly for dispersion-shifted fibers (DSFs), is by measuring the variation in λ_0 as a function of distance, which yields a function $\lambda_0(z)$. Measurements have been performed for a variety of fibers. For DSF it has been found that λ_0 can vary by about 0.5 nm, on a length scale of the order of 500 m [3]. Highly nonlinear fibers (HNLFs) have a smaller core diameter, which is harder to control; as a result, λ_0 fluctuations are relatively larger. Destructive measurements on an HNLF sample have revealed a λ_0 variation of 20 nm over 1.7 km [4]; in other work a 4 nm variation was measured nondestructively in a 300-m-long fiber [5]; improvements in fabrication techniques are reducing these numbers [6].

The impact of $\lambda_0(z)$ on a particular OPA depends a great deal on some other OPA parameters. The pump power P_0 is particularly important, in several ways. If P_0 is large, only a short fiber is needed to achieve a given gain; then the probability of encountering large λ_0 variations is reduced. In addition, the nonlinear phase shifts that enter into the phase matching are large and dominate the linear phase shifts, which are the only ones affected by dispersion. Hence high P_0 values reduce the impact of λ_0 variations in two separate ways, which combine to make the effect of high P_0 significant. As a result, one should be able to obtain near-theoretical performance with high-power pulses in short fibers; this has indeed been shown to be the case [7].

2.5 Temperature dependence of the zero-dispersion wavelength

The zero-dispersion wavelength λ_0 also exhibits temperature dependence. Its rate of change $d\lambda_0/dT$ with temperature T is of the order of 0.03 nm $°C^{-1}$ for DSF [8] and 0.06 nm $°C^{-1}$ for HNL-DSF [9, 10]. Since OPA performance depends critically on the fiber dispersion properties, it is clear that fiber OPAs will need to be protected from large temperature variations in order to operate reliably in practical circumstances.

A possibly positive aspect of the dependence of λ_0 on temperature is that this could in principle be used to control $\lambda_0(z)$. Starting from a fiber with a random distribution $\lambda_0^r(z)$, the latter could be compensated and transformed into a deterministic, desirable, distribution $\lambda_0^d(z)$ by producing the necessary temperature distribution along the fiber [10]. In this manner it would in principle be possible to tailor the shape of the gain spectrum. For example, one could obtain a flat-top gain spectrum with a one-pump OPA instead of the uneven spectrum typically obtained when there are no λ_0 variations [11].

Exposure to ultraviolet (UV) light can also be used for the same purpose. It is well known that UV irradiation of the GeO_2-doped core of silica fibers leads to a change in the refractive index. This is used extensively for writing fiber Bragg gratings (FBGs) directly into fiber cores. In addition, UV exposure also shifts λ_0: a 100 nm shift has been demonstrated [12]. Ultraviolet exposure has the advantage that the finished fiber does not require the system needed for temperature control. On the negative side, if long fibers (tens or hundreds of meters long) are needed then the exposure time could be quite long, especially if exposure has to take place through the protective plastic coating.

2.6 Fiber birefringence

The OPA theory to be presented in Chapter 3 assumes that all waves remain in the same linear state of polarization (SOP) over the entire fiber length. Depending on the type of fiber used, this may or may not be a good approximation. Here we discuss this aspect of fiber propagation and its potential impact on the performance of fiber OPAs.

2.6.1 Polarization-maintaining fibers

Polarization-maintaining fibers (PMFs) are fabricated in such a way that they exhibit different refractive indices n_x and n_y for light linearly polarized along the x- and y-directions, perpendicular to the direction of propagation. The birefringence is defined as $\Delta n = n_x - n_y$. Fibers with Δn as high as 10^{-3} have been fabricated.

A length of PMF behaves as a birefringent waveplate. In particular, if light is launched at the input in an SOP that is linear and parallel to either the x- or the y-axis, it retains that SOP over the entire length. These SOPs are known as the eigenpolarizations. Thus one way to ensure that all waves in a fiber OPA retain a common linear SOP over the entire fiber is to use a PMF and to launch all waves with the proper SOP.

This last requirement is a practical difficulty in communication applications, because communication fibers are not PMFs and do not preserve SOPs. Hence, at the end of a communication link any SOP may emerge, and it may also evolve as a function of time. Thus, to use a PMF-based OPA at the end of such a fiber, one would need to implement an active polarization control system to track the emerging SOP and to transform it into a linear SOP suitable to couple into the PMF. Such polarization-tracking systems have been developed for other applications such as coherent detection; however, their implementation is fairly complex and expensive and therefore might be difficult to justify for each OPA.

We note that PMFs can also support the transmission of waves with different orthogonal SOPs. This has been used for parametric amplification with an x-polarized pump and y-polarized signal and idler [13]. In this case the theory of Chapter 3 must be modified, as discussed in Chapter 4: γ is divided by 3 for interactions between the orthogonal components via cross-phase modulation (XPM) four-wave mixing (FWM) and $\Delta\beta$ is modified by a term proportional to Δn. This last term may be substantial, and if so it can be used to shift the gain maximum away from the pump; this can be useful for obtaining wavelength conversion between distant wavelength ranges.

2.6.2 Non-polarization-maintaining fibers

Owing to the difficulty of using PMFs in communication applications, most recent work with fiber OPAs has been performed with fibers that are not polarization maintaining.

For example, commonly available communication fibers have been used for many OPA experiments. Such fibers are designed with a circular geometry and thus ideally would exhibit no preferred azimuthal variation and no associated linear birefringence. However, the fabrication of fibers that actually behave in such an ideal manner is not possible at present: practical fibers have slight departures from the ideal circular geometry, such as slightly elliptical cores or stress due to bends, which induce the presence of linear birefringence. Furthermore, both the orientation of the local polarization axes and Δn vary randomly as a function of z. In typical fibers, Δn has a standard deviation of the order of 10^{-6}. This corresponds to a beat length of the order of 1 m, which implies that any SOP, except the local eigenpolarizations, will change substantially over 1 m. In addition, because of the randomness of orientation of the polarization axes, in a long fiber any input SOP will eventually pass through all possible SOPs. This is generally expressed by saying that the SOPs end up being uniformly distributed over the Poincaré sphere, which is a common graphical means for representing SOPs [14].

Thus propagation in non-PMFs is far more complicated than in PMFs, as it is not possible to assume that linear SOPs can exist over the entire fiber length. The analysis of OPA performance in such circumstances is presented in Chapters 4 and 6.

In communication fibers, random birefringence manifests itself by what is known as polarization-mode dispersion (PMD): an input pulse with an arbitrary SOP gives rise to two pulses, each propagating along one of the principal axes. As a result, at the output two pulses emerge, separated by a time delay τ. Because of the random nature of the birefringence, it can be shown that τ scales as the square root of the fiber length. For this reason PMD values are usually given in units of ps km$^{-1/2}$. Polarization-mode dispersion values range from about 1 to 0.01 ps km$^{-1/2}$, depending on the type of fiber and the fabrication process.

It should also be noted that since the random birefringence is weak, it can readily be modified by environmental factors such as temperature or mechanical stress. In long installed communication fibers, time-dependent temperature and stress variations are sufficient to cause the output SOP to vary randomly, on a subsecond time scale. For typical OPA work in the laboratory relatively short fibers are used, and they are not subject to strong environmental disturbances. Thus one generally finds that the output SOPs do not drift significantly over several minutes. This generally allows enough time for performing measurements before needing to readjust the SOPs.

While laboratory work is thus possible with non-PMFs this does not necessarily mean that non-PMFs could be used in commercial OPAs, because for such a use the SOPs would have to remain the same permanently. It is thus likely that for practical fiber OPAs, PMFs (or perhaps spun fibers, see below) will need to be used.

2.6.3 Spun fibers

Non-polarization-maintaining fibers always exhibit a residual amount of random linear birefringence, which causes the SOP of an input wave to vary in an unpredictable manner along the fiber length. In some situations it may be desirable to maintain the input SOP over long lengths. One solution is to use a PMF. Then, if the input SOP is linear and aligned with one of the principal axes, it remains in that state throughout the fiber.

Another possibility exists. If a non-PMF is spun about its axis at a rapid rate during fiber drawing, the resulting twist of the material will induce circular birefringence. A similar result can be obtained by taking a finished non-PMF and winding it on a spool while spinning it; however, this leads to somewhat less birefringence than spinning during drawing. In order to suppress linear birefringence to a negligible level in this manner, it is necessary to spin the fiber at a rate of 10 turns per meter, or more. The resulting circular birefringence is of the order of 10^{-6} [15–18].

The polarization eigenstates of a spun fiber are circularly polarized. If an elliptical SOP is launched at the input, the shape of the ellipse will be maintained during propagation but it will rotate at a rate proportional to the circular birefringence.

Spun fiber would be a good medium for experimenting with OPAs using circularly polarized pumps, which have advantages over linearly polarized pumps. To date, no such experiments have been reported, probably owing to the difficulty of fabricating spun fibers as compared with non-PMFs or even PMFs.

2.7 Nonlinearities

2.7.1 Third-order nonlinearity

(a) The nonlinear index n_2

The fiber nonlinearity that gives rise to parametric amplification in glass fibers is associated with the third-order susceptibility $\chi^{(3)}$. It is also known as the Kerr nonlinearity. A simple way to introduce it is via the nonlinear refractive index n_2. It can be shown that in a medium with a third-order susceptibility the presence of a strong light wave (a pump) with irradiance I modifies its refractive index according to the relation

$$n = n_0 + n_2 I, \qquad (2.23)$$

where n_0 is the weak-field refractive index and n is the index in the presence of the strong wave.

Equation (2.23) indicates that, since n is a dimensionless quantity, n_2 should have the dimensions of reciprocal irradiance, or m^2 W^{-1} in the SI system. The reader should beware of several different definitions for n_2. One definition uses $I/2$ instead of I, because I is taken as the peak value of the power per unit area, rather than the average value, as in Eq. (2.23). Other definitions use the square of the electric field (peak or r.m.s.) rather than the irradiance, which gives n_2 units of m^2 V^{-2}.

Since a change in refractive index implies a change in propagation constant and therefore a change in phase due to propagation, the presence of the pump modifies the phase of other waves passing through the same region (cross-phase modulation, or XPM) and also the pump's own phase (self-phase modulation, or SPM). By combining XPM or SPM with phase-sensitive devices such as interferometers, one can implement a variety of functions, for example all-optical switching, modulation, etc.

In glass fibers n_2 is primarily due to the anharmonic motion of bound electrons responding to an applied field. Since electrons are very light, they respond very quickly to changes in field strength: their response time is on the femtosecond time scale, i.e. about one period for visible light. As a result, the irradiance I in Eq. (2.23) can actually be considered to be instantaneous. Then, if the pump consists of two monochromatic terms with frequencies ω_1 and ω_2, I will contain terms with frequencies $\pm\omega_1 \pm \omega_2$. If another wave, with frequency ω_3, is passed through the same region, it experiences time-dependent phase modulation due to the pump. As a result, it can be shown that this wave now contains new frequency components, at $\pm\omega_1 \pm \omega_2 \pm \omega_3$. These new frequencies are said to arise from four-wave mixing (FWM), since four frequencies are involved, the original three, and a new one. (Note that in the older literature the term three-wave mixing was used to describe the same thing, in reference to the three initial waves only.)

In fiber OPAs, the basic mechanism that leads to amplification is FWM. In addition, since strong pumps are generally used, pump-induced SPM and XPM effects cannot be neglected; they play an important role in determining the growth rate as well as the gain-spectrum shape.

(b) The fiber nonlinearity coefficient γ

Just like the refractive index n, the nonlinear index n_2 is a property of an optical material. It does not depend on the structure of the optical device. To calculate the effect of n_2 in the presence of an optical beam, one must obtain the irradiance from a knowledge of the total power in the beam and of the beam profile and cross-section.

When working with optical fibers, it is more convenient to deal with the fiber nonlinearity coefficient γ, defined as

$$\gamma = \frac{\omega n_2}{c A_{\text{eff}}} = \frac{2\pi n_2}{A_{\text{eff}} \lambda}; \qquad (2.24)$$

γ contains information about the material, through n_2, and about the fiber mode, through its effective area A_{eff}. (The justification for this will be given in Chapter 3.) For a Gaussian mode, $A_{\text{eff}} = 2\pi w^2$, i.e. it is twice the area of a uniform beam with the same on-axis irradiance as the Gaussian mode and the same total power. The coefficient γ has the dimensions of $1/(\text{power} \times \text{length})$. The usefulness of γ can be seen as follows: in terms of γ the self-phase shift experienced by a wave of power P traveling along a length L of fiber is simply

$$\Phi_{\text{SPM}} = \gamma P L. \qquad (2.25)$$

Equation (2.25) shows that to obtain a large phase shift with a given power and fiber length, one needs to use a large value of γ. Equation (2.24) then shows that one should therefore use a fiber with a large n_2 as well as a small A_{eff}. A small A_{eff} value increases I for a given amount of power. Similar reasons hold for the desirability of increasing all the possible third-order nonlinear interactions.

This presentation of the third-order nonlinearity is a simplified introduction to the subject. In Chapter 3 we will investigate it at a deeper level, starting from basic electromagnetic propagation equations that include contributions due to the third-order susceptibility $\chi^{(3)}$.

2.7.2 Stimulated Brillouin scattering

Stimulated Brillouin scattering (SBS) is a high-gain nonlinear process that needs to be suppressed for fiber OPAs to operate properly. It involves an interaction between photons and phonons (acoustic waves). An incident optical wave (pump) is initially reflected by thermally excited acoustic waves. The incident and reflected optical waves form an interference pattern, which in turn reinforces the acoustic wave by means of electrostriction. This process leads to simultaneous growth of the acoustic wave and the reflected wave until a steady-state situation is reached. One defines a threshold P_{th} for SBS as the pump power beyond which the transmitted power is clamped at a constant level and the balance is reflected. If left unchecked, SBS can reflect virtually all the pump power required for driving an OPA, thus preventing its operation.

Several techniques have been demonstrated for reducing or suppressing SBS. By using air gaps along the fiber, as might be implemented with connectors, one can prevent acoustic waves from propagating over the entire fiber length, thereby increasing P_{th}. Alternatively, placing optical isolators along the fiber impedes the propagation of the reflected optical wave, with a similar effect.

Other techniques are based on the fact that the SBS gain has a narrow bandwidth ($\Delta v_B \approx 50$ MHz) and that pump frequency components falling outside that bandwidth do not generate SBS gain. As a result the Brillouin gain is decreased and P_{th} is increased. Quantitatively, this can be expressed by the relation

$$g'_B = \frac{g_B}{1 + (\Delta v_B / \Delta v)^2}, \tag{2.26}$$

where g'_B (g_B) is the Brillouin gain with (without) line broadening and Δv is the broadened linewidth. Thus, if one uses a pump with a linewidth $\Delta v = 5$ GHz, P_{th} is raised by a factor 100. This is an effective and flexible means for suppressing SBS, and it has been used in virtually all OPA experiments using CW pumps. An important consideration for OPAs is that one should avoid intensity modulation (IM) when broadening the pump, because pump intensity fluctuations will lead to temporal gain fluctuations and thus to degradation of the signal and idler quality. Hence, in OPA work, pump spectrum broadening is accomplished by means of phase or frequency dithering, which are not accompanied by IM. Various modulation schemes have been used to shape the CW pump spectrum. These include: sinusoidal phase or frequency modulation; summing several harmonic tones; pseudo-random bit sequence (PRBS) phase modulation.

It should also be noted that SBS can be suppressed by using a pulsed pump with a low duty cycle. The spectral width of such a pump is inversely proportional to the pulse width; hence, by using short pulses one can have very wide spectra and so increase P_{th}. For example, 1 ns pulses will have a spectral width of the order of 1 GHz, much larger than the SBS linewidth. Of course, the use of such pulses is not suitable for CW OPA operation. However, it is sometimes convenient for achieving high pump powers during a short time, which can be useful for demonstrating feature such as a wide OPA gain spectrum or a high OPA gain.

Another approach exploits the fact that stress shifts the location of the Brillouin peak. Thus one can design a stress distribution along a fiber that will effectively broaden the overall gain SBS spectrum for the fiber and reduce the peak SBS gain. Broadening by several hundred megahertz has been achieved in this manner, resulting in a significant increase in P_{th}.

Finally, we note that it is also possible to design the fiber core in such a way that the overlap between between optical and acoustic waves will be reduced. An increase in P_{th} by as much as 6 dB has been obtained in this manner.

Of all these SBS-suppression techniques, the advantage of the static techniques is that they do not require modulators and RF drivers to work. On the negative side, however, they may be difficult to implement and the changes that they introduce may be irreversible. Also, some of the effects used may have other undesirable consequences; for example, stress also modifies the fiber refractive index as well as its zero-dispersion wavelength.

By contrast, the active method, namely broadening of the pump spectrum by pump frequency or phase modulation, has the advantages of being compatible with any type of fiber and of leaving the fibre unchanged. For this reason it is widely used in experimental work.

2.7.3 Stimulated Raman scattering

Raman gain in fibers is described in Section 3.6. While the physical origin of stimulated Raman scattering (SRS) is quite different from that of parametric gain, from a mathematical standpoint SRS can be introduced into the basic OPA equations in a fairly simple manner, by adding an imaginary part to $\chi^{(3)}$ with suitable frequency dependence.

The parametric gain coefficient g in a well-phased-matched fiber OPA is about 2–3 times larger than the corresponding Raman gain coefficient. As a result, it can be shown that, for a broadband fiber OPA in which the gain bandwidth overlaps the Raman peak(s) due to the pump(s), the Raman contribution introduces only a minor distortion of the parametric gain spectrum. This has been shown theoretically and verified experimentally [2]. Hence the theory of Chapter 3 can be expected to provide a good first-order approximation for the gain spectrum of well-phase-matched OPAs, even when SRS is not included.

In regions where the parametric gain is weak compared with the Raman gain the reverse situation holds, and one may consider the Raman gain alone as a first approximation. This would for example be the case with a narrowband one-pump fiber OPA, with a half-gain bandwidth of say 30 nm near the pump. In that case we would expect to see the Raman gain peak at about 100 nm from the pump, i.e. not overlapping appreciably with the parametric gain spectrum.

In situations where the Raman and parametric gain contributions are comparable, it is necessary to keep both in the model to calculate the combined gain accurately. This is done in subsection 3.6.4.

For two-pump OPAs, the Raman interaction between the pumps needs to be taken into account if it leads to significant variation in pump power. In this case it is no longer possible to obtain a closed-form solution to the signal and idler equations.

2.7.4 Two-photon absorption

Just as the real part of the refractive index is modified by the passage of an intense light wave through a fiber, so is its imaginary part, which corresponds to absorption. This is mathematically described by an equation similar to Eq. (2.23), wherein the refractive indices are replaced by absorption coefficients. The form of this equation indicates that the increase in absorption can be attributed to the annihilation of pairs of photons, hence the name two-photon absorption (TPA).

To date TPA has played essentially no role in fiber OPA work. The reason is that in silica-based fibers nonlinear phase shifts of the order of π can easily be obtained at irradiances that correspond to negligible TPA. Only if one uses very high irradiances in short fibers should one be concerned with the possible impact of TPA. For example, no effects attributable to TPA were observed in a tapered silica fiber for pulses with 1 kW of peak power passing through a tapered silica fiber with a 1 μm diameter, i.e. for irradiances as high as 100 GW cm^{-2} [19]. For highly-nonlinear materials such as lead glass [20] or semiconductors [21], however, TPA can occur at much lower irradiances.

2.8 Types of fiber used for OPA work

The development of fiber-optic components for the telecommunications industry has made available a wide range of affordable components that can also be used for other applications, such as experimentation with fiber OPAs. In particular, silica-based fibers designed for optical communication links have some very attractive properties: they have a minimum loss as low as 0.15 dB km^{-1} at 1550 nm and can exhibit losses lower than 0.3 dB km^{-1} over the entire 1200–1700 nm range. With such low losses, experimentation with fibers that are tens of kilometers long can be done at a reasonable cost.

Communication fibers have a relatively simple step-index design, shown in Fig. 2.1(a). They consist of a circular core with refractive index n_{co}, and radius a of the order of a few microns, surrounded by a cladding with refractive index n_{cl} that is slightly smaller than n_{co}. As a result, they have a small numerical aperture $NA = \sqrt{n_{co}^2 - n_{cl}^2}$. In order to support a single mode, they are usually designed so that the normalized frequency $V = (NA)a\omega/c$ is of the order of 2. The resulting fundamental mode then has a mode field diameter (MFD) of the order of $2a$.

The most interesting communication fibers for OPA work are dispersion-shifted fibers (DSFs), which exhibit a ZDW located in the C-band (1535–60 nm) or close to it. In a typical design, the cladding is made of silica and the core is silica doped with a small amount (<1%) of germania (germanium dioxide). A drawback of these fibers is that their γ value is of the order of 2 W^{-1} km^{-1}, which is low. As a result, it is difficult to obtain large gain bandwidths for OPAs made from such fibers.

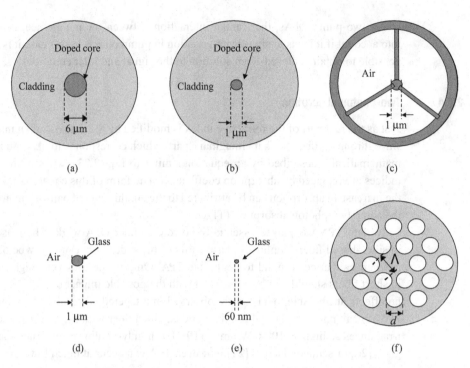

Fig. 2.1 Types of single-mode fiber: (a) step-index fiber; (b) highly nonlinear fiber (HNLF); (c) extruded fiber; (d) tapered fiber (the core is not shown as its diameter is very small); (e) nanowire; (f) microstructured fiber.

2.8.1 Highly nonlinear step-index fibers

In order to increase the γ values of DSFs, fiber manufacturers have developed a version with a smaller core size, Fig. 2.1(b) [22]. In order to maintain single-mode operation in the C-band, they have increased the numerical aperture (NA) by increasing the germania content of the core, up to about 3%. The combination of smaller core and increased germania content leads to an increase in γ by about an order of magnitude: γ values of the order of 25 W^{-1} km^{-1} have been reported. The losses are higher than for DSFs but are still reasonably low at about 0.5 dB km^{-1}. Such fibers are sometimes referred to as highly nonlinear DSFs (HNL-DSFs) or simply highly nonlinear fibers (HNLFs). The latter terminology has the disadvantage that it can be, and occasionally is, used to refer to other types of fiber with large values of γ.

The introduction of such fibers has facilitated a number of experiments, particularly demonstrations of wideband and high-gain amplifiers. They are now often the preferred medium for OPA research in the C-band.

An interesting variation on the step-index design is obtained by using a core material that consists mostly of germania, and a silica cladding [23]. Since germania has a larger n_2 value than silica, γ can be larger than for the more conventional GeO_2-doped silica fibers: fibers with $\gamma = 45$ W^{-1} km^{-1} have been made [23].

Another interesting aspect of these fibers is that the core material, obtained by blending GeO_2 and SiO_2, can in principle exhibit a zero-dispersion wavelength located anywhere between 1.74 μm (pure GeO_2) and 1.27 μm (pure SiO_2). On the negative side, the losses in these fibers are substantial, of the order of 20 dB km^{-1} at 1.85 μm for 75% GeO_2 in the core; the losses are primarily due to Rayleigh scattering.

2.8.2 Extruded fibers

Since $\gamma \propto n_2/A_{\text{eff}}$, in order to obtain a fiber with a large γ value: (i) one can use a material with a large n_2 value; (ii) one can use a guiding structure capable of supporting a mode with a small A_{eff} value.

For (i) highly nonlinear materials are needed that are suitable for making fibers by preferably more than one technique. It is also desirable that the material should exhibit low loss, but the higher n_2 is, the higher the α value that can be tolerated. For OPA work, the location of the material's ZDW is also important, although it can be shifted to some extent by suitable design of the waveguide.

For (ii) one has to design a waveguide that can provide a small A_{eff}. Above we discussed the step-index-fiber approach, which requires two different materials, one for the core and one for the cladding. Sometimes, when working with high-n_2 materials, it is difficult to find a suitable cladding material. In that case the only choice is to use air as a cladding, i.e. to make a single-material fiber [24]. One technique that has successfully been used recently is extrusion. In this approach the material is softened by heating and then pushed through a suitably shaped nozzle. It emerges as a fiber whose cross-section is a faithful replica of that of the nozzle. A design well suited for making fibers with small A_{eff} values has a very small core supported in air by a few thin blades extending to an outer cylinder that provides mechanical support for the whole structure, Fig. 2.1(c).

With this approach, A_{eff} values of the order of just a few square microns have been achieved. By using high-n_2 materials, values of γ in excess of 1000 have been obtained.

Note that, for extruded fibers, even though the core diameter is very small it may still support more than one mode, because the modes are well confined by the large index difference between the core material and the air cladding (i.e. V is large because NA is large).

2.8.3 Tapered fibers

Another approach to obtaining a small-diameter core with an air cladding is to make a so-called tapered fiber [19]. A common way of making such fibers is to start from a typical communication fiber, with a cladding diameter of 125 μm, and to heat it up past the softening point by moving a flame back and forth along a length of the order of 10 cm. Then, by slowly pulling the cold fiber ends, the central part of the fiber becomes elongated and its diameter shrinks. By this method it is possible to obtain fibers about 10 cm long having a diameter as small as 1 μm. It should be noted that when the cladding has reached that diameter, whatever remains of the original core plays essentially no part in confining the light: confinement occurs because of the large index difference between the cladding and the surrounding air, Fig. 2.1(d).

An interesting aspect of these fibers is that as a result of the fabrication process there is at each end of the small-core region a transition region where the cladding diameter increases gradually from 1 μm to 125 μm. Such a region constitutes a so-called fiber taper [25], an important component in fiber technology. Fiber tapers have the important property that when they are many wavelengths long, and are very gradual, they can transform fiber modes adiabatically, with very low loss. What this means is that, if a single mode is traveling in the core of the original fiber, after passing through a taper most of its energy has been automatically redistributed so that it is now in the mode of the 1-μm-diameter region. This reasoning works equally well in the opposite direction, for an up-taper. So altogether tapered fibers have the remarkable property that coupling in and out of a 1-μm-diameter region is not particularly difficult, because all one needs to do is couple in and out of the large taper ends, which are just standard fibers.

Tapered fibers can in principle be made out of any material that can be pulled into fibers. In particular, one could make tapered fibers out of high-n_2 materials. The combination of high n_2 and small A_{eff} that can be obtained in this manner would lead to large γ values.

In spite of their advantages, tapered fibers have not been used extensively in nonlinear optics, because of two limitations: (i) their length is limited to a few tens of centimeters; (ii) the glass is exposed to air and so the quality of the interface may deteriorate owing to environmental factors.

For OPA work, there is an additional difficulty, namely the need to have accurate control of the ZDW. Because of the small core size, however, the ZDW is very sensitive to core-radius variations, and the fabrication technique may not lend itself to making very uniform regions, with a well-defined diameter.

Because of the short lengths available, tapered fibers require high pump powers to display significant nonlinear effects. For this reason, nonlinear work with tapered fibers has been done exclusively with femtosecond lasers generating pulses with high peak power but with moderate average power [19, 26].

2.8.4 Nanowires

It is interesting to note that glass fibers with even smaller diameters have been made. The starting point is a tapered fiber, with a diameter of the order of 1 μm. It is then softened by heating, and its tip is pulled and wrapped around a small mandrel. The result is a fiber with a diameter as small as 60 nm, several centimeters long, a so-called nanowire, Fig. 2.1(e) [27]. Some very interesting experiments have been performed with nanowires, which feature coupling in and out through the ends, evanescent coupling between parallel nanowires, etc.

From the point of view of nonlinear optics, however, such fibers are not as attractive as tapered fibers, because actually only a small fraction of the power travels in the glass; most of the power travels in the air, as an evanescent wave. Hence the γ values of such fibers would be smaller that for a tapered fiber made from the same material. In addition, the coupling efficiency and available length are also less than those for tapered fibers, and dispersion control appears to be very limited. So altogether it appears that nanowires, in spite of their intriguing properties, are probably not good candidates for OPA work unless perhaps in conjunction with some other devices.

2.8.5 Microstructured fibers

In recent years much work has been done to develop new classes of fibers that have an array of holes around the core for confining the light, instead of a uniform low-index cladding as in conventional step-index fibers. This is illustrated in Fig. 2.1(f). One advantage is that such fibers can be made from a single material. They have a variety of names such as holey fibers (HFs), microstructured fibers, Bragg fibers, photonic bandgap fibers, or photonic crystal fibers (PCFs), which refer either to their physical structure or to the regime in which they operate. By changing the size, shape, and spatial distribution of the holes, in principle one can tailor most fiber properties. For fiber OPAs, the potential for achieving a small mode-field diameter (and hence for increasing γ) and for tailoring the dispersion are particularly attractive. Recently, HFs with $\gamma \approx 20$ W^{-1} km^{-1} and $\lambda_0 \approx 1550$ nm have become commercially available [28]; they can be used for OPA work in the C-band. Presently, their dispersion properties are not as good as those of conventional HNL-DSFs.

One advantage of HFs compared with step-index HNL-DSFs is that for fibers made from silica (which exhibits zero dispersion at 1.27 μm) it is possible to use the strong waveguide dispersion to shift the fiber λ_0 to *below* 1.27 μm and in fact well into the visible region, i.e. below 700 nm [29, 30]. This is in contrast with HNL-DSFs based on the classical step-index design, for which the fiber λ_0 can only be *above* 1.27 μm. Thus HFs may provide the only possible solution at the present time for making fiber OPAs that require λ_0 below 1.27 μm.

Another advantage of HFs is that they can also be endlessly single-mode, i.e. as the wavelength become shorter and shorter, the light continues to propagate in the fundamental mode and higher-order modes are not allowed to propagate. This is in contrast with step-index fibers, for which the second mode has a finite cutoff wavelength, below which it can propagate together with the fundamental mode. Endlessly single-mode fibers might present advantages over conventional fibers for making OPAs with very wideband characteristics, such as a broad gain bandwidth, extending over several hundred nanometers, or two narrow gain peaks several hundred nanometers apart [7]. In such situations, if the second mode could propagate on the short-wavelength side it would cause loss for the fundamental mode and could significantly alter the OPA's performance.

The large number of degrees of freedom available with these fibers indicates that dispersion could be tailored by appropriate design. So far a number of design strategies have been proposed for flattening the chromatic dispersion curve of HFs [31–36], and some have been demonstrated [37–39].

A complication arises with HFs designed to have a small A_{eff}. The coefficient γ has a fairly strong dependence on wavelength, because as the wavelength increases the mode has more overlap with air holes, which do not contribute to the nonlinearity. It has been calculated that, for a certain type of PCF, γ varies by 12.5% over a 60 nm wavelength range [40]. Considering that in some OPA applications the wavelength range can be as large as 400 nm, it appears that γ could vary considerably from one end of the gain spectrum to the other. While this may not prevent OPAs from providing a reasonable performance, it will have an impact on the shape of the gain spectrum, which will differ from that obtained by assuming that γ is constant over the entire range, as is currently done in most analytic solutions and even in numerical simulations. In such a situation it might be desirable to modify the basic model to include the wavelength dependence of γ, in order to obtain good agreement between theory and experiment.

2.8.6 Coupling to small-core fibers

In many practical situations (particularly for communication work) it is desirable to use communication-type single-mode fibers wherever possible, since they are commonly available and inexpensive. Also many types of standard components are compatible with them (couplers, connectors, isolators) and come equipped with pigtails that are also made from such fibers.

For this reason, fiber OPA work with DSFs is straightforward since the OPA fiber itself is of the same type as the rest of the system. As a result the OPA can be spliced into a system with very low insertion loss, of the order of 0.1 dB per splice.

Using a step-index HNLF is not quite as straightforward as using a DSF, because the HNLF has an MFD that is smaller by a factor 2–3 than that of a communication fiber. If a connector is used to butt-couple these two fibers, with no adjustment of the MFDs, the MFD mismatch leads to a loss that can be as large as 10 dB and is generally unacceptable. The mode mismatch can be alleviated in two ways.

1. **Thermally expanded core (TEC)** This refers to a method that uses heat to cause the germania in a fiber core to diffuse slowly into the surrounding cladding, thereby increasing the MFD. (Note that the TEC method does not modify the cladding diameter.) By using this method at the end of an HNL-DSF, one can in principle enlarge the mode of the latter until it matches that of a DSF. Inside the HNL-DSF the core shape changes slowly over a distance that may be from millimeters to centimeters, determined by the structure of the heating system, and the mode makes a slow transition from the HNL-DSF mode to that of the DSF. Hence such a section is called a mode transformer. When the transition takes place over a distance large compared with the wavelength, it is said to be adiabatic. An adiabatic transition can be almost perfect, in the sense that it transforms the shape of the mode with very little loss. Then splicing between a DSF and an HNL-DSF with TEC is essentially the same as between two DSFs and results in low loss. So overall the TEC method can in principle provide a low-loss connection between a DSF and an HNL-DSF. Note that this method can provide low loss even when a connector is used, if a good mode match is achieved.

   On the negative side, the TEC method is not simple to implement in most laboratories. To alleviate this, several companies offer to manufacture to order TEC fibers for customers, who need only to specify the two MFDs that need to be matched; the company then provides a suitable mode transformer. This approach may be convenient for manufacturing commercial products.

2. **Fusion splice** For those who do not wish to deal with TEC, an alternative seems to work just as well for typical fibers. It simply relies on making a fusion splice between the two dissimilar fibers without TEC preparation at all. It has been found in practice that by suitable setting of the discharge parameters, one can end up with a splice loss as low as 0.05 dB. To accomplish this, however, it is necessary to monitor the power of a reference laser transmitted through the splice. Using a basic discharge current and duration, one keeps repeating the same discharge. With some practice, one finds that the loss keeps decreasing, perhaps at a fairly slow rate, and eventually reaches a low level. It may take up to 50 discharges to accomplish this, but in general it can be done. While this procedure has not been thoroughly investigated, the general consensus is that the repeated discharges in effect create a TEC core in both fibers and eventually make the MFDs in the two fibers essentially the same at the junction, thereby leading to very low mode-mismatch loss.

The couplings described above, between a DSF and a step-index HNL-DSF, are relatively straightforward because the two fibers are not very dissimilar. They are basically made from the same silica material and are both of step-index type; hence they differ primarily through the core dimensions. The fact they are both made from silica means that they can easily be spliced to each other. The step-index structure makes it possible for the core to diffuse into the cladding and therefore to equalize the MFDs.

With other types of nonlinear fiber, one or more of these two basic features is or are lost, which makes low-loss coupling much more difficult. First, if the nonlinear fiber is not made from silica then fusion splicing to a silica-based fiber is all but impossible, owing to differences in softening point (an exception is BiO_2 fibers). Second, if the fiber structure is not of the step-index type then neither mode-matching method described above is feasible and so one must find another method; this is true for the various extruded fibers and HFs discussed above.

Microscope objective lenses can be used for coupling into small-core fibers; generally, however, the coupling efficiency is not very high. Another approach, more compatible with coupling to communication fibers, is to use an auxiliary piece of fiber, a mode transformer made either by TEC or by tapering. If TEC is used, one can start from a small-core fiber such as HNL-DSF and expand its MFD on one side to match that of the communication fiber. The small MFD on the other side will then be a better match for that of the highly nonlinear fiber. Since splicing is not possible, one must align the two small cores very accurately. If the fibers used all have a common diameter, such as 125 μm, accurate alignment by means of a V-groove may be possible.

By using this technique, a coupling loss of just 0.5 dB has been obtained for coupling from a communication fiber to an extruded TeO_2 fiber with core diameter 2 μm.

References

1. "Efficient conversion of light over a wide spectral range by four-photon mixing in a multimode graded-index fiber," Hill, K. O., Johnson, D. C., Kawasaki, B. S. *Appl. Opt.*; 1981; vol. 20, pp. 1075–9.

2. "200-nm-bandwidth fiber optical amplifier combining parametric and Raman gain," Ho, M.-C., Uesaka, K., Marhic, M. E., Akasaka, Y., Kazovsky, L. G. *J. Lightwave Technol.*; 2001; vol. 19, pp. 977–81.
3. "Nondestructive position-resolved measurement of the zero-dispersion wavelength in an optical fiber," Eiselt, M., Jopson, R. M., Stolen, R. H. *J. Lightwave Technol.*; 1997; vol. 15, pp. 135–43.
4. "Recent advances in the design and experimental implementation of fiber optical parametric amplifiers," Marhic, M. E., Wong, K. K., Kalogerakis, G., Kazovsky, L. G. In *Proc. Conf. on Passive Components and Fiber-based Devices*, APOC 2004, Beijing, China; *Proc. SPIE*; vol. 5623, pp. 691–704.
5. "Zero-dispersion wavelength mapping in short single-mode optical fibers using parametric amplification," Mussot, A., Lantz, E., Durecu-Legrand, A., Simonneau, C., Bayart, D., Sylvestre, T., Maillotte, H. *IEEE Photon. Technol. Lett.*; 2006; vol. 18, pp. 22–4.
6. "Broadband wavelength conversion over 193-nm by HNL-DSF improving higher-order dispersion performance," Hirano, M., Nakanishi, T., Okuno, T., Onishi, M. In *Proc. 31st European Conf. on Optical Communication*, September 2005, Glasgow, UK; vol. 6, pp. 43–4.
7. "Wide-band tuning of the gain spectra of one-pump fiber optical parametric amplifiers," Marhic, M. E., Wong, K. K. Y., Kazovsky, L. G. *IEEE J. Selected Topics in Quantum Electron.*; 2004; vol. 10, pp. 1133–41.
8. "Shifts in zero dispersion wavelength due to pressure, temperature and strain in dispersion shifted single-mode fibres," Byron, K. C., Bedgood, M. A., Finney, A., McGauran, C., Savory, S., Watson, I. *Electron. Lett.*; 1992; vol. 28, pp. 1712–4.
9. "Increase of the SBS threshold in a short highly nonlinear fiber by applying a temperature distribution," Hansryd, J., Dross, F., Westlund, M., Andrekson, P. A., Knudsen, S. N. *J. Lightwave Technol.*; 2001; vol. 19, pp. 1691–7.
10. "Temperature control of the gain spectrum of fiber optical parametric amplifiers," Wong, K. K. Y., Marhic, M. E., Kazovsky, L. G. *Optics Express*; 2005; vol. 13, pp. 4666–73.
11. "Broadband and flat parametric amplifiers with a multisection dispersion-tailored nonlinear fiber arrangement," Provino, L., Mussot, A., Lantz, E., Sylvestre, T., Maillotte, H. *J. Opt. Soc. Amer. B*; 2003; vol. 20, pp. 1532–7.
12. "UV processing of highly nonlinear fibers for enhanced supercontinuum generation," Westbrook, P. S., Nicholson, J. W., Feder, K., Yablon, A. D. In *Proc. Optical Fiber Communication Conf.*, February 2004, Los Angeles CA; postdeadline paper PDP27.
13. "Phase matching in birefringent fibers," Stolen, R. H., Bosch, M. A., Lin, C. *Opt. Lett.*; 1981; vol. 6, pp. 213–5.
14. *Principles of Optics*, Born, M., Wolf, E. Pergamon, Oxford, 1970.
15. "Polarization optics of twisted single-mode fibers," Ulrich, R., Simon, A. *Appl. Opt.*; 1979; vol. 18, pp. 2241–51.
16. "80 Gb/s to 10 Gb/s polarization-insensitive demultiplexing with circularly polarized spun fiber in a two-wavelength nonlinear optical loop mirror," Lou, J. W., Jepsen, K. S., Nolan, D. A., Tarcza, S. H., Bouton, W. J., Evans, A. F., Islam, M. N. *IEEE Photon. Technol. Lett.*; 2000; vol. 12, pp. 1701–3.
17. "Simplified phenomenological model for randomly birefringent strongly spun fibers," Galtarossa, A., Palmieri, L., Schenato, L. *Opt. Lett.*; 2006; vol. 31, pp. 2275–7.
18. "Polarization-insensitive asymmetric four-wave mixing using circularly polarized pumps in a twisted fiber," Tanemura, T., Katoh, K., Kikuchi, K. *Optics Express*; 2005; vol. 13, pp. 7497–505.

19. "Enhanced self-phase modulation in tapered fibers," Dumais, P., Gonthier, F., Lacroix, S., Bures, J., Villeneuve, A., Wigley, P. G. J., Stegeman, G. I. *Opt. Lett.*; 1993; vol. 18, pp. 1996–8.
20. "Two-photon absorption as a limitation to all-optical switching," Mizrahi, V., DeLong, K. W., Stegeman, G. I., Staifi, M. A., Andrejco, M. J. *Opt. Lett.*; 1989; vol. 14, pp. 1140–2.
21. "Large nonlinear phase shifts in low-loss $Al_xGa_{1-x}As$ waveguides near half-gap," Ho, S. T., Soccolich, C. E., Islam, M. N., Hobson, W. S., Levi, A. F. J., Slusher, R. E. *Applied Phys. Lett.*; 1991; vol. 59, pp. 2558–60.
22. "Highly nonlinear optical fiber for all optical processing applications," Holmes, M. J., Williams, D. L., Manning, R. J. *IEEE Photon. Technol. Lett.*; 1995; vol. 7, pp. 1045–7.
23. "Four-wave mixing with large Stokes shifts in heavily Ge-doped silica fibers," Yatsenko, Y. P., Pryamikov, A. D., Mashinsky, V. M., Likhachev, M. E., Mavritsky, A. O., Dianov, E. M., Guryanov, A. N., Khopin, V. F., Salgansky, M. Y. *Opt. Lett.*; 2005; vol. 30, pp. 1932–4.
24. "The fabrication and modelling of non-silica microstructured optical fibres," Hewak, D. W., West, Y. D., Broderick, N. G. R., Monro, T. M., Richardson, D. J. In *Proc. Optical Fiber Communication Conf., Technical Digest*, March 2001, Anaheim CA; vol. 2, pp. TuC4-1–3.
25. "Launching into single mode optical fibre waveguide," Stern, J. R., Dyott, R. B. In *Proc. Conf. on Trunk Telecommunications by Guided Waves*, September 1970, London UK; pp. 191–96.
26. "Soliton self-frequency shift in a short tapered air–silica microstructure fiber," Liu, X., Xu, C., Knox, W. H., Chandalia, J. K., Eggleton, B. J., Kosinski, S. G., Windeler, R. S. *Opt. Lett.*; 2001; vol. 26, pp. 358–60.
27. "Subwavelength-diameter silica wires for low-loss optical wave guiding," Tong, L., Gattass, R. R., Ashcom, J. B., He, S., Lou, J., Shen, M., Maxwell, I., Mazur, E. *Nature*; 2003; vol. 426, pp. 816–9.
28. www.crystal-fibre.com.
29. "Visible continuum generation in air–silica microstructure optical fibers with anomalous dispersion at 800 nm," Ranka, J. K., Windeler, R. S., Stentz, A. J. *Opt. Lett.*; 2000; vol. 25, pp. 25–7.
30. "Anomalous dispersion in photonic crystal fiber," Knight, J. C., Arriaga, J., Birks, T. A., Ortigosa-Blanch, A., Wadsworth, W. J., Russell, P. S. J *IEEE Photon. Technol. Lett.*, 2000; vol. 12, pp. 807–9.
31. "Designing the properties of dispersion-flattened photonic crystal fibers," Ferrando, A., Silvestre, E., Andres, P. *Optics Express*; 2001; vol. 9, pp. 687–97.
32. "Designing a photonic crystal fibre with flattened chromatic dispersion," Ferrando, A., Silvestre, E., Miret, J. J., Monsoriu, J. A., Andres, M. V., Russell, P. S. *Electron. Lett.*; 1999; vol. 35, pp. 325–7.
33. "Nearly zero ultraflattened dispersion in photonic crystal fibers," Ferrando, A., Silvestre, E., Miret, J. J., Andres, P. *Opt. Lett.*; 2000; vol. 25, pp. 790–2.
34. "A novel ultraflattened dispersion photonic crystal fiber," Wu, T. L., Chao, C. H. *IEEE Photon. Technol. Lett.*; 2005; vol. 17, pp. 67–9.
35. "An efficient approach for calculating the dispersions of photonic-crystal fibers: design of the nearly zero ultra-flattened dispersion," Wu, T. L., Chao, C. H. *J. Lightwave Technol.*; 2005; pp. 2055–61.
36. "Chromatic dispersion control in photonic crystal fibers: application to ultra-flattened dispersion," Saitoh, K., Koshiba, M., Hasegawa, T., Sasaoka, E. *Optics Express*; 2003; vol. 11, pp. 843–52.

37. "Demonstration of ultra-flattened dispersion in photonic crystal fibers," Reeves, W., Knight, J., Russell, P., Roberts, P. *Optics Express*; 2002; vol. 10, pp. 609–13.
38. "Dispersion-flattened photonic crystal fibers at 1550 nm," Reeves, W., Knight, J., Russell, P., Roberts, P., Mangan, B. In *Proc. Optical Fiber Communication Conf. Technical Digest*, March 2003, Atlanta GA; vol. 2, pp. 696–7.
39. "Fully dispersion controlled triangular-core nonlinear photonic crystal fiber," Hansen, K. P., Folkenberg, J. R., Peucheret, C., Bjarklev, A. In *Proc. Optical Fiber Communication Conf.*, Postdeadline Papers, March 2003, Atlanta GA; vol. 3, pp. PD2-1-3.
40. "Impact of the wavelength dependence of the mode field on the nonlinearity coefficient of PCFs," Hainberger, R., Watanabe, S. *IEEE Photon. Technol. Lett.*; 2005; vol. 17, pp. 70–2.

3 Scalar OPA theory

3.1 Introduction

In this chapter we introduce the basic types of fiber optical parametric amplifier (OPA) and simple models to study their characteristics. We will confine ourselves to situations in which all the waves are launched into the fiber with the same linear state of polarization (SOP) and remain in that state along the entire fiber. This allows us to consider a single component of the electric field and hence to write scalar equations for it. We first set up the basic OPA equations starting from Maxwell's equations for the case of nonlinear polarization. In the process we derive an expression for the fiber nonlinearity coefficient γ in terms of the waveguide properties and those of the interacting modes. We then proceed with the solution of the OPA equations in a variety of situations for which exact solutions are known. The solutions in the absence of loss and pump depletion are relatively simple, being expressible in terms of exponentials, or alternatively in terms of sinh and cosh functions. They are used extensively in practice as a first approximation to calculating the gain spectra of various fiber OPAs. Solutions in other regimes are more complicated; some involve Bessel functions, others involve Jacobian elliptic functions, etc. While it is more difficult to grasp their properties, they can be useful computational tools for obtaining accurate results when the exponential solutions are not applicable.

We have made an attempt to present a fairly exhaustive list of known solutions, in order to provide a compact source for such solutions for researchers who need to evaluate OPA performance. We have also presented complete derivations, so that the interested reader can follow them step by step and thereby gain physical and mathematical insight into the origin of the solutions. Those who do not have the time or the inclination to follow the proofs can of course skip the details and go directly to the final results. The most commonly used formulas have been placed in Appendix 5 for easy reference.

3.2 Types of fiber OPA

We begin by introducing the various types of fiber OPA and related devices. We also describe some of their fundamental properties resulting from energy conservation, as well as the quantum features of four-wave mixing (FWM) interactions.

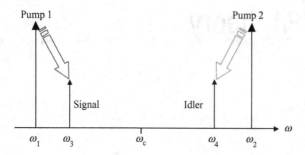

Fig. 3.1 Frequency assignments for a two-pump OPA.

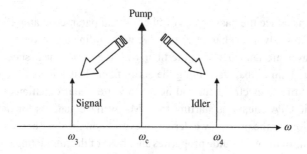

Fig. 3.2 Frequency assignments for a one-pump OPA.

3.2.1 Two-pump and one-pump OPAs

We first consider a two-pump fiber OPA. At the fiber input we introduce pumps, with angular frequencies ω_1 and ω_2, and a signal at ω_3. Generally the pumps have much higher power than the signal, and this is represented by the lengths of the vertical arrows in Fig. 3.1. At the fiber output, a new wave emerges along with the first three. It is called the idler, and its frequency ω_4 is such that

$$\omega_1 + \omega_2 = \omega_3 + \omega_4, \tag{3.1}$$

i.e. ω_4 and ω_3 are symmetric with respect to the center frequency $\omega_c = (\omega_1 + \omega_2)/2 = (\omega_3 + \omega_4)/2$, which is halfway between the two pump frequencies.

(In reality, many other frequency components are also generated in this process, but in a suitably designed OPA they should be much smaller because of phase mismatch; we will thus neglect them as a first approximation, to simplify our model.)

If we let the pump frequency spacing go to zero, Fig. 3.1 is replaced by Fig. 3.2. The pump frequency now coincides with the center frequency ω_c, halfway between signal and idler, i.e. $\omega_3 + \omega_4 = 2\omega_p = 2\omega_c$.

(This notation for the four frequencies is now fairly standard in the field, and we will adhere to it throughout most of the book. In some situations we will use other conventions, but we will make it clear when we do.)

3.2.2 Photon interpretation: Manley–Rowe relations

If we multiply both sides of Eq. (3.1) by $h/2\pi$, where h is Planck's constant, and let $\nu_k = \omega_k/2\pi$ be the optical frequency of the kth wave, $k = 1$–4, we obtain $h\nu_1 + h\nu_2 = h\nu_3 + h\nu_4$. Since $h\nu_k$ is the energy of one photon of the kth wave, we can interpret this relation as expressing the conservation of photon energy and photon number: each nonlinear interaction is such that each pump loses one photon and the signal and idler each gain one photon. (For a one-pump OPA the conclusion is similar except that the pump loses photons in pairs.)

This conclusion, arrived here by a simple argument, can in fact be shown to be correct by studying OPAs at the quantum-mechanical level (see Chapter 9). This simple photon description has some immediate consequences: since the signal and idler simultaneously gain photons, it follows that, for OPAs with a large gain, the signal and idler waves will have virtually identical temporal photon distributions. Hence, OPAs can generate pairs of waves that are virtually identical copies of each other (except for differing wavelengths), down to the level of individual photons. This remarkable property can be exploited for a variety of interesting classical and quantum-mechanical effects (see Chapter 9).

Another important consequence of this photon energy conservation rule is that it introduces basic relations between the rates at which the various waves grow or decay. Let us consider a fiber in which the loss and Raman gain can be neglected, so that the only way in which the three waves can exchange energy is via the mechanism described above. Let $P_k(z)$, $k = 1$–4, denote the power of the kth wave at a distance z from the fiber input. Then the photon picture implies that we must have the equations

$$\frac{dP_1}{dz} = \frac{dP_2}{dz} = -\frac{dP_3}{dz} = -\frac{dP_4}{dz}. \tag{3.2}$$

By combination and integration, this yields the relations

$$P_3(z) - P_4(z) = P_3(0) - P_4(0), \tag{3.3}$$

$$P_1(z) - P_2(z) = P_1(0) - P_2(0), \tag{3.4}$$

$$P_1(z) + P_3(z) = P_1(0) + P_3(0), \tag{3.5}$$

$$P_2(z) + P_4(z) = P_2(0) + P_4(0), \tag{3.6}$$

$$P_1(z) + P_4(z) = P_1(0) + P_4(0), \tag{3.7}$$

$$P_2(z) + P_3(z) = P_2(0) + P_3(0), \tag{3.8}$$

which reveal the existence of six related invariants for propagation along the fiber. Relations of this type are known as Manley–Rowe relations [1] and play an important role in many areas of nonlinear optics.

It is important to note that Eq. (3.1) must hold regardless of the local properties of the fiber, such as dispersion and birefringence. Therefore Eq. (3.1), as well as its consequences, Eqs. (3.3)–(3.8), must hold in any type of fiber, even those exhibiting random longitudinal variations of dispersion and birefringence.

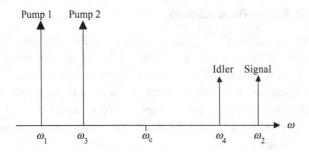

Fig. 3.3 Frequency assignments for wavelength exchange.

3.2.3 Wavelength exchange

An interesting variation on the two-pump OPA arrangement is obtained by switching the roles (that is the relative magnitudes) of the waves at ω_2 and ω_3, as shown in Fig. 3.3: the second pump is now at ω_3, while the signal is at ω_2. Now we have one pump and the signal symmetric with respect to ω_c, and the other pump and the idler also symmetric with respect to ω_c. Because this change does not alter the relative values of the frequencies themselves, Eq. (3.1) is still valid.

In terms of photon energies, we now have $h\nu_1 + h\nu_3 = h\nu_2 + h\nu_4$. This can be interpreted by saying that the pumps exchange photons between themselves, and so do the signal and idler. As a result, the signal and idler cannot receive energy from the pumps leading to gain. The most that can happen is that all the signal (idler) photons end up at the idler (signal), i.e. that there is an exchange of photons between signal and idler.

From a communication standpoint, this arrangement has the remarkable property that if the signal and idler are present at the input and are modulated by different signals, at the output the modulation can be exchanged between the two wavelengths. For this reason, when this phenomenon occurs we refer to it as *wavelength exchange*. Actually this term does not capture the richness of the phenomenon because the powers of the signal and idler are also exchanged, along with their actual quantum states. Perhaps it would be better to refer to this arrangement as "photon exchange."

(For simplicity, at times we will classify wavelength exchange under the general heading of optical parametric amplification in the rest of the text, although this does not mean that it amplifies with a gain greater than unity.)

3.2.4 Degenerate OPAs

The case of a fully degenerate OPA is obtained by considering a one-pump OPA and letting the signal and idler frequencies become equal to the pump frequency. Since all the waves are now indistinguishable in the nonlinear fiber, it is necessary to introduce some other means to combine them before they pass through the fiber and to separate them afterwards. This is generally done by using some type of interferometer, where the signal and pump can be fed into different fiber inputs; the output waves can be similarly separated.

By definition, degenerate OPAs have little or no bandwidth and so can only have a limited role in optical communication. They can be used, however, for manipulating quantum states of light; in particular they can generate "squeezed" states, which are finding applications in high-precision interferometric sensing.

3.2.5 Polarization aspects

So far we have used the frequency assignments of pump(s), signal, and idler to classify parametric processes in fibers into four basic categories.

However, light propagating in fibers is also characterized by its state of polarization. In principle the three or four waves that interact may have distinct SOPs, which will influence device performance. Thus within each of the four basic classes there will be a number of OPA types, characterized by different SOPs for the waves. The simplest, and probably the most useful, type of OPA is one in which all waves are in the same state of linear polarization, everywhere along the fiber. For these reasons, we will consider such a system as we begin our detailed investigations of the OPA equations. After that we will investigate types of OPA that correspond to different SOPs for the waves and compare their properties with those of linearly polarized OPAs; this will be done in the next chapter.

3.3 Derivation of the OPA equations and γ

We start with Maxwell's equations in a non-magnetic material,

$$\nabla \times \vec{E} = -\mu_0 \frac{\partial \vec{H}}{\partial t}, \tag{3.9}$$

$$\nabla \times \vec{H} = \varepsilon_0 \frac{\partial \vec{E}}{\partial t} + \vec{J}, \tag{3.10}$$

where $\vec{E}$ and $\vec{H}$ are the electric and magnetic field vectors, $\vec{J}$ is the current density vector, given by $\vec{J} = \partial \vec{P}/\partial t$, and $\vec{P}$ is the polarization density vector.

Taking the curl of Eq. (3.9) and making use of the identity

$$\nabla \times \nabla \times \vec{E} = \nabla(\nabla \cdot \vec{E}) - \nabla^2 \vec{E} = -\nabla^2 \vec{E}$$

in a charge-free medium, which we assume, we obtain

$$\nabla^2 \vec{E} - \frac{1}{c^2} \frac{\partial^2 \vec{E}}{\partial t^2} = \mu_0 \frac{\partial^2 \vec{P}}{\partial t^2}. \tag{3.11}$$

We now make the assumption that the vectors $\vec{E}$ and $\vec{P}$ are parallel to a fixed direction everywhere; therefore we only need to consider their scalar components along that direction, which we denote by E and P.

The polarization density P is related to E by

$$P = \varepsilon_0 \left(\chi^{(1)} E + \chi^{(3)} E^3 \right), \tag{3.12}$$

where $\chi^{(m)}$, $m = 1, 3$, is the mth-order susceptibility; $\chi^{(1)}$ governs the linear properties of the medium, i.e. the dispersion, while $\chi^{(3)}$ is responsible for the third-order nonlinear effects on which fiber OPAs are based. For now we can think of $\chi^{(3)}$ as being a constant, i.e. independent of frequency.

Let us first consider the case $\chi^{(3)} = 0$. When E consists of a single frequency component at ω, i.e. when

$$E(x, y, z, t) = E_s(x, y, z)e^{-i\omega t},$$

Eq. (3.11) leads to Helmholtz' equation

$$\nabla^2 E_s + K^2 E_s = \left(\frac{\partial^2}{\partial x^2} + \frac{\partial^2}{\partial y^2} + \frac{\partial^2}{\partial z^2}\right) E_s + K^2 E_s = 0, \quad (3.13)$$

where $K = n\omega/c$ is the (local) propagation constant (wavevector) and $n = \sqrt{1 + \chi^{(1)}}$ is the refractive index at ω. Note that, in a fiber, n must be a function of transverse position as this is necessary to confine the mode.

For a cylindrical waveguide, such as the fibers in which we are interested, n is a function only of the transverse coordinates x and y and not of the distance z along the fiber. In this case the solutions of Eq. (3.13), which are the fiber modes, are of the form

$$E_s(x, y, z) = \psi(x, y)e^{i\beta z}, \quad (3.14)$$

where β is the longitudinal propagation constant; $\psi(x, y)$ represents the transverse mode profile and is the solution of

$$\left(\frac{\partial^2}{\partial x^2} + \frac{\partial^2}{\partial y^2}\right) \psi + (K^2 - \beta^2)\psi = 0. \quad (3.15)$$

The dependence of β on ω is obtained by solving Eq. (3.15) in detail. The complete mode field dependence on space and time is then

$$E(x, y, z, t) = \psi(x, y)e^{i\beta z - i\omega t}. \quad (3.16)$$

We now consider the case where $\chi^{(3)} \neq 0$ and E consists of four frequency components, which satisfy Eq. (3.1). We write the total real electric field as

$$E = \frac{1}{2}\sum_{k=1}^{4}[B_k(z)\psi_k(x, y)e^{i\beta_k z - i\omega_k t} + \text{c.c.}], \quad (3.17)$$

where $B_k(z)$ is the slowly varying envelope of the mode at ω_k and is the main quantity of interest in the following. By contrast, $\psi_k(x, y)$ is known and therefore we will eventually remove it from our calculations. The equations for the $B_k(z)$ are obtained by substituting Eq. (3.17) into Eq. (3.11) and keeping only those terms on both sides that are at the frequency ω_k. As an example, for $B_1(z)$ we obtain

$$(\nabla^2 + K_1^2)\left[\psi_1 B_1 e^{(i\beta_1 z - i\omega_1 t)}\right]$$
$$= -\frac{3\chi^{(3)}\omega_1^2}{4c^2}\{[\psi_1^2 B_1 B_1^* + 2(\psi_2^2 B_2 B_2^* + \psi_3^2 B_3 B_3^* + \psi_4^2 B_4 B_4^*)]\psi_1 B_1 e^{i\beta_1 z - i\omega_1 t}$$
$$+ 2\psi_2\psi_3\psi_4 B_2^* B_3 B_4 e^{i(-\beta_2+\beta_3+\beta_4)z - i\omega_1 t}\}, \quad (3.18)$$

where we have assumed for simplicity that $\chi^{(3)}$ is independent of frequency. This can be rewritten as

$$-i\psi_1 \frac{dB_1}{dz} = \frac{3\chi^{(3)}\omega_1}{8c\bar{n}_1} \left\{ \left[\psi_1^2 B_1 B_1^* + 2\left(\psi_2^2 B_2 B_2^* + \psi_3^2 B_3 B_3^* + \psi_4^2 B_4 B_4^*\right) \right] \psi_1 B_1 \right.$$
$$\left. + 2\psi_2\psi_3\psi_4 B_2^* B_3 B_4 e^{i\Delta\beta z} \right\}, \tag{3.19}$$

where $\bar{n}_1 = \beta_1 c/\omega_1$ is the mode's effective index of refraction, and

$$\Delta\beta = \beta_3 + \beta_4 - \beta_1 - \beta_2 \tag{3.20}$$

is the propagation constant (wavevector) mismatch.

To arrive at this form, we have assumed that

$$\left| \frac{d^2 B_1}{dz^2} \right| \ll 2\beta_1 \left| \frac{dB_1}{dz} \right|,$$

which can be shown to be equivalent to the statement that the envelope of the wave varies very little over one wavelength. This is the slowly varying envelope approximation (SVEA), which is well satisfied in many cases of practical interest.

We now want to eliminate the x and y dependences from Eq. (3.19), to obtain an equation for A_1 that depends only on z. To do this we select the light propagating in the lowest-order (or other desired) transverse mode, by making use of the orthogonality of the transverse modes. In lossless waveguides these modes are orthogonal, in the sense that

$$\int\int \vec{E}^{(t,a)} \times \left[\vec{H}^{(t,b)}\right]^* dx dy = 0 \quad \text{if} \quad \beta_a \neq \beta_b, \tag{3.21}$$

where $\vec{E}^{(t,a)}$ and $\vec{H}^{(t,b)}$ are respectively the transverse electric and magnetic fields associated with transverse modes a and b, and β_a and β_b are the corresponding longitudinal propagation constants [2]. This form should be used in the most general case, where the field has two transverse components and the magnetic components are not simply related to the electric components. This could for example be the case for fibers designed to have a fundamental mode with mode field diameter (MFD) of the order of 1 μm. We will perform a detailed calculation involving arbitrary polarization states in Chapter 4.

For now, however, we will assume that we are dealing with larger MFDs, for which the transverse electric field can be assumed to be primarily linearly polarized along the x-axis while the magnetic field is linearly polarized along the y-axis. In this case Maxwell's equations show that

$$H_y = -\frac{\beta}{\omega\mu_0} E_x. \tag{3.22}$$

Hence the orthogonality condition (3.21) is now equivalent to

$$\int\int E^{(t,a)} \left(E^{(t,b)}\right)^* dx dy = \int\int \psi^{(a)} \left(\psi^{(b)}\right)^* dx dy = 0. \tag{3.23}$$

Thus if we multiply both sides of Eq. (3.19) by $\psi_1^*(x, y)$, integrate with respect to x and y, and divide by $D_1 = \int\int |\psi_1|^2 dx dy$ we can eliminate x and y and select on both sides the complex field amplitude field for the transverse mode of interest. This procedure transforms Eq. (3.19) into

$$-i\frac{dB_1}{dz} = \left[\delta_{1111} B_1 B_1^* + 2\left(\delta_{1212} B_2 B_2^* + \delta_{1313} B_3 B_3^* + \delta_{1414} B_4 B_4^*\right)\right] B_1$$
$$+ 2\delta_{1234} B_2^* B_3 B_4 e^{i\Delta\beta z}, \tag{3.24}$$

where

$$\delta_{jklm} = \frac{3\omega_j}{8c\bar{n}_j} \frac{\iint \chi^{(3)} \psi_j^* \psi_k^* \psi_l \psi_m \, dx\, dy}{D_j} \tag{3.25}$$

and

$$D_j = \iint |\psi_j|^2 dx\, dy. \tag{3.26}$$

When studying the nonlinear optics of guided waves, it is customary to introduce scaled mode field amplitudes A_k such that $A_k A_k^* = P_k$, the average power in the mode at ω_k. Imposing this condition introduces changes in the coefficients in Eq. (3.24), which can be calculated as follows.

The irradiance (power per unit area) at x and y for the field in the mode at ω_k is $I_k = c\varepsilon_0 \bar{n}_k \langle E_k^2 \rangle$, where the angle brackets represent a time average and $\bar{n}$ is the effective refractive index for mode k. Since

$$E_k^2 = \tfrac{1}{4}\left[B_k \psi_k e^{i\beta_k z - i\omega_k t} + \text{c.c.}\right]^2$$
$$= \tfrac{1}{4}\left[2 B_k B_k^* \psi_k \psi_k^* + B_k^2 \psi_k^2 e^{2i\beta_k z - 2i\omega_k t} + \left(B_k^* \psi_k^*\right)^2 e^{-2i\beta_k z + 2i\omega_k t}\right], \tag{3.27}$$

we find that

$$\langle E_k^2 \rangle = \tfrac{1}{2} B_k B_k^* \psi_k \psi_k^* \tag{3.28}$$

and

$$P_k = \iint I_k \, dx\, dy = \tfrac{1}{2} c\varepsilon_0 \bar{n}_k B_k B_k^* \iint \psi_k \psi_k^* \, dx\, dy$$
$$= \tfrac{1}{2} c\varepsilon_0 \bar{n}_k D_k B_k B_k^*$$
$$= A_k A_k^*. \tag{3.29}$$

Hence we have

$$B_k = A_k \sqrt{\frac{2}{c\varepsilon_0 \bar{n}_k D_k}}. \tag{3.30}$$

Substituting into Eq. (3.24) we find that

$$-i\frac{dA_1}{dz} = \left[\gamma_{1111} A_1 A_1^* + 2\left(\gamma_{1212} A_2 A_2^* + \gamma_{1313} A_3 A_3^* + \gamma_{1414} A_4 A_4^*\right)\right] A_1$$
$$+ 2\gamma_{1234} A_2^* A_3 A_4 e^{i\Delta\beta z}, \tag{3.31}$$

where

$$\gamma_{jklm} = \frac{3\mu_0 \omega_j}{4} \frac{\int\int \chi^{(3)} \psi_j^* \psi_k^* \psi_l \psi_m \, dx\, dy}{\sqrt{\bar{n}_j \bar{n}_k \bar{n}_l \bar{n}_m} D_j D_k D_l D_m}. \tag{3.32}$$

If $\chi^{(3)}$ has a significant frequency dependence, i.e. if $\chi^{(3)} = \chi^{(3)}(-\omega_j; -\omega_k, \omega_l, \omega_m)$ [3], then the coefficients γ_{jklm} are all different. Even if $\chi^{(3)}$ is independent of frequency, if the frequencies are sufficiently different that their mode profiles are different then the γ_{jklm} are still different.

When the frequency spacings are not very large, however, it is often a good approximation to assume that $\chi^{(3)}$ is independent of frequency and that all four mode profiles are essentially the same. If in addition $\chi^{(3)}$ depends only weakly on position, it can be pulled out of the integral. Then all the γ_{jklm} have the same value, which can be written as

$$\gamma = \frac{3\mu_0 \omega \chi^{(3)}}{8\bar{n}^2 A_{\text{eff}}}, \tag{3.33}$$

where ω is a nominal or average value for the angular frequency; A_{eff} is the effective area, defined as

$$A_{\text{eff}} = \frac{\left[\int\int (\psi^*\psi)\, dx\, dy\right]^2}{\int\int (\psi^*\psi)^2 dx\, dy}.$$

As discussed in Chapter 2, A_{eff} is generally of the same order as the area of the core.

It can be shown that this expression for γ is equivalent to that given in [4], provided that an appropriate choice is made for the nonlinear index n_2 in [4]. The parameter most commonly used to describe nonlinear effects in fibers is γ. For work with fiber OPAs, it is convenient to express γ in W^{-1} km^{-1}. Typical communication fibers have $\gamma \approx 2$ W^{-1} km^{-1} and HNL-DSFs have $\gamma \approx 20$ W^{-1} km^{-1}.

Assuming a single value for all the γ_{jklm} greatly simplifies the form of equations such as Eq. (3.31), and therefore this assumption is made most of the time. However, one should bear in mind that when investigating OPAs or wavelength converters operating over very wide wavelength ranges, it may lead to inaccurate results.

At any rate, when a single value of γ can be used it can be shown that for a two-pump OPA the propagation equations for the slowly varying envelopes A_k of the four waves are of the form

$$\frac{dA_l}{dz} = i\gamma \left[|A_l|^2 A_l + 2\sum_{j \neq l=1}^{4} |A_j|^2 A_l + 2 A_m A_n A_k^* e^{i\varepsilon \Delta \beta z} \right], \quad l = 1\text{--}4. \tag{3.34}$$

The three terms on the right-hand side of Eq. (3.34) respectively correspond to:

1. the interaction of one wave with itself, which leads to self-phase modulation (SPM);
2. the interaction between two waves, which leads to cross-phase modulation (XPM);
3. the interaction between four distinct waves, i.e. non-degenerate four-wave mixing (FWM).

The integers $k, l, m,$ and n are such that in Eq. (3.34):

for $\varepsilon = 1$, if $l = 1$ or 2 then $k = 3 - l, m = 3, n = 4;$ \hfill (3.35)

for $\varepsilon = -1$, if $l = 3$ or 4 then $k = 7 - l, m = 1, n = 2.$

As before, $\Delta\beta = \beta_3 + \beta_4 - \beta_1 - \beta_2$.

One-pump OPA

For a one-pump OPA, we remove the second pump and consider the interactions between only three distinct waves. Their frequencies satisfy $\omega_3 + \omega_4 = 2\omega_1$. Following the same procedure as above, it can be shown that Eq. (3.34) is now replaced by

$$\frac{dA_1}{dz} = i\gamma \left[|A_1|^2 A_1 + 2 \sum_{j\neq 1}^{4} |A_j|^2 A_1 + 2A_3 A_4 A_1^* e^{i\Delta\beta z} \right], \quad (3.36)$$

$$\frac{dA_l}{dz} = i\gamma \left[|A_l|^2 A_l + 2 \sum_{j\neq l=1}^{4} |A_j|^2 A_l + (A_1)^2 A_k^* e^{-i\Delta\beta z} \right],$$

$$l = 3, 4, \quad k = 7 - l, \quad (3.37)$$

with $\Delta\beta = \beta_3 + \beta_4 - 2\beta_1$. The first two terms of Eqs. (3.36) and (3.37) are similar to those in Eq. (3.34). The third term of Eq. (3.36) corresponds to nondegenerate FWM, but that in Eq. (3.37) corresponds to degenerate FWM.

3.4 Scaling laws

Above, we have used the variables A_k, $\Delta\beta$, and z, as they lead to results that are directly interpretable in terms of the actual system parameters. In physics, however, it is often useful to use dimensionless variables as these can bring out the similarities between different systems. We can do this for fiber OPAs by introducing the new dimensionless variables [5]

$$a_k = \frac{A_k}{\sqrt{P_0}}, \quad k = 1\text{--}4, \quad \zeta = \gamma P_0 z, \quad S = \frac{\Delta\beta}{\gamma P_0}, \quad (3.38)$$

where P_0 is the total power of the signal, pump(s), and idler. With these, Eq. (3.34) for a two-pump OPA becomes

$$\frac{da_k}{d\zeta} = i \left(2 - |a_k|^2\right) a_k + 2i a_m a_n a_l^* e^{i\varepsilon S\zeta}, \quad k = 1\text{--}4, \quad (3.39)$$

where we have used the fact that $\sum_{k=1}^{4} |a_k|^2 = 1$. A similar set of equations would hold for a one-pump OPA, and so the following will hold for either type of OPA.

Equation (3.39) is a function only of the dimensionless variables a_k, ζ, and S. Let us assume that a particular solution exists, of the form

$$a_k = f_k(S, \zeta), \quad k = 1\text{--}4. \quad (3.40)$$

We can show that this dimensionless solution can in principle correspond to an infinite family of physical realizations, represented by the original dimensional variables A_k, $\Delta\beta$, and z. To see this, we consider two OPAs, respectively denoted by unprimed and primed quantities, and assume that the following relations exist:

$$a_k = \frac{A_k}{\sqrt{P_0}} = \frac{A'_k}{\sqrt{P'_0}}, \qquad (3.41)$$

$$S = \frac{\Delta\beta}{\gamma P_0} = \frac{\Delta\beta'}{\gamma' P'_0}, \qquad (3.42)$$

$$\zeta = \gamma P_0 z = \gamma' P'_0 z'. \qquad (3.43)$$

Clearly, if the unprimed quantities correspond to a solution of Eq. (3.43) then so do the primed quantities. We will say that two OPAs satisfying these conditions are similar.

An important consequence of Eq. (3.39) is that similar OPAs have exactly the same gain, at the respective distances z and z'. If we define the signal field gain as $h_3(z) = a_3(S, \zeta)/a_3(S, 0) = A_3(S, \zeta)/A_3(S, 0)$, and similarly for the idler field gain, it can be shown that

$$h_3(z) = h'_3(z') \quad \text{and} \quad h_4(z) = h'_4(z'). \qquad (3.44)$$

Since this is true for any z, two equivalent OPAs have exactly the same gain profile along the fibers, within a linear scaling factor of the distance along the fiber. In particular, they have the same output gain provided that the fiber lengths L and L' satisfy Eq. (3.43), or $L'/L = \gamma P_0/(\gamma' P'_0)$.

It should be noted that this conclusion is valid even if pump depletion occurs. In this case, however, the input signal levels are not arbitrary but themselves should be scaled according to Eq. (3.41). Nevertheless, when pump depletion is negligible, i.e. when we are in the linear-gain region, the gain is independent of signal input power; therefore the gain profiles can be similar, regardless of the input signal levels.

There are a number of ways in which Eqs. (3.42) and (3.43) can be satisfied mathematically. To see this, we rewrite these two equations as

$$\frac{\Delta\beta}{\Delta\beta'} = \frac{\gamma P_0}{\gamma' P'_0} = \frac{z'}{z}. \qquad (3.45)$$

For two different fibers, with different (positive) ratios γ/γ' and $\Delta\beta/\Delta\beta'$ at a given frequency, we can always adjust pump powers and fiber lengths so that Eq. (3.45) is satisfied. This can be interpreted as follows: if we choose P_0/P'_0 so that Eq. (3.41) is satisfied then the gain profiles along the two will be the *same* if the distances along the fibers are themselves scaled according to Eq. (3.45). Since only the ratio P_0/P'_0 is fixed by Eq. (3.45), we can find many operating conditions that will correspond to such similar field distributions in the fibers.

While these scaling laws are of theoretical interest, they are not particularly interesting in practice, for the following reason. If two different fibers are compared, the first equality of Eq. (3.45) will generally be satisfied by pairs of frequency spacings $\Delta\omega_s$ and $\Delta\omega'_s$ with no simple relation between them, because of the different dispersion properties of the fibers. Under these circumstances the scaling laws will be useful only for obtaining identical gain profiles as a function of z at a given frequency, not for obtaining related spectral shapes in the frequency domain at fixed z.

In Chapter 5 we will show that under certain circumstances it is also possible to find OPAs that have similar gain spectra, i.e. gain spectra that are actually identical within a suitable scaling of the frequency axis.

It should be noted that the above scaling laws are not applicable unless the four waves considered in Section 3.2 are the only ones involved in the nonlinear interaction. In practice these four waves will normally generate additional waves by FWM. If such waves are not negligible, the set of FWM equations has to be enlarged beyond the original four and correspondingly additional $\Delta\beta$'s must be introduced. These new $\Delta\beta$'s will be different from the original ones and therefore it is unlikely that proper scaling of all $\Delta\beta$'s with pump power can be achieved simultaneously; this makes it virtually impossible to find two OPAs with similar gain spectra.

3.5   Solution of the two-pump OPA equations

In this section we proceed with the solution of the four basic OPA equations arrived at in the preceding section. We will first consider the general case where pump depletion is allowed, which leads to solutions in terms of Jacobian elliptic functions. We then proceed with the simpler case where pump depletion is negligible, which leads to simpler solutions, in terms of exponential functions. We will also consider the case of loss, wavelength exchange, and a six-wave model useful under particular circumstances.

3.5.1   Pump depletion

We will assume that the pump powers may vary with distance owing to depletion. Since it is then possible that the signal and idler powers will become comparable with the pump powers, it is not possible to neglect any terms in Eq. (3.34). We have the set of propagation equations

$$-i\frac{dA_l}{dZ} = |A_l|^2 A_l + 2 \sum_{j \neq l=1}^{4} |A_j|^2 A_l + 2 A_m A_n A_k^* e^{i\varepsilon \Delta\beta Z/\gamma}, \qquad l = 1\text{--}4, \qquad (3.46)$$

where $Z = \gamma z$. The solution of this set of equations was obtained by Chen and Snyder [6], following closely the formalism developed in [1]. Here we present a shortened version of this solution. It is simplified by our use of γ, and our assumption that it is independent of frequency. In addition, throughout the derivation we will use as unknowns the powers of the waves, instead of their square roots as is customary in this field [6, 1]. We also do not normalize the powers. This has several advantages: (i) the powers have a straightforward physical meaning; (ii) we can keep track of the dimensions of various terms; (iii) since powers are generally the quantities of interest in practice, and are always calculated at the end of the derivation, it may be preferable to use them throughout rather than to introduce temporary intermediate variables; (iv) we save on notation and conversions; (v) the derivation is slightly shorter; (vi) programming of the solution on a computer is also clarified and shortened by avoiding the use of multiple related variables and conversion between them.

3.5 Solution of the two-pump OPA equations

Let us define θ_k as the phase of A_k, so that $A_k = \sqrt{P_k}\, e^{i\theta_k}$. Multiplying Eq. (3.46) by A_l^*, we obtain

$$-i\frac{dA_l}{dZ}A_l^* = P_l(2P_0 - P_l) + 2\sqrt{P_1 P_2 P_3 P_4}\, e^{i\varepsilon\theta}, \quad l = 1\text{--}4, \tag{3.47}$$

where $\theta = \Delta\beta z + \theta_3 + \theta_4 - \theta_1 - \theta_2$. Equation (3.47) leads to

$$\frac{dA_l}{dZ}A_l^* + \frac{dA_l}{dZ}A_l^* = \frac{dP_l}{dZ} = 2i\sqrt{P_1 P_2 P_3 P_4}\left(e^{i\varepsilon\theta} - e^{-i\varepsilon\theta}\right)$$

$$= -4\varepsilon\sqrt{P_1 P_2 P_3 P_4}\sin\theta \tag{3.48}$$

or

$$\frac{dP_1}{dZ} = \frac{dP_2}{dZ} = -\frac{dP_3}{dZ} = -\frac{dP_4}{dZ} = -4\sqrt{P_1 P_2 P_3 P_4}\sin\theta. \tag{3.49}$$

Equation (3.49) implies that the powers of the four waves are of the form

$$\begin{aligned}P_1 &= P_1(0) - x, \quad P_2 = P_2(0) - x,\\ P_3 &= P_3(0) + x, \quad P_4 = P_4(0) + x,\end{aligned} \tag{3.50}$$

i.e. that they all go up or down by x. This is the same conclusion as was reached in subsection 3.2.2 by looking at the interaction in terms of individual photons. Because of its fundamental role, we will use x as one of the unknowns in solving the equations.

From Eq. (3.47) we also obtain

$$i\left(\frac{dA_l^*}{dZ}A_l - \frac{dA_l}{dZ}A_l^*\right) = 2P_l(2P_0 - P_l) + 4\sqrt{P_1 P_2 P_3 P_4}\cos\theta,$$
$$l = 1\text{--}4, \tag{3.51}$$

which leads to

$$\frac{d\theta_l}{dZ} = 2P_0 - P_l + \frac{2}{P_l}\sqrt{P_1 P_2 P_3 P_4}\cos\theta. \tag{3.52}$$

Then

$$\frac{d\theta}{dZ} = \frac{\Delta\beta}{\gamma} + P_1 + P_2 - P_3 - P_4$$
$$+ 2\sqrt{P_1 P_2 P_3 P_4}\left(P_3^{-1} + P_4^{-1} - P_1^{-1} - P_2^{-1}\right)\cos\theta. \tag{3.53}$$

We also have

$$\frac{d\theta}{dZ} = \frac{d\theta}{dx}\frac{dx}{dZ} = \frac{d\theta}{dx} 4\sqrt{P_1 P_2 P_3 P_4}\sin\theta, \tag{3.54}$$

and hence

$$\frac{d\theta}{dZ} = -4\sqrt{P_1 P_2 P_3 P_4}\frac{d(\cos\theta)}{dx}. \tag{3.55}$$

Substituting in Eq. (3.53) and using the fact that $dP_l/dx = -\varepsilon$ leads to

$$\frac{\Delta\beta}{\gamma} + P_1 + P_2 - P_3 - P_4 + 4\frac{d}{dx}\left(\sqrt{P_1 P_2 P_3 P_4}\cos\theta\right) = 0. \tag{3.56}$$

Now substituting Eq. (3.50) into Eq. (3.56), and integrating both sides with respect to x, we obtain

$$\left(\frac{\Delta\beta}{\gamma} + \Delta P_0\right)x - 2x^2 + 4\sqrt{P_1 P_2 P_3 P_4}\cos\theta = K_2, \quad (3.57)$$

where $\Delta P_0 = P_{10} + P_{20} - P_{30} - P_{40}$ and P_{l0} is the initial power of the lth wave. The definition of the constant K_2 is slightly different from that of C in [3]. Since it is independent of x, K_2 can be evaluated at $x = 0$, which yields

$$K_2 = 4\sqrt{P_{10} P_{20} P_{30} P_{40}}\cos(\theta_{30} + \theta_{40} - \theta_{10} - \theta_{20}), \quad (3.58)$$

where θ_{l0} is the initial phase of the lth wave. We can now eliminate θ in Eq. (3.49) by using Eq. (3.57) to obtain an equation relating only x and Z. We have

$$\frac{dx}{dZ} = 4\sqrt{P_1 P_2 P_3 P_4}\sin\theta = 4\sqrt{P_1 P_2 P_3 P_4(1 - \cos^2\theta)} = 2\sqrt{h(x)}, \quad (3.59)$$

where $h(x)$ is a fourth-order polynomial in x:

$$h(x) = 4(P_{10} - x)(P_{20} - x)(P_{30} + x)(P_{40} + x)$$
$$- \frac{1}{4}\left[K_2 - \left(\frac{\Delta\beta}{\gamma} + \Delta P_0\right)x + 2x^2\right]^2. \quad (3.60)$$

This can be rewritten as

$$dZ = \frac{dx}{2\sqrt{h(x)}}. \quad (3.61)$$

The solution $x(Z)$ of Eq. (3.61) is known to be expressible in terms of Jacobian elliptic functions [7]. It is obtained in terms of the four roots of $h(x)$, listed in increasing order, i.e. $\eta_1 < \eta_2 < \eta_3 < \eta_4$. Note that if there is no idler at the input then $P_{40} = 0$ and $x = 0$ is a root of $h(x)$.

Since the coefficient of x^4 in $h(x)$ is $C_0^2 = 3$, x oscillates between η_2 and η_3. Specifically,

$$x(Z) = \eta_1 + (\eta_2 - \eta_1)\left[1 - \eta\operatorname{sn}^2\left(\frac{Z + Z_0}{Z_c}, k\right)\right]^{-1}, \quad (3.62)$$

where sn denotes a Jacobian elliptic function,

$$Z_c^{-1} = |C_0|\left[(\eta_3 - \eta_1)(\eta_4 - \eta_2)\right]^{1/2},$$

$$\eta = \frac{\eta_3 - \eta_2}{\eta_3 - \eta_1} \quad Z_0 = Z_c F\left(\sin^{-1}\left[\frac{P_{40} - \eta_2}{\eta(P_{40} - \eta_1)}\right], k\right), \quad (3.63)$$

and $F(\phi, k)$ is a standard elliptical integral [7].

Complete pump depletion
As an application, let us see whether it is possible to obtain complete pump depletion with a two-pump fiber OPA. This means that, at a certain distance, we would like the entire pump power to be transferred to the signal and idler (in equal amounts, as per described

in subsection 3.2.2). Then necessarily $(P_{10} + P_{20})/2$ must be a root of $h(x) = 0$; hence

$$16 \left(\frac{P_{10} - P_{20}}{2} \right) \left(\frac{P_{20} - P_{10}}{2} \right) \left(P_{30} + \frac{P_{10} + P_{20}}{2} \right)$$
$$= \left(\frac{P_{10} + P_{20}}{2} \right) \left(P_{10} + P_{20} - \frac{\Delta \beta}{\gamma} - \Delta P_0 \right)^2. \quad (3.64)$$

Since the left-hand side and the right-hand side have opposite signs, Eq. (3.64) can be satisfied only if they both vanish. For the left-hand side this implies that $P_{10} = P_{20}$, which is easy to achieve. For the right-hand side also to vanish, we must have $\Delta \beta = \gamma P_{30}$; whether this can be achieved in practice depends on fiber dispersion, signal and pump wavelengths, etc.

If these conditions are satisfied, we find that $\eta_3 = \eta_4 = P_{10}$. Also, $k = 1$, which implies that sn becomes tanh; then $x(Z)$ is no longer periodic but approaches P_{10} asymptotically. Thus, while complete depletion is in principle possible, mathematically it is achieved only at an infinite distance. In this case we find that $\{\eta_1, \eta_3, \eta_4\} = \{-4P_{30}/3, P_{10}, P_{10}\}$ and that x can be written as

$$x(Z) = -\frac{4P_{30}}{3} \left\{ 1 - \left[1 - \frac{3P_{10}}{3P_{10} + 4P_{30}} \tanh^2 \left(Z\sqrt{(3P_{10} + 4P_{30})P_{10}} \right) \right]^{-1} \right\}.$$

This expression is useful in that it shows that if Z is a few times larger than $Z_c = 1/\sqrt{(3P_{10} + 4P_{30})P_{10}}$, nearly complete depletion can be obtained.

An example that could be of interest for high-power wavelength conversion has the following parameters: $\gamma = 20$ W^{-1} km^{-1}, $P_{10} = P_{20} = 2$ W, and $P_{30} = 1$ W. This corresponds to $z_c = Z_c/\gamma \approx 11$ m. So we expect that nearly complete pump depletion can be obtained after a few tens of meters. Figure 3.4 shows the graph of $x(z) = P_4(z)$ for this case. We see that indeed the idler power becomes very close to the initial power of one pump after 50 m. In fact, if a somewhat smaller conversion efficiency is acceptable then one can use an even shorter fiber: for example, 90% conversion efficiency is achieved in about 26 m.

3.5.2 Cubic Cardano solution

We assume that $P_{40} = 0$. Then $g(x) = h(x)/x$ is a third-order polynomial. It can be written in the form $g(x) = ax^3 + bx^2 + cx + d$, where

$$a = 3,$$
$$b = 3(P_{30} - P_{10} - P_{20}) + \frac{\Delta \beta}{\gamma},$$
$$c = 4(P_{10}P_{20} - P_{10}P_{30} - P_{20}P_{30}) - \frac{1}{4}\left(\frac{\Delta \beta}{\gamma} + P_{10} + P_{20} - P_{30} \right)^2,$$
$$d = 4P_{10}P_{20}P_{30}. \quad (3.65)$$

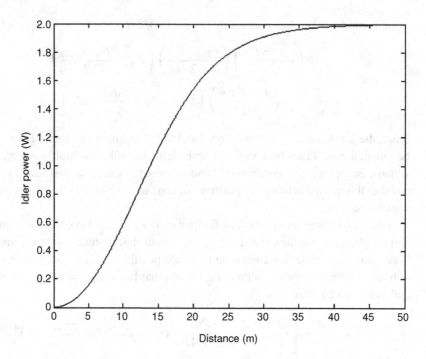

Fig. 3.4 Plot of idler power versus distance. The idler power saturates when it reaches nearly 50% of the total initial pump power.

Letting $b' = b/a$, $c' = c/a$, $d' = d/a$, and $y = x + w = x - b'/3$, we obtain

$$g(x) = l(y) = a(y^3 + 3py - 2q), \quad (3.66)$$

where

$$p = \frac{1}{3}(3w^2 + 2b'w + c') \quad \text{and} \quad q = -\frac{1}{2}(w^3 + b'w^2 + c'w + d'). \quad (3.67)$$

The three roots of $l(y) = 0$ are then given by Cardano's formula [8]

$$y_1 = -\frac{p}{\Delta} + \Delta, \quad y_{2,3} = \frac{p}{2\Delta}(1 \pm i\sqrt{3}) - \frac{\Delta}{2}(1 \mp i\sqrt{3}), \quad (3.68)$$

where

$$\Delta = \left(q + \sqrt{p^3 + q^2}\right)^{1/3}.$$

The three roots of $g(x) = 0$ are given by $x_1 = y_1 - b'/3$, $x_2 = y_2 - b'/3$, $x_3 = y_3 - b'/3$. Finally, $\{\eta_1, \eta_3, \eta_4\}$ is obtained by listing x_1, x_2, x_3 in increasing order.

Of interest are the following.

1. Two of the roots, η_3 and η_4, coincide when $p^3 + q^2 = 0$. It can be shown mathematically that this corresponds to $\Delta\beta = \Delta\beta_0$, i.e. to a $\Delta\beta$ value suitable for maximum pump depletion. This makes physical sense because η_3 reaches its maximum value when it equals η_4, which must correspond to maximum pump depletion.

2. When the OPA is well optimized for gain flatness, the sn function is very close to 1 and therefore $P_4 \approx \eta_3$. This can be a very useful approximation, because the obtaining η_3 involves only elementary functions and avoids the calculation of sn altogether.

3.5.3 No pump depletion

It is often desirable to operate fiber OPAs in a regime where the signal and idler are kept small compared with the pumps, in order to ensure nearly constant gain as a function of signal level. This also avoids pump depletion and therefore possible cross-gain modulation (XGM) in WDM systems. In this regime, it is thus appropriate to assume that the pumps are not depleted and that their powers remain constant. We will study this case by going back to the basic equations, simplifying them, and solving the new set of equations. This is simpler than trying to take the limit of the Jacobian elliptic solution. (In addition, the solution given in subsection 3.5.1 does not have a closed form for phases, but the linearized solution does.)

(a) Pump equations

To set up the pump equations, we assume that the signal and idler remain much smaller than the pumps. Hence the pump powers are not depleted and we let $P_k(z) = P_k(0) = P_k$, $k = 1, 2$. The pumps interact only through the first two terms of Eq. (3.34), i.e. through SPM and XPM. We obtain

$$\frac{dA_k}{dz} = i\gamma(P_k + 2P_l)A_k = ip_k A_k, \qquad k = 1, 2, \quad l = 3 - k, \qquad (3.69)$$

where $p_k = \gamma(P_k + 2P_l)$. The solution of Eq. (3.69) is

$$A_k(z) = A_k(0)e^{ip_k z}, \qquad k = 1, 2. \qquad (3.70)$$

(b) Signal and idler equations

Signal and idler SPM and XPM between the signal and idler are neglected. Then the equations begin with two XPM terms due to the two pumps, which are similar to the second term on the right-hand side of Eq. (3.69). In addition, there is an FWM term coupling signal and idler.

The equations for A_k take the form

$$\frac{dA_k}{dz} = i\gamma(2P_1 + 2P_2)A_k + 2i\gamma A_1 A_2 A_l^* e^{-i\Delta\beta z}$$

$$= 2i\gamma(P_1 + P_2)A_k + ir_k A_l^* e^{i(p_1+p_2-\Delta\beta)z}$$

$$= ip_k A_k + ir_k e^{iq_k z} A_l^*, \qquad (3.71)$$

where

$$p_k = 2\gamma(P_1 + P_2) = 2\gamma P_0, \qquad q_k = p_1 + p_2 - \Delta\beta,$$
$$r_k = 2\gamma A_1(0)A_2(0), \qquad k = 3, 4, \quad l = 7 - k. \qquad (3.72)$$

The total pump power is $P_0 = P_1 + P_2$ and r_k is the amplitude of the FWM coupling coefficient between the four waves. Since $r_3 = r_4$, we simply denote both by r.

It is useful to note for future reference that p_k, $k = 1$–4, is a real, positive, linear function of the pump powers. It has an important physical interpretation as the rate of increase of the phase shift due to the pump SPM and/or XPM experienced by the kth wave. It has the dimensions of a propagation constant, and we can expect that it should add to (or subtract from) the linear propagation constant β_k.

Equation (3.71) has the same form as Eq. (A1.5) in Appendix 1, and so we can use the general solution provided there directly. To do so we calculate $\kappa_k = p_k + p_l^* - q_k = p_3 + p_4 - p_1 - p_2 + \Delta\beta$, which has the same value for $k = 3$ and 4; hence we simply denote it by κ. The quantity can be interpreted as the total propagation constant or wavevector mismatch, $\kappa = \Delta\beta + \Delta\beta_{NL}$, by introducing $\Delta\beta_{NL} = p_3 + p_4 - p_1 - p_2 = \gamma P_0$, as the nonlinear wavevector mismatch.

Following Appendix 1, we then make the change of variable

$$C_k = A_k \exp\left[i\left(\frac{\kappa}{2} - p_k\right)z\right], \quad k = 3, 4, \quad l = 7 - k. \tag{3.73}$$

This leads to an equation for C_k:

$$\frac{d^2 C_k}{dz^2} - g^2 C_k = 0, \quad k = 3, 4, \tag{3.74}$$

where g is the parametric gain coefficient and is given by

$$g^2 = r^2 - \left(\frac{\kappa}{2}\right)^2. \tag{3.75}$$

The output fields are then obtained as linear combinations of e^{gz} and e^{-gz}. Making use of the initial conditions, we find that

$$C_k(z) = A_k(0)\cosh(gz) + \frac{i}{g}\left[\frac{\kappa}{2}A_k(0) + r A_l^*(0)\right]\sinh(gz),$$
$$k = 3, 4, \quad l = 7 - k. \tag{3.76}$$

This result is similar to that first arrived at in [9] for a one-pump OPA.

Let us now consider the important case where there is no idler at the input, i.e. $A_4(0) = 0$. In this case the power gain is independent of the initial phases of the pumps and the signal. One may then assume for simplicity that all initial phases are equal to zero, so that $A_k(0) = \sqrt{P_k}$, $k = 1, 2$; r is then real. We will make this assumption for the rest of this subsection. Then the signal and idler gains are respectively

$$h_3(z) = \frac{A_3(z)}{A_3(0)} = \left[\cosh(gz) + i\frac{\kappa}{2g}\sinh(gz)\right]\exp\left[i\left(p_3 - \frac{\kappa}{2}\right)z\right], \tag{3.77}$$

$$h_4(z) = \frac{A_4(z)}{A_3^*(0)} = i\frac{r}{g}\sinh(gz)\exp\left[i\left(p_4 - \frac{\kappa}{2}\right)z\right]. \tag{3.78}$$

Introducing the signal-to-idler power conversion efficiency $G_i(z) = |h_4(z)|^2$ and the signal power gain $G_s(z) = |h_3(z)|^2$, we have

$$G_i(z) = \left|\frac{r}{g}\sinh(gz)\right|^2 \tag{3.79}$$

and
$$G_s(z) = G_i(z) + 1. \tag{3.80}$$

Equation (3.80) is simply a consequence of the fact that the signal and idler simultaneously gain photons (subsection 3.2.2.). We note that, since $g_{\max} = r$, the maximum values of G_i and G_s always have the simple forms

$$G_{i,\max}(z) = |\sinh(rz)|^2 \quad \text{and} \quad G_{s,\max}(z) = |\cosh(rz)|^2. \tag{3.81}$$

When the gain is large, we have approximately $G_{s,\max}(z) \approx G_{i,\max}(z) \approx e^{2rz}/4$. Expressing the gains in decibels, we have $G_{s,\max}^{dB}(z) \approx G_{i,\max}^{dB}(z) \approx 17.2(\gamma\sqrt{P_1 P_2}z) - 6$. For a given total pump power $P_0 = P_1 + P_2$, the gains are a maximum for $P_1 = P_2 = P_0/2$; then

$$G_{s,\max}^{dB}(z) \approx G_{i,\max}^{dB}(z) \approx 8.64\Phi - 6, \tag{3.82}$$

where $\Phi = \gamma P_0 z$ is a measure of the pump nonlinear phase shift. This is a convenient expression, which allows us to estimate quickly the maximum gain of a fiber OPA. We note that the gain is a function only of the product $\gamma P_0 z$; hence we can trade fiber nonlinearity, length, and pump power to obtain a particular gain. As an example, if $\Phi = 3$ then the maximum gain is approximately 20 dB; this could be achieved with a 1-km-long dispersion-shifted fiber (DSF) with $\gamma \approx 2$ W^{-1} km^{-1} and a 1 W pump or, equivalently, with a 100-m-long HNL-DSF with $\gamma \approx 20$ W^{-1} km^{-1} and the same pump power.

Equations (3.75), (3.79), and (3.80) show that the small-signal gain spectrum of a fiber OPA depends only on three quantities, namely r, $\Delta\beta_{NL}$, and $\Delta\beta$. The wavevector mismatch $\Delta\beta$ is determined by the dispersion of the type of fiber used and plays a crucial role in determining the shape of the gain spectrum; its effect will be studied in Chapter 5. The nonlinear wavevector mismatch $\Delta\beta_{NL}$ determines at which wavelengths κ vanishes, and hence where maximum gain can be achieved. The latter is solely determined by r; r and $\Delta\beta_{NL}$ can be calculated without any knowledge of fiber dispersion since they depend only on the pump powers and γ.

We now consider the accuracy of the results obtained by neglecting pump depletion. Figure 3.5 shows plots of the idler output power of a two-pump OPA versus distance, calculated with and without pump depletion. The parameters used are the same as those for Fig. 3.4. We see that when the idler power reaches about 1 W, i.e. 50% of the input power of one pump, the model without pump depletion gives an idler power less than 10% larger than that predicted by the more accurate model including pump depletion. Hence the model that does not include pump depletion is in fact accurate over a large range of pump depletion, the accuracy being excellent at low idler output powers.

3.5.4 No pump depletion, loss

When the fiber loss in an OPA is fairly large, the normal expressions for OPA gain, which are obtained by neglecting fiber loss, become inapplicable. One must then go back to first principles to derive gain expressions applicable in the case of lossy fibers. Here we present a derivation that provides a solution for either one- or two-pump OPAs.

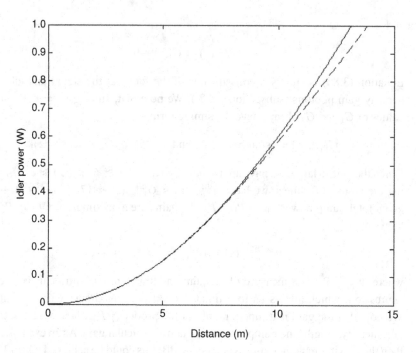

Fig. 3.5 Plots of idler power versus distance, with pump depletion either included (broken line) or not included (solid line).

The basic equations for one- and two-pump fiber OPAs are obtained by considering FWM interactions between three or four waves, respectively. We assume that the pump(s) are not depleted by the interaction but suffer only exponential attenuation due to distributed loss. The initial system of equations then reduces to two coupled equations for the signal and the idler. Eliminating one unknown leads to a second-order ordinary differential equation (ODE), which can be solved in terms of confluent hypergeometric functions [10].

We assume that the fiber nonlinearity and attenuation coefficients, γ and α, are frequency independent. For simplicity we take $P_{10} = P_{20}$ for the two-pump case (the solution is slightly more complicated as the pump powers are arbitrary). Then, for either one- or two-pump OPAs, the FWM equations for the signal ($k = 3$) and the idler ($k = 4$) are of the form

$$\frac{dA_k}{dz} = ipA_k + ir \exp\left(i \int_{\xi=0}^{z} q\,d\xi\right) A_l^*, \qquad k = 3, 4, \quad l = 7 - k, \qquad (3.83)$$

where A_k represents a complex slowly varying envelope, z is the distance along the fiber, and the asterisk indicates complex conjugation. The quantities p, q, and r are given by

$$p = i\alpha/2 + 2\gamma P_0 e^{-\alpha z}, \qquad q = i\alpha - \Delta\beta + (1 + n_p)\gamma P_0 e^{-\alpha z},$$
$$r = 2\gamma\sqrt{P_{10}P_{20}}, \qquad (3.84)$$

where P_{10} and P_{20} are the initial pump powers, $P_0 = P_{10} + P_{20}$, n_p is the number of pumps, and $\Delta\beta$ is the wavevector mismatch. These expressions are applicable for either one- or two-pump OPAs, provided that the convention $P_{10} = P_{20} = P_0/2$ is used for one-pump OPAs.

Following Appendix 1, we let

$$A_k = C_k \exp\left[\int_{\xi=0}^{z} \left(p - \frac{\kappa}{2}\right) d\xi\right]. \tag{3.85}$$

This leads to

$$\frac{d^2 C_k}{dz^2} - g^2(z) C_k = 0, \tag{3.86}$$

where $g^2(z) = r^2 e^{-2\alpha z} - \kappa^2/4 + i\kappa'/2$, $\kappa = p + p^* - q$, and the prime stands for a derivative with respect to z. It is found that $g^2(z)$ is of the form $g^2(z) = u e^{-2\alpha z} + v e^{-\alpha z} + w$, where

$$u = \gamma^2 \left[4 P_{10} P_{20} - \tfrac{1}{4}(3 - n_p)^2 P_0^2\right], \tag{3.87}$$
$$v = -(3 - n_p)\Delta\beta\gamma P_0/2, \qquad w = \tfrac{1}{4}(\alpha + i\Delta\beta)^2.$$

The initial conditions for C_k are

$$C_k(0) = A_k(0),$$
$$C'_k(0) = s(0) A_k(0)/2 + r A'_i(0).$$

Letting $x = e^{-\alpha z}$, we obtain

$$\frac{d^2 C_k}{dx^2} + \frac{1}{x}\frac{dC_k}{dx} - \left(\frac{u}{\alpha^2} + \frac{v}{\alpha^2 x} + \frac{w}{\alpha^2 x^2}\right) C_k = 0, \quad k-3,4. \tag{3.88}$$

Now letting $C_k = Y_k/\sqrt{x}$ yields

$$\frac{d^2 Y_k}{dx^2} + \left[-\frac{u}{\alpha^2} - \frac{v}{\alpha^2 x} + \left(\frac{1}{4} - \frac{w}{\alpha^2}\right)\frac{1}{x^2}\right] Y_k = 0. \tag{3.89}$$

Writing $x = Z/Z_0$, where $Z_0 = Z(0)$ is a constant, gives

$$\frac{d^2 Y_k}{dZ^2} + \left[-\frac{u}{\alpha^2 Z_0^2} - \frac{v}{\alpha^2 Z_0 Z} + \left(\frac{1}{4} - \frac{w}{\alpha^2}\right)\frac{1}{Z^2}\right] Y_k = 0. \tag{3.90}$$

Making the substitutions

$$Z_0 = \frac{2\sqrt{u}}{\alpha}, \quad \kappa = -\frac{v}{\alpha^2 Z_0}, \quad \mu = \frac{w^{1/2}}{\alpha}, \tag{3.91}$$

we finally obtain

$$\frac{d^2 Y_k}{dZ^2} + \left[-\frac{1}{4} + \frac{\kappa}{Z} + \left(\frac{1}{4} - \mu^2\right)\frac{1}{Z^2}\right] Y_k = 0, \tag{3.92}$$

which is Whittaker's equation [7].

Two independent solutions are the Whittaker confluent hypergeometric functions $M_{\kappa,\mu}(Z)$ and $M_{\kappa,\mu}(Z)$. We must then have $Y_k(Z) = U_k M_{\kappa,\mu}(Z) + V_k W_{\kappa,\mu}(Z)$, where U_k and V_k are constants to be determined from the initial conditions, which are

$$Y_k(Z_0) = C_k(0)$$
$$Y'_k(Z_0) = \frac{1}{Z_0}\left[\frac{1}{2} C_k(0) - \frac{1}{\alpha} C'_k(0)\right]. \tag{3.93}$$

The solution for the idler field ($k = 4$) is relatively simple if no idler is present at the input. Then the equations for U_4 and V_4 are

$$Y_4(Z_0) = U_4 M_{\kappa,\mu}(Z_0) + V_4 W_{\kappa,\mu}(Z_0) = 0, \tag{3.94}$$

$$Y_4'(Z_0) = U_4 M'_{\kappa,\mu}(Z_0) + V_4 W'_{\kappa,\mu}(Z_0) = -\frac{r}{2\sqrt{u}} A_3^*(0). \tag{3.95}$$

The solution is

$$U_4 = \frac{r}{2\sqrt{u}} A_3^*(0) \frac{W_{\kappa,\mu}(Z_0)}{W_{\text{Whit}}(Z_0)},$$

$$V_4 = -\frac{r}{2\sqrt{u}} A_3^*(0) \frac{M_{\kappa,\mu}(Z_0)}{W_{\text{Whit}}(Z_0)},$$

where $W_{\text{Whit}}(Z_0)$ is the Wronskian of the Whittaker functions $M_{\kappa,\mu}(Z)$ and $M_{\kappa,\mu}(Z)$ at Z_0, i.e.

$$W_{\text{Whit}}(Z_0) = M_{\kappa,\mu}(Z_0) W'_{\kappa,\mu}(Z_0) - W_{\kappa,\mu}(Z_0) M'_{\kappa,\mu}(Z_0).$$

It can be shown that $W_{\text{Whit}}(Z_0) = -\Gamma(b)/\Gamma(a)$, where $a = 1/2 + \mu - \kappa$ and $b = 1 + 2\mu$. We can then write

$$Y_4(Z) = \frac{r}{\alpha Z_0} \frac{\Gamma(a)}{\Gamma(b)} \left[W_{\kappa,\mu}(Z) M_{\kappa,\mu}(Z_0) - W_{\kappa,\mu}(Z_0) M_{\kappa,\mu}(Z) \right] A_3^*(0). \tag{3.96}$$

The idler power gain (or conversion efficiency) can now be calculated as

$$G_i(z) = \left| \frac{A_4(z)}{A_3^*(0)} \right|^2 = e^{-\alpha z} \left| \frac{Y_4(Z)}{A_3^*(0)} \right|^2$$

$$= e^{-\alpha z} \left| \frac{r}{\alpha Z_0} \frac{\Gamma(a)}{\Gamma(b)} \left[W_{\kappa,\mu}(Z) M_{\kappa,\mu}(Z_0) - W_{\kappa,\mu}(Z_0) M_{\kappa,\mu}(Z) \right] \right|^2. \tag{3.97}$$

An alternative expression can be obtained in terms of the Kummer functions M and U [7]. It is

$$G_i(z) = e^{-\alpha z} \left| \frac{r}{\alpha} \frac{\Gamma(a)}{\Gamma(b)} e^{-(Z+Z_0+\alpha b z)/2} Z_0^{b-1} \right.$$

$$\left. \times \left[U(a,b,Z) M(a,b,Z_0) - U(a,b,Z_0) M(a,b,Z) \right] \right|^2. \tag{3.98}$$

The advantage of this form for G_i is that it is expressed only in terms of M and U, which are confluent hypergeometric functions available in mathematical programs such as Mathematica [11] or Matlab [12]. By using the relationship between M and U, one can also write an expression for G_i entirely in terms of M (this is advantageous for taking certain limits, since the properties of M are better known than those of U). This yields

$$G_i(z) = \left| \frac{\pi r}{\alpha} \frac{e^{-(Z+Z_0)/2}}{\sin(\pi b) \Gamma(b) \Gamma(d)} \right.$$

$$\left. \times \left[M(c,d,Z) M(a,b,Z_0) x^{(3-b)/2} - M(c,d,Z_0) M(a,b,Z) x^{(1+b)/2} \right] \right|^2 \tag{3.99}$$

where $c = 1 + a - b$ and $d = 2 - b$.

The expression for $A_3(z)$ is more complicated than that for $A_4(z)$. However, the signal power gain $G_s(z) = |A_3(z)/A_3(0)|^2$ can be obtained simply from $G_i(z)$, since it can be shown from Eq. (3.83) that $G_s(z) = G_i(z) + e^{-\alpha z}$.

The advantage of these closed-form solutions is that they can be used for rapidly obtaining the shape of the gain spectrum under specified experimental conditions. This permits a rapid optimization in terms of pump power, gain bandwidth, etc.

3.5.5 Wavelength exchange

Two-pump optical parametric amplification and wavelength exchange are simply particular cases of four-wave mixing that involve only two pairs of waves, the frequencies of each pair being symmetric to those of the other pair with respect to the center frequency ω_c. Therefore the general formalism of subsection 3.5.1 is applicable to both two-pump OPA and wavelength exchange. A distinction between the two cases occurs only when specific assignments are given to pump, signal, and idler frequencies, i.e. when we specify which two input waves (the pumps) are considered to be much larger than the other two (the signal and idler). Therefore all the mathematical derivations up to Eq. (3.68) are common to both situations.

To be able to reuse the notation of subsection 3.5.1 we leave the frequency symmetries the same, as in Fig. 3.1. But we let ω_1 and ω_3 correspond to the pumps, ω_2 to the signal, and ω_4 to the idler. (*Note that this departs from our usual notation for the frequencies.*)

We can now use the results of subsection 3.5.1 to justify the statements that we made about wavelength exchange in subsection 3.2.3. Equation (3.50) gives

$$P_2(z) + P_4(z) = P_2(0) + P_4(0) \quad \text{and} \quad P_1(z) + P_3(z) = P_1(0) + P_3(0). \quad (3.100)$$

This states that the total pump power does not change along the fiber and neither does the sum of the signal power and idler power. This can be interpreted by saying that there is no exchange of power between the two pumps as a whole, and between the signal and idler as a whole. Hence the two pumps exchange power only between themselves, and so do the signal and idler.

To demonstrate that all the power from the signal can move to the idler wavelength, we must show that complete depletion of the signal can take place. We assume that there is no idler power at the input, i.e. that $P_{40} = 0$. Then necessarily P_{20} must be a root of $h(x) = 0$. This leads to the phase-matching condition

$$\Delta\beta = \gamma(P_{30} - P_{10} - 3P_{20}). \quad (3.101)$$

In practice, we can easily arrange to have equal pump powers, i.e. $P_{30} = P_{10}$. Furthermore, in communication applications the signal power is generally small, and therefore it can be considered as negligible in Eq. (3.101). Under these circumstances, Eq. (3.101) simply reduces to $\Delta\beta \approx 0$, which implies that *linear* phase matching is sufficient to obtain a complete transfer of signal power to the idler.

In general the sn functions are periodic, with half-wave symmetry, and so total power transfer will take place after half a period.

If the signal and idler are launched simultaneously then in this model it is difficult to separate the power transferred from the signal from the original idler power and thus to justify the notion of an exchange between signal and idler.

Therefore, to do this we proceed to the situation where the pump powers are much larger than those of the signal and idler, which allows us to assume that pump depletion is negligible and to proceed as in subsection 3.5.3.

The pumps are affected only by SPM and XPM, and their fields can be written as

$$A_1(z) = A_{10} \exp[i\gamma(P_{10} + 2P_{30})z], \quad (3.102)$$

$$A_3(z) = A_{30} \exp[i\gamma(P_{30} + 2P_{10})z]. \quad (3.103)$$

The signal and idler equations are then

$$\begin{aligned}
\frac{dA_2}{dz} &= i\gamma(P_{10} + P_{30})A_2 + 2i\gamma A_3 A_4 A_1^* e^{i\Delta\beta z} \\
&= i\gamma P_{p0} A_2 + 2i\gamma A_{30} A_{10}^* A_4 \exp[i(\Delta\beta + \gamma P_{10} - \gamma P_{30})z], \\
\frac{dA_4}{dz} &= i\gamma(P_{10} + P_{30})A_4 + 2i\gamma A_1 A_2 A_3^* e^{-i\Delta\beta z} \\
&= i\gamma P_{p0} A_4 + 2i\gamma A_{10} A_{30} A_2^* \exp[-i(\Delta\beta + \gamma P_{10} - \gamma P_{30})z],
\end{aligned} \quad (3.104)$$

where $P_{p0} = P_{10} + P_{30}$ is the total pump power, and $\Delta\beta = \beta_3 + \beta_4 - \beta_1 - \beta_2$. These equations have the form of Eq. (A1.5) in Appendix 1, with coefficients $p_2 = p_4 = \gamma P_{p0}$, $r_2 = 2\gamma A_{10}^* A_{30}$, $r_4 = 2\gamma A_{30}^* A_{10}$, and $q_2 = -q_4 = \Delta\beta + \gamma P_{10} - \gamma P_{30}$. From these we find the parametric gain coefficients through their squares,

$$g_2^2 = g_4^2 = g^2 = -4\gamma^2 P_{10} P_{30} - [\Delta\beta + \gamma(P_{10} - P_{30})]^2/4. \quad (3.105)$$

If we assume that $P_{10} = P_{30}$ and $\Delta\beta = 0$, we simply have $g = i\gamma P_{p0}$. As shown in Appendix 1, the complex amplitudes of signal and idler (to within phase factors) are then related by a unitary transfer matrix, i.e. we have

$$\begin{bmatrix} A_2(z) \\ A_4(z) \end{bmatrix} = \begin{bmatrix} \cos(\gamma P_{p0} z) & \sin(\gamma P_{p0} z) \\ -\sin(\gamma P_{p0} z) & \cos(\gamma P_{p0} z) \end{bmatrix} \begin{bmatrix} A_2(0) \\ A_4(0) \end{bmatrix}. \quad (3.106)$$

As a result, at the distance $L_{ex} = \pi/(2\gamma P_{p0})$ we have $A_2(L_{ex}) = A_4(0)$ and $A_4(L_{ex}) = -A_2(0)$, which shows that the two initial fields are exchanged and so are the corresponding powers. This now provides full justification for the exchange terminology used for naming this phenomenon [13]. Additional insights into its physical nature will be provided in Chapter 9.

3.5.6 Six-wave model

In the preceding we restricted our attention to the interaction between four waves only, namely, the two pumps, the signal, and the idler. In practice this can be well approximated

3.5 Solution of the two-pump OPA equations

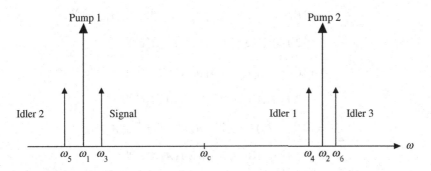

Fig. 3.6 Frequency assignments for a two-pump OPA, with two sidebands around each pump.

by placing the pumps and the signal appropriately with respect to each other and to the ZDW. Under these circumstances, these four waves carry most of the power over the entire fiber and other waves arising from four-wave mixing are poorly phase matched and remain at negligible levels.

Under certain circumstances, however, some of these other waves may be well phase matched and so may reach levels comparable with the signal and the idler. In such situations one must take these waves into account to obtain an adequate description of the situation. An example of such a situation occurs with a two-pump OPA when the signal is close in frequency to one of the pumps, as shown in Fig. 3.6. Then two new waves appear, located at ω_5 and ω_6. Their origin can be understood as follows.

Since the signal is close to pump 1, efficient four-wave mixing (FWM) between these two waves generates the wave at ω_5, which is symmetric to the signal with respect to the pump, i.e. we have $\omega_3 + \omega_5 = 2\omega_1$. Similarly, since idler 1 is close to pump 2, efficient FWM between these two waves generates the wave at ω_6, which is symmetric to idler 1 with respect to pump 2, i.e. we have $\omega_4 + \omega_6 = 2\omega_2$.

Since $\omega_3 + \omega_4 = 2\omega_c = \omega_1 + \omega_2$ we also have $\omega_5 + \omega_6 = 2\omega_c$, which shows that the two new idlers are themselves coupled by the two-pump OPA process. Furthermore, we also have $\omega_4 - \omega_5 \omega_6 - \omega_3 = \omega_2 - \omega_1$, which indicates that the signal and the third idler are coupled by a wavelength-exchange type of interaction, as are the first and second idlers. Because of all these tight couplings when the signal is close to a pump, the signal and the three idlers can grow together, with gains that are similar.

The six waves are coupled by the following system of equations:

$$-i\frac{dA_1}{dZ} = (2P_0 - P_1)A_1 + 2A_3A_4A_2^*e^{i\Delta\beta_{3412}z} + 2A_5A_6A_2^*e^{i\Delta\beta_{5612}z}$$
$$+ 2A_2A_3A_6^*e^{i\Delta\beta_{2316}z} + 2A_2A_5A_4^*e^{i\Delta\beta_{2514}z} + 2A_3A_5A_1^*e^{i\Delta\beta_{3511}z},$$

$$-i\frac{dA_2}{dZ} = (2P_0 - P_2)A_2 + 2A_3A_4A_1^*e^{i\Delta\beta_{3412}z} + 2A_5A_6A_1^*e^{i\Delta\beta_{5612}z}$$
$$+ 2A_1A_6A_3^*e^{i\Delta\beta_{1623}z} + 2A_1A_4A_5^*e^{i\Delta\beta_{1425}z} + 2A_4A_6A_2^*e^{i\Delta\beta_{4622}z},$$

$$-i\frac{dA_3}{dZ} = (2P_0 - P_3)A_3 + 2A_1A_2A_4^* e^{i\Delta\beta_{1234}z} + A_1^2 A_5^* e^{i\Delta\beta_{1135}z}$$
$$+ 2A_1A_6A_2^* e^{i\Delta\beta_{1623}z} + 2A_5A_6A_4^* e^{i\Delta\beta_{5634}z},$$

$$-i\frac{dA_4}{dZ} = (2P_0 - P_4)A_4 + 2A_1A_2A_3^* e^{i\Delta\beta_{1234}z} + A_2^2 A_6^* e^{i\Delta\beta_{2246}z}$$
$$+ 2A_2A_5A_1^* e^{i\Delta\beta_{2514}z} + 2A_5A_6A_3^* e^{i\Delta\beta_{5634}z},$$

$$-i\frac{dA_5}{dZ} = (2P_0 - P_5)A_5 + 2A_1A_2A_6^* e^{i\Delta\beta_{1256}z} + A_1^2 A_3^* e^{i\Delta\beta_{1135}z}$$
$$+ 2A_1A_4A_2^* e^{i\Delta\beta_{1425}z} + 2A_3A_4A_6^* e^{i\Delta\beta_{3456}z},$$

$$-i\frac{dA_6}{dZ} = (2P_0 - P_6)A_6 + 2A_1A_2A_5^* e^{i\Delta\beta_{1256}z} + A_2^2 A_4^* e^{i\Delta\beta_{2246}z}$$
$$+ 2A_2A_3A_1^* e^{i\Delta\beta_{2316}z} + 2A_3A_4A_5^* e^{i\Delta\beta_{3456}z}, \tag{3.107}$$

where $\Delta\beta_{klmn} = \beta_k + \beta_l - \beta_m - \beta_n$. From these equations we can derive equations for the powers:

$$\frac{dP_1}{dZ} = 2i A_3 A_4 A_2^* A_1^* e^{i\Delta\beta_{3412}z} + 2i A_5 A_6 A_2^* A_1^* e^{i\Delta\beta_{5612}z} + 2i A_2 A_3 A_6^* A_1^* e^{i\Delta\beta_{2316}z}$$
$$+ 2i A_2 A_5 A_4^* A_1^* e^{i\Delta\beta_{2514}z} + 2i A_3 A_5 (A_1^*)^2 e^{i\Delta\beta_{3511}z} + \text{c.c.}$$

$$\frac{dP_2}{dZ} = 2i A_3 A_4 A_1^* A_2^* e^{i\Delta\beta_{3412}z} + 2i A_5 A_6 A_1^* A_2^* e^{i\Delta\beta_{5612}z} + 2i A_1 A_6 A_3^* A_2^* e^{i\Delta\beta_{1623}z}$$
$$+ 2i A_1 A_4 A_5^* A_2^* e^{i\Delta\beta_{1425}z} + 2i A_4 A_6 (A_2^*)^2 e^{i\Delta\beta_{4622}z} + \text{c.c.}$$

$$\frac{dP_3}{dZ} = 2i A_1 A_2 A_4^* A_3^* e^{i\Delta\beta_{1234}z} + i A_1^2 A_5^* A_3^* e^{i\Delta\beta_{1135}z}$$
$$+ 2i A_1 A_6 A_2^* A_3^* e^{i\Delta\beta_{1623}z} + 2i A_5 A_6 A_4^* A_3^* e^{i\Delta\beta_{5634}z} + \text{c.c.},$$

$$\frac{dP_4}{dZ} = 2i A_1 A_2 A_3^* A_4^* e^{i\Delta\beta_{1234}z} + i A_2^2 A_6^* A_4^* e^{i\Delta\beta_{2246}z}$$
$$+ 2i A_2 A_5 A_1^* A_4^* e^{i\Delta\beta_{2514}z} + 2i A_5 A_6 A_3^* A_4^* e^{i\Delta\beta_{5634}z} + \text{c.c.},$$

$$\frac{dP_5}{dZ} = 2i A_1 A_2 A_6^* A_5^* e^{i\Delta\beta_{1256}z} + i A_1^2 A_3^* A_5^* e^{i\Delta\beta_{1135}z}$$
$$+ 2i A_1 A_4 A_2^* A_5^* e^{i\Delta\beta_{1425}z} + 2i A_3 A_4 A_6^* A_5^* e^{i\Delta\beta_{3456}z} + \text{c.c.},$$

$$\frac{dP_6}{dZ} = 2i A_1 A_2 A_5^* A_6^* e^{i\Delta\beta_{1256}z} + i A_2^2 A_4^* A_6^* e^{i\Delta\beta_{2246}z}$$
$$+ 2i A_2 A_3 A_1^* A_6^* e^{i\Delta\beta_{2316}z} + 2i A_3 A_4 A_5^* A_6^* e^{i\Delta\beta_{3456}z} + \text{c.c.} \tag{3.108}$$

From these expressions it can be shown that

$$\frac{d}{dz}(P_1 + P_3 + P_5) = \frac{d}{dz}(P_2 + P_4 + P_6) = 0, \tag{3.109}$$

which indicates that the total power of each pump and its associated sidebands remains constant.

In addition, we have the following relationship between the powers of the sidebands:

$$\frac{d}{dz}(P_3 - P_5) = \frac{d}{dz}(P_4 - P_6). \tag{3.110}$$

We also note that combining Eqs. (3.109) and (3.110) leads to

$$\frac{d}{dz}(P_1 + 2P_3) = \frac{d}{dz}(P_2 + 2P_4) \quad \text{and} \quad \frac{d}{dz}(P_1 + 2P_5) = \frac{d}{dz}(P_2 + 2P_6).$$

Most of the eight distinct wavevector mismatches in Eq. (3.107) cannot be expressed as even functions of $\omega_3 - \omega_c$, and therefore the signal gain spectrum generally will not be symmetric about ω_c. This is in contrast with the gain spectra predicted by the four-wave model, which are always symmetric. The asymmetries here will be most pronounced near the pumps, where the two extra idlers have amplitudes comparable with those of the signal and the first idler. This can readily be observed in Fig. 5.9.

We now consider only the small-signal case, because there is no known solution for the case of pump depletion. When the pumps are much stronger than the sidebands we can assume that their propagation is independent of the sidebands. Hence the pump fields are calculated in exactly the same way as for a two-pump OPA with undepleted pumps, Eq. (3.69). The pump amplitudes remain constant, and they experience phase shifts φ_k due to SPM and XPM. Specifically, we have

$$A_k(z) = A_{k0} \exp\left[i\gamma(P_{k0} + 2P_{l0})z\right] = A_{k0} \exp(i\varphi_k),$$
$$k = 1, 2, \quad l = 3 - k. \tag{3.111}$$

The four sidebands are then governed by the remaining four equations, in which we substitute the pump fields. It is also legitimate to neglect the last term in each of these equations, as they involve only sideband fields and are thus very small compared with the other terms.

The resulting equations have been studied in detail [14, 15], and it is known that solutions can be found by assuming that the sidebands grow exponentially, i.e. that

$$A_k(z) = A_k(0) \exp^{i\lambda_k z}, \quad k = 3\text{–}6, \tag{3.112}$$

where the λ_k are unknown propagation constants (wavevectors). This leads to a fourth-order polynomial equation, which in general does not have simple closed-form solutions and must be solved numerically.

Some simple solutions have been found for particular limiting cases [14, 15]. Here we will limit ourselves to the case where the signal is very close to the pump, and therefore all the nonlinear interactions are well phase matched. Thus all the exponentials in Eq. (3.107) are replaced by unity, and we obtain

$$-i\frac{dA_3}{dZ} = 2P_0 A_3 + 2A_1 A_2 A_4^* + A_1^2 A_5^* + 2A_1 A_6 A_2^*,$$
$$-i\frac{dA_4}{dZ} = 2P_0 A_4 + 2A_1 A_2 A_3^* + A_2^2 A_6^* + 2A_2 A_5 A_1^*,$$
$$-i\frac{dA_5}{dZ} = 2P_0 A_5 + 2A_1 A_2 A_6^* + A_1^2 A_3^* + 2A_1 A_4 A_2^*,$$
$$-i\frac{dA_6}{dZ} = 2P_0 A_6 + 2A_1 A_2 A_5^* + A_2^2 A_4^* + 2A_2 A_3 A_1^*. \tag{3.113}$$

Scalar OPA theory

Substituting the expressions for the pump fields we obtain

$$-i\frac{dA_3}{dZ} = 2P_0 A_3 + 2A_{10}A_{20}A_4^* e^{i(\varphi_1+\varphi_2)} + A_{10}^2 A_5^* e^{2i\varphi_1}$$
$$+ 2A_{10}A_{20}^* A_6 e^{i(\varphi_1-\varphi_2)},$$
$$-i\frac{dA_4}{dZ} = 2P_0 A_4 + 2A_{10}A_{20}A_3^* e^{i(\varphi_1+\varphi_2)} + A_{20}^2 A_6^* e^{2i\varphi_2}$$
$$+ 2A_{10}^* A_{20} A_5 e^{i(\varphi_2-\varphi_1)},$$
$$-i\frac{dA_5}{dZ} = 2P_0 A_5 + 2A_{10}A_{20}A_6^* e^{i(\varphi_1+\varphi_2)} + A_{10}^2 A_3^* e^{2i\varphi_1}$$
$$+ 2A_{10}A_{20}^* A_4 e^{i(\varphi_1-\varphi_2)},$$
$$-i\frac{dA_6}{dZ} = 2P_0 A_6 + 2A_{10}A_{20}A_5^* e^{i(\varphi_1+\varphi_2)} + A_{20}^2 A_4^* e^{2i\varphi_2}$$
$$+ 2A_{10}^* A_{20} A_3 e^{i(\varphi_2-\varphi_1)}. \tag{3.114}$$

The exponentials can be eliminated by making the changes of variable $A_3 = B_3 e^{i\varphi_1}$, $A_4 = B_4 e^{i\varphi_2}$, $A_5 = B_5 e^{i\varphi_1}$, and $A_6 = B_6 e^{i\varphi_2}$. Upon complex conjugating the second and third equations, we obtain

$$\frac{d}{dz}\begin{bmatrix} B_3 \\ B_4^* \\ B_5^* \\ B_6 \end{bmatrix} = i\gamma \begin{bmatrix} A_{10}A_{10}^* & 2A_{10}A_{20} & A_{10}^2 & 2A_{10}A_{20}^* \\ -2A_{10}^* A_{20}^* & -A_{20}A_{20}^* & -2A_{10}A_{20}^* & -A_{20}^{*2} \\ -A_{10}^{*2} & -2A_{10}^* A_{20} & -A_{10}A_{10}^* & -2A_{10}^* A_{20}^* \\ 2A_{10}^* A_{20} & A_{20}^2 & 2A_{10}A_{20} & A_{20}A_{20}^* \end{bmatrix} \begin{bmatrix} B_3 \\ B_4^* \\ B_5^* \\ B_6 \end{bmatrix}$$

$$= N \begin{bmatrix} B_3 \\ B_4^* \\ B_5^* \\ B_6 \end{bmatrix}, \tag{3.115}$$

which has the form $d\tilde{B}/dz = N\tilde{B}$, where $\tilde{B}$ is the column vector in Eq. (3.115). Since N is a constant matrix, the solution is

$$\tilde{B}(z) = \exp(Nz)\tilde{B}(0) = \sum_{m=1}^{\infty} \frac{(Nz)^m}{m!} \tilde{B}(0). \tag{3.116}$$

It can be shown that $N^2 = 0$. Therefore Eq. (3.116) reduces to

$$\tilde{B}(z) = (I + Nz)\tilde{B}(0). \tag{3.117}$$

This shows that in this situation the OPA has a field gain that is linear in z (and a power gain that is quadratic in z), for all four sidebands. This is in contrast with the four-wave model for a two-pump OPA, which generally predicts an exponential gain near the pumps. This linear gain is similar to that which exists near the pump in a one-pump OPA.

3.5 Solution of the two-pump OPA equations

Let us now assume for simplicity that the two pumps have zero initial phases, so that $A_{10} = \sqrt{P_{10}}$ and $A_{20} = \sqrt{P_{20}}$. Then N can be rewritten as

$$N = i\gamma \begin{bmatrix} P_{10} & 2\sqrt{P_{10}P_{20}} & P_{10} & 2\sqrt{P_{10}P_{20}} \\ -2\sqrt{P_{10}P_{20}} & -P_{20} & -2\sqrt{P_{10}P_{20}} & -P_{20} \\ -P_{10} & -2\sqrt{P_{10}P_{20}} & -P_{10} & -2\sqrt{P_{10}P_{20}} \\ 2\sqrt{P_{10}P_{20}} & P_{20} & 2\sqrt{P_{10}P_{20}} & P_{20} \end{bmatrix}. \quad (3.118)$$

In the common situation where only a signal at ω_3 is present at the input, we obtain

$$\begin{aligned} B_3(z) &= (1 + i\gamma P_{10}z) B_3(0), \\ B_4(z) &= 2i\gamma \sqrt{P_{10}P_{20}} z B_3^*(0), \\ B_5(z) &= i\gamma P_{10}z B_3^*(0), \\ B_6(z) &= 2i\gamma \sqrt{P_{10}P_{20}} z B_3(0). \end{aligned} \quad (3.119)$$

From Eq. (3.119) we obtain the expressions for the power gains:

$$\begin{aligned} P_3/P_{30} &- 1 + (\gamma P_{10}z)^2, & P_4/P_{30} &- (2\gamma \sqrt{P_{10}P_{20}}z)^2, \\ P_5/P_{30} &= (\gamma P_{10}z)^2, & P_6/P_{30} &= (2\gamma \sqrt{P_{10}P_{20}}z)^2. \end{aligned} \quad (3.120)$$

These relations show that the sidebands of the first pump grow at equal rates, and that this rate is the same as if the second pump were absent. The sidebands of the second pump also grow at equal rates, but these growth rates are generally different for the two pumps. We also have $P_3 - P_5 = P_{30}$ and $P_4 - P_6 = 0$, so that Eq. (3.110) is satisfied.

The different growth rates of the pump sidebands can also be understood as follows. The signal and pump 1 together can be viewed as pump 1 with intensity modulation (IM). As the waves propagate, this IM does not change but generates imaginary sidebands for the two pumps, which can be interpreted as SPM for pump 1 and XPM for pump 2. The factor 2 in the second and fourth lines of Eq. (3.119) corresponds to the ratio of the XPM and the SPM for parallel linear SOPs.

From Eq. (3.119) we can also write the expressions for the B_k as

$$B_3 = A_{30} + i\gamma z \left(P_{10} I_{10} + 2\sqrt{P_{10}P_{20}} I_{20} \right), \quad (3.121)$$

$$B_4 = A_{40} + i\gamma z \left(P_{20} I_{02}^* + 2\sqrt{P_{10}P_{20}} I_{10}^* \right), \quad (3.122)$$

$$B_5 = A_{50} + i\gamma z \left(P_{10} I_{10}^* + 2\sqrt{P_{10}P_{20}} I_{20}^* \right), \quad (3.123)$$

$$B_6 = A_{60} + i\gamma z \left(P_{20} I_{20} + 2\sqrt{P_{10}P_{20}} I_{10} \right), \quad (3.124)$$

where $I_{10} = A_{30} + A_{50}^*$ and $I_{20} = A_{40}^* + A_{60}$ are constants. We note that $B_3 + B_5^* = I_{10}$ and $B_4 + B_6^* = I_{20}$; this shows that, for each pump, the sum of one sideband amplitude and the complex conjugate of the other sideband amplitude is an invariant.

An interesting consequence is that if initially pump 1 has frequency modulation (FM) and pump 2 has no modulation then $A_{30} + A_{50}^* = I_{10} = 0$ and $B_3 + B_5^* = I_{10} = 0$, which implies that pump 1 can have only FM at any point along the fiber. Pump 2 acquires no modulation at all.

Also, if initially pump 1 has amplitude modulation (AM) and pump 2 has no modulation then pump 1 keeps the same AM but acquires growing phase modulation (PM), by SPM. Pump 2 acquires FM only by means of XPM. So, all these results are in agreement with our expectations concerning SPM and XPM.

The presence of the low-gain regions near the pumps restricts somewhat the range over which the gain spectra predicted by the four-wave model are applicable. When the pump spacing is large, the impact is rather limited [16]. However, for smaller spacings the gain dips can be significant. Their widths are similar to those of the one-pump OPA gain spectra associated with each pump. Examples of these spectra are shown in Chapter 5.

We note that the idlers at ω_4 and ω_5 are complex conjugates of the signal. However, the idler at ω_6 is proportional to the signal itself; this can be used for obtaining wavelength conversion without phase conjugation or spectral inversion.

A number of interesting phenomena occur when the pump spacing becomes even smaller. Then significant FWM can take place between the two pumps, which can create one or more pairs of additional pumps. All these pumps then divide the spectrum into bands of equal width. If a signal is introduced in one of these bands then one idler is created in the same band and two in all other bands. This type of arrangement can therefore be used for generating multiple copies of a signal, or multicasting [17]. Of course, the six-wave model is no longer sufficient to analyze this situation, the complexity of which indicates that it is probably best to use the nonlinear Schrödinger equation (NLSE) to investigate it (see Chapter 6).

3.6   Theory for one-pump OPA

3.6.1   Pump depletion

Pump depletion in one-pump OPAs was first investigated by Chen and Snyder [6]. Cappellini and Trillo then investigated the stability of the solutions [5]. The treatment presented here parallels that of subsection 3.5.1. We thus refer the reader to that subsection for further details. The starting point is three equations obtained in Section 3.3:

$$-i\frac{dA_1}{dZ} = |A_1|^2 A_1 + 2\sum_{k=3}^{4} |A_k|^2 A_1 + 2A_3 A_4 A_1^* e^{i\Delta\beta Z/\gamma}, \tag{3.125}$$

$$-i\frac{dA_l}{dZ} = |A_l|^2 A_l + 2\sum_{j\neq l=1}^{4} |A_j|^2 A_l + (A_1)^2 A_k^* e^{-i\Delta\beta Z/\gamma},$$

$$l = 3, 4, \quad k = 7 - l, \tag{3.126}$$

where $\Delta\beta = \beta_3 + \beta_4 - 2\beta_1$. Multiplying Eq. (3.125) by A_1^* and Eq. (3.126) by A_l^* leads to

$$-i\frac{dA_1}{dZ}A_1^* = P_1(2P_0 - P_1) + 2P_1\sqrt{P_3 P_4}e^{i\theta}, \tag{3.127}$$

$$-i\frac{dA_l}{dZ}A_l^* = P_l(2P_0 - P_l) + P_1\sqrt{P_3 P_4}e^{-i\theta}, \quad l = 3, 4, \tag{3.128}$$

where $\theta = \Delta\beta Z/\gamma + \theta_3 + \theta_4 - 2\theta_1$ and $P_0 = P_1 + P_3 + P_4$. Equations (3.127) and (3.128) lead to

$$\frac{dP_1}{dZ} = -2\frac{dP_3}{dZ} = -2\frac{dP_4}{dZ} = -4P_1\sqrt{P_3 P_4} \sin\theta. \quad (3.129)$$

Equation (3.129) implies that the powers of the three waves have the forms $P_1 = P_1(0) - 2x$, $P_3 = P_3(0) + x$, and $P_4 = P_4(0) + x$; thus when the pump releases two photons, the signal and the idler each gain one photon.

Equation (3.127) also yields

$$i\left(\frac{dA_1^*}{dZ}A_1 - \frac{dA_1}{dZ}A_1^*\right) = 2P_1\frac{d\phi_1}{dz} = 2P_1(2P_0 - P_1) + 4P_1\sqrt{P_3 P_4} \cos\theta. \quad (3.130)$$

Hence

$$\frac{d\theta_1}{dZ} = 2P_0 - P_1 + 2\sqrt{P_3 P_4} \cos\theta. \quad (3.131)$$

Similarly, we obtain

$$\frac{d\theta_l}{dZ} = 2P_0 - P_l + \frac{P_1}{P_l}\sqrt{P_3 P_4} \cos\theta, \quad l = 3, 4. \quad (3.132)$$

Then

$$\frac{d\theta}{dZ} = \frac{d}{dZ}(\theta_3 + \theta_4 - 2\theta_1) + \frac{\Delta\beta}{\gamma}$$
$$= \frac{\Delta\beta}{\gamma} + 2P_1 - P_3 - P_4 + P_1\sqrt{P_3 P_4}\left(P_3^{-1} + P_4^{-1} - 4P_1^{-1}\right)\cos\theta. \quad (3.133)$$

We also have

$$\frac{d\theta}{dZ} = \frac{d\theta}{dx}\frac{dx}{dZ} = \frac{d\theta}{dx}2P_1\sqrt{P_3 P_4} \sin\theta = -2P_1\sqrt{P_3 P_4}\frac{d(\cos\theta)}{dx}. \quad (3.134)$$

Substituting in Eq. (3.133), and using the facts that $dP_1/dx = -2$ and $dP_3/dx = dP_4/dx = 1$ leads to

$$\frac{\Delta\beta}{\gamma} + 2P_1 - P_3 - P_4 + 2\frac{d}{dx}\left(P_1\sqrt{P_3 P_4} \cos\theta\right) = 0. \quad (3.135)$$

Integrating both sides with respect to x, we obtain

$$\left(\frac{\Delta\beta}{\gamma} + \Delta P_0\right)x - 3x^2 + 4\sqrt{P_1 P_2 P_3 P_4} \cos\theta = K_1, \quad (3.136)$$

where $\Delta P_0 = 2P_{10} - P_{30} - P_{40}$ and K_1 is a constant given by

$$K_1 = 2P_{10}\sqrt{P_{30} P_{40}} \cos(\theta_{30} + \theta_{40} - 2\theta_{10}). \quad (3.137)$$

Eliminating θ in Eq. (3.129) by using Eq. (3.136), we obtain

$$\frac{dx}{dZ} = 2P_1\sqrt{P_3 P_4}\sin\theta = 2P_1\sqrt{P_3 P_4(1-\cos^2\theta)} = 2\sqrt{h(x)}, \qquad (3.138)$$

where

$$h(x) = (P_{10} - 2x)^2 (P_{30} + x)(P_{40} + x)$$

$$-\frac{1}{4}\left[K_1 - \left(\frac{\Delta\beta}{\gamma} + \Delta P_0\right)x + 3x^2\right]^2. \qquad (3.139)$$

We find that $C_0^2 = 7/4$. The solution for $x(Z)$ then proceeds as in subsection 3.5.1, but with the constants

$$\Delta P_0 = 2P_{10} - P_{30} - P_{40},$$
$$K_1 = 2P_{10}\sqrt{P_{30}P_{40}}\cos(\theta_{30} + \theta_{40} - 2\theta_{10}),$$
$$C_0 = \sqrt{7}/2.$$

Equations (3.62) and (3.63) can be used as before but with appropriate values for the roots of $h(x)$.

Let us consider the possibility of obtaining complete pump depletion with a one-pump fiber OPA. Then necessarily $P_{10}/2$ must be a root of $h(x) = 0$. Let us also assume that there is no idler at the input, so that $K_1 = 0$. Then we must have

$$-\left(\frac{\Delta\beta}{\gamma} + 2P_{10} - P_{30}\right) + \frac{3}{2}P_{10} = 0,$$

or $\Delta\beta = \gamma(P_{30} - P_{10}/2)$.

With no idler at the input, $x = 0$ must be a root of $h(x)$, and so $g(x) = h(x)/x$ is a third-order polynomial of the form $g(x) = ax^3 + bx^2 + cx + d$, where

$$a = \frac{7}{4}, \qquad b = \frac{5}{2}P_{30} - P_{10} + \frac{3}{2}\frac{\Delta\beta}{\gamma},$$

$$c = P_{10}^2 - 4P_{10}P_{30} - \frac{1}{4}\left(\frac{\Delta\beta}{\gamma} + 2P_{10} - P_{30}\right)^2, \qquad d = P_{10}^2 P_{30}.$$

The roots of $g(x)$ can then be found by means of the Cardano solution, as in subsection 3.5.2.

3.6.2 No pump depletion

We now neglect the XPM and FWM due to the signal and idler in Eq. (3.125), which then reduces to

$$\frac{dA_1}{dz} = i\gamma P_1 A_1 = ip_1 A_1, \qquad (3.140)$$

where $p_1 = \gamma P_1$. Equation (3.140) has the solution

$$A_1(z) = A_1(0)e^{ip_1 z}. \qquad (3.141)$$

3.6 Theory for one-pump OPA

Similarly, neglecting the signal and idler SPM and XPM in Eq. (3.126) leads to

$$\frac{dA_k}{dz} = ip_k A_k + ir_k A_l^* \exp i(2p_1 - \Delta\beta)z, \qquad k = 3, 4, \quad l = 7 - k, \quad (3.142)$$

$$p_k = 2\gamma P_1, \qquad q_k = 2p_1 - \Delta\beta, \qquad r_k = \gamma [A_1(0)]^2, \qquad (3.143)$$

$$k = 3, 4, \quad l = 7 - k,$$

where $\Delta\beta = \beta_3 + \beta_4 - 2\beta_1$.

We note that Eq. (3.142) has the same form as Eq. (3.71) for the two-pump OPA, provided that we let $p_2 = p_1$ and $\beta_2 = \beta_1$. It follows that the solution of two-pump OPA equations presented in Eqs. (3.71)–(3.82) holds for a one-pump OPA as well, with the appropriate expressions for r and $\Delta\beta$.

For a one-pump OPA we have $g_{\max} = \gamma P_1$ and $\Delta\beta_{NL} = 2\gamma P_1$, while for a two-pump OPA with $P_1 = P_2 = P_0/2$ we have $g_{\max} = \gamma P_0$ and $\Delta\beta_{NL} = \gamma P_0$. Hence a one-pump OPA and a two-pump OPA with the same total pump power have the same maximum gain, but it is obtained for different values of $\Delta\beta = -\Delta\beta_{NL}$.

Although they have the same maximum gain for equal pump power, one- and two-pump OPAs generally have different gain spectra, owing to their different expressions for $\Delta\beta_{NL}$ and $\Delta\beta$. The shape of their respective gain spectra will be studied in detail in Chapter 5.

3.6.3 Loss, no pump depletion

In subsection 3.5.4. we calculated the gain in the presence of fiber loss for both the one- and two-pump cases. In the case of a one-pump OPA it can be shown that $u = 0$, $Z_0 = 0$, and κ is infinite. Then $M(a, b, Z)$ can be expressed in terms of modified Bessel functions I by taking suitable limits [7].

An alternative approach, which directly leads to the Bessel functions J, is obtained by going back to Eq. (3.83) and letting $n_p = 1$ as well as $P_{10} = P_{20} = P_0/2$ [18]. Then

$$p = i\frac{\alpha}{2} + 2\gamma P_0 e^{-\alpha z}, \qquad q = i\alpha + 2\gamma P_0 e^{-\alpha z} - \Delta\beta,$$

$$r = \gamma P_0, \qquad \kappa = -i\alpha + \Delta\beta + 2\gamma P_0 e^{-\alpha z}. \qquad (3.144)$$

Equation (3.87) then reduces to $u = 0$, $v = -\Delta\beta\gamma P_0$, and $w = (\alpha + i\Delta\beta)^2/4$, so that

$$g^2(z) = (\alpha + i\Delta\beta)^2/4 - \Delta\beta\gamma P_0 e^{-\alpha z}. \qquad (3.145)$$

Making the change of variable

$$Z = \frac{2}{\alpha} \left(\Delta\beta\gamma P_0 e^{-\alpha z}\right)^{1/2},$$

Equation (3.86) becomes

$$Z^2 \frac{d^2 D_k(Z)}{dZ^2} + Z \frac{dD_k(Z)}{dZ} + (Z^2 - \nu^2) D_k(Z) = 0, \qquad (3.146)$$

where $D_k(Z) = C_k(z)$ and $\nu = 1 + i\Delta\beta/\alpha$. Equation (3.146) is Bessel's equation [7], hence $D_k(Z)$ must have the form $D_k(Z) = U_k J_\nu(Z) + V_k J_{-\nu}(Z)$ where U_k and V_k are constants to be determined from the initial conditions, which are $D_k(Z_0) = C_k(0) = A_k(0)$ and

$$D_k'(Z_0) = -\frac{2}{\alpha Z_0}\left[\frac{s}{2}C_k(0) + rC_l^*(0)\right] = -\frac{2}{\alpha Z_0}\left[\frac{s}{2}A_k(0) + rA_l^*(0)\right]. \quad (3.147)$$

The solution for the idler field ($k = 4$) is relatively simple if no idler is present at the input, i.e. $A_4(0) = 0$. Then the equations for U_4 and V_4 are

$$D_4(Z_0) = U_4 J_\nu(Z_0) + V_4 J_{-\nu}(Z_0) = 0, \quad (3.148)$$

$$D_4'(Z_0) = U_4 J_\nu'(Z_0) + V_4 J_{-\nu}'(Z_0) = -\frac{2r}{\alpha Z_0}A_3^*(0). \quad (3.149)$$

The solution is

$$U_4 = \frac{2r}{\alpha Z_0}A_3^*(0)\frac{J_{-\nu}(Z_0)}{W_{\text{Bess}}(Z_0)},$$

$$V_4 = -\frac{2r}{\alpha Z_0}A_3^*(0)\frac{J_\nu(Z_0)}{W_{\text{Bess}}(Z_0)},$$

where $W_{\text{Bess}}(Z_0)$ is the Wronskian of the Bessel functions $J_\nu(Z)$ and $J_{-\nu}(Z)$ at Z_0, i.e.

$$W_{\text{Bess}}(Z_0) = J_\nu(Z_0)J_{-\nu}'(Z_0) - J_{-\nu}(Z_0)J_\nu'(Z_0) = -\frac{2\sin(\pi\nu)}{\pi Z_0}. \quad (3.150)$$

We can then write

$$D_4(Z) = \frac{i\pi\gamma P_0}{\alpha \sin(\pi\nu)}[J_\nu(Z_0)J_{-\nu}(Z) - J_{-\nu}(Z_0)J_\nu(Z)]A_3^*(0), \quad (3.151)$$

$$A_4(z) = D_4(Z)\exp\left[\int_{\xi=0}^z \left(\frac{q+p+p^*}{2}\right)d\xi\right]$$

$$= \frac{i\pi\gamma P_0}{\alpha \sin(\pi\nu)}\exp\left[-\alpha z - \frac{i\Delta\beta z}{2} + i\gamma P_0\left(\frac{1-e^{-\alpha z}}{\alpha}\right)\right]$$

$$\times[J_\nu(Z_0)J_{-\nu}(Z) - J_{-\nu}(Z_0)J_\nu(Z)]A_3^*(0). \quad (3.152)$$

The idler power gain (or conversion efficiency) can then be calculated as

$$G_4(z) = \left|\frac{A_4(z)}{A_3^*(0)}\right|^2$$

$$= e^{-2\alpha z}\left|\frac{\pi\gamma P_0}{\alpha \sin(\pi\nu)}[J_\nu(Z_0)J_{-\nu}(Z) - J_{-\nu}(Z_0)J_\nu(Z)]\right|^2. \quad (3.153)$$

To calculate the signal gain directly, it is necessary to go back to Eq. (3.147) and write down the initial conditions for the signal, which are more complicated than for the idler, since the signal is present at the input. They are

$$D_3(Z_0) = C_3(0) = A_3(0), \tag{3.154}$$

$$D_3'(Z_0) = -\frac{2}{Z_0}\left[\frac{\nu+i\varepsilon}{2}C_3(0) + \frac{r}{\alpha}C_4^*(0)\right] = -\frac{(\nu+i\varepsilon)A_3(0)}{Z_0}, \tag{3.155}$$

where $\varepsilon = 2\gamma P_{10}/\alpha$. Hence the equations for U_3 and V_3 are

$$D_3(Z_0) = U_3 J_\nu(Z_0) + V_3 J_{-\nu}(Z_0) = A_3(0), \tag{3.156}$$

$$D_3'(Z_0) = U_3 J_\nu'(Z_0) + V_3 J_{-\nu}'(Z_0) = -\frac{(\nu+i\varepsilon)A_3(0)}{Z_0}. \tag{3.157}$$

The solution of these equations leads to the signal power gain

$$G_3(z) = \left|\frac{A_3(z)}{A_3(0)}\right|^2$$

$$= e^{-2\alpha z}\left|\frac{\pi}{2\sin(\pi\nu)}\right|^2 \Big|[(\nu+i\varepsilon)J_{-\nu}(Z_0) + Z_0 J_{-\nu}'(Z_0)]J_\nu(Z)$$

$$- [(\nu+i\varepsilon)J_\nu(Z_0) + \alpha Z_0 J_\nu'(Z_0)]J_{-\nu}(Z)\Big|^2. \tag{3.158}$$

Expressing the derivatives of Bessel functions in terms of Bessel functions leads to the alternative form

$$G_3(z) = e^{-2\alpha z}\left|\frac{\pi}{2\sin(\pi\nu)}\right|^2 \Big|[i\varepsilon J_{-\nu}(Z_0) - Z_0 J_{-\nu+1}(Z_0)]J_\nu(Z)$$

$$- [i\varepsilon J_\nu(Z_0) + Z_0 J_{\nu-1}(Z_0)]J_{-\nu}(Z)\Big|^2. \tag{3.159}$$

As in the two-pump case, G_3 and G_4 must satisfy the relation

$$G_3(z) = G_4(z) + e^{-\alpha z}. \tag{3.160}$$

Substituting Eqs. (3.153) and (3.159) into Eq. (3.160) leads to a non-trivial identity between Bessel functions.

3.6.4 Raman gain, no pump depletion

Until now we considered that the fiber nonlinearity governing parametric interactions is purely electronic in origin and that it can be described by a single real constant γ. This is only an approximation, because in fibers stimulated Raman scattering (SRS) is always present to some extent. Here we are going to include SRS in the basic OPA equations and see under what circumstances it significantly modifies the gain expressions that we obtained previously.

Stimulated Raman scattering is characterized by the third-order susceptibility $\chi_R^{(3)}(\Omega)$, where $\Omega = \omega_3 - \omega_1$ and ω_1 and ω_3 are the pump and signal frequencies; respectively; $\chi_R^{(3)}(\Omega)$ is proportional to the Fourier transform $f(t)$ of the time-domain response in SRS which is normalized in such a way that $\int_{-\infty}^{\infty} f(t)dt = 1$ [19]. In silica fibers $f(t)$ has a width of the order of tens of femtoseconds, which leads to a Raman gain spectrum several terahertz wide (tens of nanometers). When $f(t)$ is real, $\chi_R^{(3)}(\Omega)$ has Hermitian symmetry, i.e. $\chi_R^{(3)}(-\Omega) = [\chi_R^{(3)}(\Omega)]^*$.

The part of the nonlinearity that is purely electronic in origin is described by $\chi_e^{(3)}$; it can be assumed to be constant, because the electronic response in the time domain is nearly instantaneous. It is convenient to define the ratio of these two susceptibilities as

$$\delta(\Omega) = \frac{\chi_R^{(3)}(\Omega)}{\chi_e^{(3)}}. \tag{3.161}$$

A number of authors have investigated aspects of the interaction between the parametric gain and the Raman gain, using different basic equations [9, 20–25]. Here we are going to follow the model of [21], which has recently been used for studying the Raman contribution to the OPA noise figure [23]. Assuming that the pump is not affected by the signal and idler, the pump equation is

$$\frac{dA_1}{dz} = i\gamma_e[1 + \delta(0)]P_{10}A_1, \tag{3.162}$$

where γ_e is the fiber nonlinearity coefficient, which accounts only for the electronic nonlinearity. The solution is

$$A_1(z) = A_1(0) \exp\{i\gamma_e[1 + \delta(0)]P_{10}z\}. \tag{3.163}$$

When the pump signal is traveling in isolation its power cannot change, and this implies that $\delta(0)$ must be real.

It is important to note that Eq. (3.163) indicates that the pump experiences only SPM, with effective nonlinearity coefficient $\gamma = \gamma_e[1 + \delta(0)]$. In practice, the fiber nonlinearity coefficient is measured by performing SPM or XPM measurements, which necessarily yields γ as the result, not γ_e. In fact measuring γ_e alone is not a simple matter.

The signal and idler equations are

$$\frac{dA_3}{dz} = i\gamma_e[2 + \delta(0) + \delta(\Omega)]P_{10}A_3$$
$$+ i\gamma_e[1 + \delta(\Omega)]P_{10}A_4^* \exp[i(-\Delta\beta + 2\gamma P_{10})z], \tag{3.164}$$

$$\frac{dA_4}{dz} = i\gamma_e[2 + \delta(0) + \delta(-\Omega)]P_{10}A_4$$
$$+ i\gamma_e[1 + \delta(-\Omega)]P_{10}A_3^* \exp[i(-\Delta\beta + 2\gamma P_{10})z], \tag{3.165}$$

where $\Delta\beta$ is the usual wavevector mismatch. Equations (3.164) and (3.165) constitute a particular case of the general set of coupled equations studied in Appendix 1 for which the coefficients are given by

$$p_3 = \gamma_e[2 + \delta(0) + \delta(\Omega)]P_{10}, \tag{3.166}$$
$$q_3 = 2\gamma P_{10} - \Delta\beta, \tag{3.167}$$
$$r_3 = \gamma_e[1 + \delta(\Omega)]P_{10}, \tag{3.168}$$
$$p_4 = \gamma_e[2 + \delta(0) + \delta(-\Omega)]P_{10}, \tag{3.169}$$
$$q_4 = 2\gamma P_{10} - \Delta\beta, \tag{3.170}$$
$$r_4 = \gamma_e[1 + \delta(-\Omega)]P_{10}. \tag{3.171}$$

Therefore the solution of Eqs. (3.164) and (3.165) can be found simply by substituting these coefficients into the general solution given by Eq. (A1.22) of Appendix 1.

3.6.5 Loss, Raman gain, no pump depletion

In the presence of distributed fiber loss and Raman gain, the one-pump OPA equations need to be modified.

(a) Pump equation

The pump power is reduced only by the fiber loss and therefore varies as $P_1(z) = P_{10} e^{-\alpha z}$. The pump-field equation then becomes

$$\frac{dA_1}{dz} = \left(-\frac{\alpha}{2} + i\gamma P_1\right) A_1 = \left[-\frac{\alpha}{2} + i\gamma P_{10} e^{-\alpha z}\right] A_1 \quad (3.172)$$

which has the solution

$$A_1(z) = \sqrt{P_{10}} \exp\left(-\frac{\alpha z}{2} + i\gamma P_{10} L_{\text{eff}}\right), \quad (3.173)$$

where

$$L_{\text{eff}} = \int_{\xi=0}^{z} e^{-\alpha\xi} d\xi = \frac{1 - e^{-\alpha z}}{\alpha}.$$

(b) Signal and idler equations

Using Eq. (3.173) we can write down the signal and idler equations:

$$\frac{dA_3}{dz} = -\frac{\alpha}{2} A_3 + 2i\gamma_{33} P_{10} \exp(-\alpha z) A_3$$
$$+ i\gamma_{34} P_{10} \exp(-\alpha z - i\Delta\beta z + 2i\gamma P_{10} L_{\text{eff}}) A_4^*, \quad (3.174)$$

$$\frac{dA_4}{dz} = -\frac{\alpha}{2} A_4 + 2i\gamma_{44} P_{10} \exp(-\alpha z) A_4$$
$$+ i\gamma_{43} P_{10} \exp(-\alpha z - i\Delta\beta z + 2i\gamma P_{10} L_{\text{eff}}) A_3^*, \quad (3.175)$$

where $\gamma = \gamma_e[1 + \delta(0)]$, $\gamma_{33} = \gamma_e[2 + \delta(0) + \delta(\Omega)]/2$, $\gamma_{34} = \gamma_e[1 + \delta(\Omega)]$, $\gamma_{44} = \gamma_e[2 + \delta(0) + \delta(-\Omega)]/2$, and $\gamma_{43} = \gamma_e[1 + \delta(-\Omega)]$.

We see that Eqs. (3.174) and (3.175) are of the same general form as Eq. (A1.5) in Appendix 1, with

$$p_k = \frac{i\alpha}{2} + 2\gamma_{kk} P_{10} e^{-\alpha z}, \qquad q_k = i\alpha - \Delta\beta + 2\gamma P_{10} e^{-\alpha z}, \qquad r_k = \gamma_{kl} P_{10} \quad (3.176)$$

$k = 3, 4$, $l = 7 - k$. From these we obtain

$$\kappa_k = -i\alpha + \Delta\beta + (\gamma_{kl} + \gamma_{lk}^*) P_{10} e^{-\alpha z},$$
$$\kappa_k' = -\alpha(\gamma_{kl} + \gamma_{lk}^*) P_{10} e^{-\alpha z}, \quad (3.177)$$

where we have made use of the fact that $2\gamma_{kk} - \gamma = \gamma_{kl}$. Following Appendix 1 we find that

$$4g_k^2 = (\alpha + i\Delta\beta)^2 - (\gamma_{kl} - \gamma_{lk}^*)^2 P_{10}^2 e^{-2\alpha z} - 2\Delta\beta(\gamma_{kl} + \gamma_{lk}^*) P_{10} e^{-\alpha z}. \quad (3.178)$$

We see that g_k^2 is a second-order polynomial in $e^{-\alpha z}$, as in subsection 3.5.4. From this we conclude that the gain can be expressed in terms of confluent hypergeometric functions. Because of the complex nature of such functions, we do not write down the full solution here.

The solution is somewhat simpler if it can be assumed that $\delta(\Omega)$ has Hermitian symmetry. Then we have

$$\gamma_{kl} = \gamma_{lk}^*, \tag{3.179}$$

so that g_k^2 reduces to

$$g_k^2 = (\alpha + i\Delta\beta)^2/4 - \Delta\beta\gamma_{kl}P_{10}e^{-\alpha z}. \tag{3.180}$$

We see that g_k^2 has the same form as in Eq. (3.145), except that γ is replaced by γ_{kl}. Thus the solution will be similar, involving Bessel functions. However, because there is no longer a single value for γ the appropriate coefficients must be used at each step, and they are in general different for the signal and idler equations.

We now make separate changes of variable when solving for the signal or the idler:

$$Z_k = \frac{2}{\alpha}\left[\Delta\beta\gamma_{kl}P_{10}e^{-\alpha z}\right]^{1/2} \quad \text{and} \quad D_k(Z_k) = C_k(z), \quad k = 3, 4. \tag{3.181}$$

The initial conditions for $D_k(Z_k)$ at $z = 0$, i.e. at

$$Z_{k0} = \frac{2}{\alpha}(\Delta\beta\gamma_{kl}P_{10})^{1/2},$$

are

$$D_k(Z_{k0}) = C_k(0) = A_k(0),$$

$$D_k'(Z_{k0}) = -\frac{2}{Z_{k0}}\left[\frac{\nu + i\varepsilon_k}{2}C_k(0) + \frac{r_k}{\alpha}C_l^*(0)\right]$$

$$= -\frac{2}{Z_{k0}}\left[\frac{\nu + i\varepsilon_k}{2}A_k(0) + \frac{r_k}{\alpha}A_l^*(0)\right], \tag{3.182}$$

where $\varepsilon_k = 2\gamma_{kl}P_{10}$ and $\nu = 1 + i\Delta\beta/\alpha$ is the same as in the Raman-free case. Assuming as usual that there is no idler at the input, the solution then proceeds as in subsection 3.6.3, from Eq. (3.149) to Eq. (3.159). The only difference in the calculations of the idler conversion efficiency $G_4(z)$ and the signal gain $G_3(z)$ is that the proper subscripts should be used. The result is

$$G_4(z) = e^{-2\alpha z}\left|\frac{\pi\gamma_{43}P_0}{\alpha\sin(\pi\nu)}[J_\nu(Z_{40})J_{-\nu}(Z_4) - J_{-\nu}(Z_{40})J_\nu(Z_4)]\right|^2. \tag{3.183}$$

Similarly, the signal power gain is

$$G_3(z) = e^{-2\alpha z}\left|\frac{\pi}{2\sin(\pi\nu)}\right|^2$$

$$\times \left|[i\varepsilon_3 J_{-\nu}(Z_{30}) - Z_{30}J_{-\nu+1}(Z_{30})]J_\nu(Z_3)\right.$$

$$\left. - [i\varepsilon_3 J_\nu(Z_{30}) + Z_0 J_{\nu-1}(Z_{30})]J_{-\nu}(Z_3)\right|^2. \tag{3.184}$$

A major difference between this derivation and that in subsection 3.6.3 is that in the presence of Raman gain the relation between signal gain and idler conversion efficiency used in subsection 3.6.3, namely

$$G_3(z) = G_4(z) + e^{-\alpha z},$$

is no longer valid. The reason is that the Raman gain introduces an asymmetry between signal and idler, so that they no longer gain photons at the same rate.

3.7 Case of no dispersion with loss

In the preceding sections we have studied the solutions of the OPA equations for various numbers of waves: three waves for the one-pump OPA and either four or six waves for the two-pump OPA and wavelength exchange. In all these cases, when there is no loss and no dispersion all the $\Delta\beta$'s vanish, and the basic equations are all of the form

$$\frac{dA_n}{dz} = i\gamma \left(2 \sum_k A_k A_k^* - A_n A_n^* \right) A_n + i\gamma \sum_{k,l,m} D_{klm} A_k A_l A_m^*, \qquad (3.185)$$

where the summations contain a number of terms and $D_{klm} = 2\,(1)$ if the corresponding FWM term is degenerate (non-degenerate).

In the presence of loss, the latter being assumed the same at all wavelengths, Eqs. (3.185) are replaced by

$$\frac{dA_n}{dz} = -\frac{\alpha}{2} A_n + i\gamma \left(2 \sum_k A_k A_k^* - A_n A_n^* \right) A_n + i\gamma \sum_{k,l,m} D_{klm} A_k A_l A_m^*. \qquad (3.186)$$

Making the transformation

$$B_k = A_k e^{\alpha z/2}, \qquad \text{for all } k, \qquad (3.187)$$

Eq. (3.186) can be rewritten as

$$\frac{dB_n}{dL_{\text{eff}}} = i\gamma \left(2 \sum_k B_k B_k^* - B_n B_n^* \right) B_n + i\gamma \sum_{k,l,m} D_{klm} B_k B_l B_m^*, \qquad (3.188)$$

where

$$L_{\text{eff}} = \frac{1 - e^{-\alpha z}}{\alpha}. \qquad (3.189)$$

We see that Eq. (3.188) has exactly the same form as Eq. (3.185). Also, Eq. (3.187) shows that B_k and A_k have the same initial conditions at $z = 0$. As a result, the solution of Eq. (3.188) for B_k is the same of that of Eq. (3.185) for A_k (with the variable z replaced by L_{eff}). Therefore, if we know the solution for a particular situation in the absence of loss, we can immediately infer the solution in the presence of loss, by making use of the exponential transformations of Eqs. (3.187) and (3.189). More explicitly, if the solution for A_k in the absence of loss is $A_k^0(z)$ then in the presence of loss it is

$$A_k(z) = e^{-\alpha z/2} A_k^0 \left(\frac{1 - e^{-\alpha z}}{\alpha} \right). \quad (3.190)$$

This transformation holds when dispersion does not play a role in the interaction between a number of waves. For four waves, this implies that $\Delta\beta = 0$; the latter holds automatically if the waves all have the same frequency (or nearly so) but can also occur with distinct frequencies, particularly for wavelength exchange. For more than four waves, dispersion effects can be neglected only if all have nearly the same frequency or if the fiber dispersion coefficients actually vanish.

Important applications are:

1. one-pump fiber OPAs when the signal is close to the pump;
2. degenerate fiber OPAs, i.e. fiber OPAs for which the signal and pump frequencies are identical (see Section 3.8);
3. wavelength exchange.

The case of wavelength exchange is particularly interesting. Applying Eq. (3.190) to Eq. (3.106), we see that the interaction is still governed by sine and cosine functions and so the signal power still vanishes at particular distances. This indicates that complete exchange can still be obtained in spite of the loss. Of course 'exchange' here has a somewhat different meaning because of the attenuation experienced by the waves. Hence we can no longer look at this as lossless photon-by-photon exchange, but we can still talk about the exchange of signal modulation in a classical sense.

Another interesting aspect of this is that the same reasoning also holds if α is negative, i.e. if there is some other gain mechanism in the fiber. In particular, if signal and idler are close together, on the long-wavelength side of the pumps, they will experience essentially the same Raman gain and so an exchange of signal modulation will still be achievable with good extinction ratios.

3.8   Solutions for degenerate OPAs

The completely degenerate case, where signal, pump, and idler have the same frequency, as well as the same SOP, is very interesting. The main reason is that it can be shown that under ideal circumstances its quantum-limited noise figure can be 0 dB, as compared with 3 dB in the case of a non-degenerate OPA with only signal at the input (see Chapter 9). This remarkable feature could in principle be used to lessen the degradation of a signal as it propagates along an optical transmission line and experiences periodic amplification to restore the signal weakened by transmission fiber attenuation.

A related feature is that a degenerate OPA can also generate squeezed light (i.e. light whose amplitude fluctuations in one quadrature are smaller than for a coherent state) or vacuum fluctuations. This feature also offers interesting prospects for improving the characteristics of communication systems or of very accurate optical sensors, such as those used for detecting gravitational waves [27].

It is thus of interest to investigate the gain properties of such OPAs from a classical standpoint. It is tempting to obtain an expression for the gain of a degenerate OPA by taking the limit of the gain of a nondegenerate one-pump OPA, by replacing ω_3 and ω_1 by ω_p. However, there are two problems with doing so.

1. In the basic equations for the non-degenerate OPA there are terms corresponding to both SPM and XPM, with coefficients that differ by a factor 2. But when all frequencies become the same, the use of XPM is no longer justified and therefore one should revisit the basic equations, which will lead to a modified gain expression.
2. When all three waves have the same frequency and the same SOP in the fiber, they can no longer be distinguished by using these characteristics. From a practical standpoint, this also means that one can no longer use the same components as in nondegenerate OPAs to combine and separate the waves before and after the OPA. Specifically, components such as WDM couplers cannot be used.

3.8.1 Mach–Zehnder interferometer

The standard way to get around the problem described above is to use an interferometer, for example a Mach–Zehnder interferometer (MZI) [28]. As we will show below, one can then inject the pump and the signal separately into the two input ports and obtain an amplified signal at one of the outputs (when the MZI is properly balanced). The pump, slightly depleted by the power gained by the signal, emerges from the second output. There is no longer a separate idler to speak of in this case. Clearly this situation is more complex than for a non-degenerate OPA, since we must now consider the simultaneous propagation of signal and idler in the two arms of the MZI, not just in a single fiber.

Let us then consider the MZI of Fig. 3.7. We assume that all the fields have the same frequency ω. Hence any electric field E can be written in terms of its complex amplitude A, as follows.

$$E(t) = \frac{1}{2}[A \exp(-i\omega t) + A^* \exp(i\omega t)]. \qquad (3.191)$$

Because the time dependences are the same they can be suppressed, and we can use equations that contain only the various A's.

The pump field A_{p0} enters the lower input port while the signal field A_{s0} enters the upper one. These fields get mixed by the first 2×2 coupler, which we assume to be lossless and to divide each input equally between its two outputs. Hence we assume that the fields at the upper and lower outputs of the first coupler are respectively given by

$$A_u = \frac{A_{p0} + A_{s0}}{\sqrt{2}} \quad \text{and} \quad A_\ell = \frac{A_{p0} - A_{s0}}{\sqrt{2}}. \qquad (3.192)$$

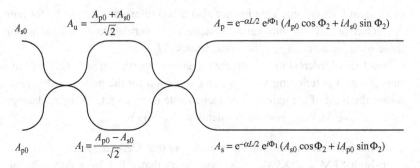

Fig. 3.7 Schematic of the nonlinear Mach–Zehnder interferometer with nonlinear arms.

The two arms of the MZI have the same length L and are made of nonlinear fiber, with nonlinearity coefficient γ and power attenuation constant α. The field in each arm experiences attenuation and a nonlinear phase shift due to SPM only. Thus the fields at the inputs of the second 2×2 coupler are

$$A'_u = A_u \exp\left(-\frac{\alpha L}{2} + i\gamma L_{\text{eff}} A_u A_u^*\right),$$
$$A'_1 = A_1 \exp\left(-\frac{\alpha L}{2} + i\gamma L_{\text{eff}} A_1 A_1^*\right).$$
(3.193)

The second 2×2 coupler is identical to the first, and so its upper output is given by

$$\begin{aligned}A_p &= \frac{A'_u + A'_1}{\sqrt{2}} \\ &= \frac{1}{\sqrt{2}} e^{-\alpha L/2} \left[A_u \exp(i\gamma L_{\text{eff}} A_u A_u^*) + A_1 \exp(i\gamma L_{\text{eff}} A_1 A_1^*) \right] \\ &= \frac{1}{2} e^{-\alpha L/2} \left\{ (A_{p0} + A_{s0}) \exp\left[\frac{i\gamma L_{\text{eff}}}{2}(A_{p0} + A_{s0})(A_{p0} + A_{s0})^*\right] \right. \\ &\quad \left. + (A_{p0} - A_{s0}) \exp\left[\frac{i\gamma L_{\text{eff}}}{2}(A_{p0} - A_{s0})(A_{p0} - A_{s0})^*\right] \right\}, \\ &= e^{-\alpha L/2} e^{i\Phi_1} (A_{p0} \cos \Phi_2 + i A_{s0} \sin \Phi_2)\end{aligned}$$
(3.194)

where

$$\Phi_1 = \frac{\gamma L_{\text{eff}}}{2}(A_{p0} A_{p0}^* + A_{s0} A_{s0}^*) \quad \text{and} \quad \Phi_2 = \frac{\gamma L_{\text{eff}}}{2}(A_{p0} A_{s0}^* + A_{s0} A_{p0}^*).$$
(3.195)

Since the signal and pump play similar roles in Fig. 3.7, the expression for A_s (which emerges from the lower output) is obtained by exchanging s and p in Eq. (3.194), i.e.

$$A_s = e^{-\alpha L/2} e^{i\Phi_1} (A_{s0} \cos \Phi_2 + i A_{p0} \sin \Phi_2).$$
(3.196)

We note that, taken together, Eqs. (3.194) and (3.196) can be rewritten as

$$\begin{bmatrix} A_s \\ A_p \end{bmatrix} = e^{-\alpha L/2} e^{i\Phi_1} U \begin{bmatrix} A_{s0} \\ A_{p0} \end{bmatrix},$$
(3.197)

where U is the unitary matrix

$$U = \begin{bmatrix} \cos\Phi_2 & i\sin\Phi_2 \\ i\sin\Phi_2 & \cos\Phi_2 \end{bmatrix}. \qquad (3.198)$$

It is interesting to note that this type of matrix is the same as that representing a lossless passive 2×2 coupler with an arbitrary coupling ratio. However, the coupling ratio of the MZI depends on A_{s0} and A_{p0} and therefore on the signal and pump powers as well as on their relative phase difference.

We see that when the pump is off $\Phi_1 = \Phi_2 = 0$ and therefore $A_s = e^{-\alpha L/2} A_{s0}$; thus the MZI is well balanced and all the attenuated signal power emerges from the lower output. When the pump is on the output signal is modified, and we define the signal gain of the MZI as

$$h_s = \frac{A_s}{A_{s0}} e^{-i\Phi_1} = e^{-\alpha L/2} \left(\cos\Phi_2 + i\sin\Phi_2 \frac{A_{p0}}{A_{s0}} \right). \qquad (3.199)$$

The signal power gain $G_s = h_s h_s^*$ is then given by

$$G_s = e^{-\alpha L} \left(\cos^2\Phi_2 + \sin^2\Phi_2 \frac{P_{p0}}{P_{s0}} - 2\sin\Phi_2 \cos\Phi_2 \sqrt{\frac{P_{p0}}{P_{s0}}} \sin\Phi_0 \right), \qquad (3.200)$$

where $P_{p0} = A_{p0} A_{p0}^*$ and $P_{s0} = A_{s0} A_{s0}^*$ are respectively the signal and pump input powers and $\Phi_0 = \Phi_{p0} - \Phi_{s0}$ is the phase difference between the pump and signal at the input.

(Note that we have removed $e^{-i\Phi_1}$ from the definition of h_s, because it has unit magnitude and thus does not affect the power gain. However, if Φ_1 is large and has rapid variations then $e^{-i\Phi_1}$ will correspond to large phase variations and therefore spectral broadening; in that case, it might be useful to reintroduce it into the definition of h_s.)

Equation (3.200) is exact, regardless of signal and pump powers, initial phases, fiber length, etc. An important particular case is that of a small signal and a pump power sufficient to provide moderate gain. We then assume that $\Phi_2 \ll \pi$ and linearize Eq. (3.199). This yields

$$h_s \approx e^{-\alpha L/2} \left(1 + i\Phi_2 \frac{A_{p0}}{A_{s0}} \right)$$
$$= e^{-\alpha L/2} \left(1 + i\Phi e^{i\Phi_0} \cos\Phi_0 \right) \qquad (3.201)$$

and

$$G_s \approx e^{-\alpha L} \left[1 + (\Phi\cos\Phi_0)^2 - 2\Phi\cos\Phi_0 \sin\Phi_0 \right], \qquad (3.202)$$

where $\Phi = \gamma P_{p0} L_{\text{eff}}$ is the pump's self-phase shift (the shift if all its power passed through a single fiber). Equation (3.202) shows that $G_s \approx e^{-\alpha L}$ if $\Phi_0 = \pi/2$, so that there is essentially no signal gain (or attenuation other than fiber loss) in that case. If $\Phi_0 = 0$, however, then $G_s \approx e^{-\alpha z}(1 + \Phi^2)$, which can be large: for example, if $\Phi = 3$ and $\alpha = 0$, $G_s \approx 10$ (i.e. 10 dB), which is substantial.

Closer analysis reveals that the gain extrema are reached for values of Φ_0 that are slightly away from 0 and $\pi/2$. The maximum gain is close to $e^{-\alpha z}(1+\Phi^2)$ but the minimum gain is much smaller than $e^{-\alpha z}$. To show this, let us study the dependence of G_s on Φ_0, as follows. The locations of the gain extrema are obtained by setting

$$\frac{dG_s}{d\theta_0} = e^{-\alpha z}\left[-\Phi^2 \sin(2\Phi_0) - 2\Phi \cos(2\Phi_0)\right] = 0, \quad (3.203)$$

which leads to

$$\tan(2\Phi_0) = -\frac{2}{\Phi},$$

$$\cos(2\Phi_0) = \pm\frac{1}{\sqrt{1+4/\Phi^2}}, \quad \sin(2\Phi_0) = \mp\frac{1}{\sqrt{1+\Phi^2/4}}. \quad (3.204)$$

Note that the two possible values of Φ_0 are exactly $\pi/2$ apart. They are said to correspond to two signal quadratures, respectively having maximum and minimum gain.

Let us now assume for simplicity that $\alpha = 0$. The gain extrema are then given by

$$G_\pm = 1 + \frac{\Phi^2}{2}\left[1 \pm \sqrt{1+\frac{4}{\Phi^2}}\right]. \quad (3.205)$$

Here G_+ corresponds to the quadrature that is nearly in phase with the pump while G_- corresponds to the quadrature that is nearly $\pi/2$ out of phase with it. It can be shown that $G_+ G_- = 1$ and that $G_+ \approx 1 + \Phi^2$. Hence we have $G_- \approx 1/(1+\Phi^2)$, which can be very small compared with 1. This is interesting, because it means that for that particular value of Φ_0 the signal is attenuated rather than amplified by the pump. Since this attenuation is also applicable for noise in phase with that particular signal, this phenomenon in principle provides a means for attenuating noise below the input level. If the input signal is a coherent state or a vacuum state with minimum fluctuations set by the uncertainty principle at the vacuum level, the output fluctuations will be reduced below that level and are said to be squeezed, as mentioned earlier. According to the preceding analysis, fluctuations can be squeezed by a factor equal to the reciprocal of the maximum gain; in practice this could correspond to tens of decibels, since such gains can readily be obtained in fiber OPAs.

If $\alpha \neq 0$, G_+ and G_- are obtained by multiplying the preceding values by $e^{-\alpha z}$. However, the discussion concerning squeezing becomes considerably more complicated, because the presence of loss in a fiber can be shown to introduce additional field fluctuations, distributed along the fiber length. Hence one cannot simply multiply the input vacuum fluctuation power by G_- to obtain the output power of the fluctuations of the squeezed quadrature.

3.8.2 Sagnac interferometer

The MZI architecture shown in Fig. 3.7 has a major drawback. If implemented by means of fibers, the upper and lower arms may experience changes in optical path length due to variations in temperature, stress, acoustic vibrations, etc. Therefore the difference in their optical path lengths may fluctuate randomly over time. This would be unacceptable, of course, because the desired expression for the outputs, given by Eq. (3.197), holds only if the correct phase relationship exists between the two arms. Thus, while the MZI is a convenient tool for understanding how the amplified signal can be extracted with an interferometer, it is not used in practice because of this stability issue.

One way to avoid this instability is to use a Sagnac interferometer (SI), which simply consists of a 2×2 coupler with its two outputs connected to each other, forming a loop. Feeding light into the inputs then creates waves that travel in opposite directions within the loop. The equations work in a manner very similar to those for the MZI, and one can again separate the amplified signal. The main advantage of the SI is that, because the counter-propagating waves travel along exactly the same path, any phase shifts due to environmental changes are the same in both directions and therefore have no net effect on the output signal. For this reason, SIs have been used for demonstrating amplification or squeezing in degenerate fiber OPAs.

A possible drawback of the SI architecture is that the counter-propagating waves influence each other. This is not present in the MZI, where the evolution of the field in each arm takes place independently of what happens in the other arm. In the SI, a wave traveling in one direction experiences a phase shift due to the wave traveling in the other direction. If the powers are not well balanced, this causes a phase shift difference between the arms and alters the interference conditions at the output. Thus accurate operation of an SI may require a 2×2 coupler with an adjustable coupling ratio as well as careful control of splice losses within the loop, which can create power imbalance if they are not symmetrically located.

3.9 Conclusion

While the set of known solutions in scalar OPA theory is fairly extensive, it is limited in the sense that the solutions are obtained for OPAs with idealized properties, such as a single SOP, constant ZDW along their length, etc. Practical OPAs, however, may have significantly non-ideal properties and thus their performance may be quite different from what one would expect from the analytic solutions. In such situations it is necessary to refine the model. In some situations it is possible to make use of analytic solutions over short segments, and to combine them. In other situations it may be necessary to resort to purely numerical solutions of the basic propagation equations; this will be described in Chapter 6.

References

1. "Interactions between light waves in a non-linear dielectric," Armstrong, J. A., Bloembergen, N., Ducuing, J., Pershan, P. S. *Phys. Rev.*; 1962; vol. 127, pp. 1918–39.
2. "Theory of dielectric waveguides," Kogelnik, H. In *Integrated Optics*, Tamir, T., ed. Springer, Berlin, Topics in Applied Physics; 1979; p. 37.

3. *The Elements of Nonlinear Optics*, Butcher, P. N., Cotter, D. Cambridge University Press, Cambridge; 1993; p. 21.
4. *Nonlinear Fiber Optics*, Agrawal, G. P. Academic, Boston; 1989.
5. "Third-order three-wave mixing in single-mode fibers: exact solutions and spatial instability effects," Cappellini, G., Trillo, S. *J. Opt. Soc. Amer. B*; 1991; vol. 8, pp. 824–37.
6. "Four-photon parametric mixing in optical fibers: effect of pump depletion," Chen, Y., Snyder, A. W. *Opt. Lett.*; 1989; vol. 14, pp. 87–9.
7. *Handbook of Mathematical Functions*, Abramowitz, M., Stegun, I., eds. National Bureau of Standards, Washington DC, Applied Mathematics Series; 1964; vol. 55.
8. "Tunable fiber parametric wavelength converter with 900 mW of CW output power at 1665 nm," Marhic, M. E., Williams, G. M., Goldberg, L., Delavaux, J. M. P. In *Proc. Conf. Photonics West*, San Jose CA, January 2006; *Proc. SPIE*; 2006; vol. 6103, pp. 165–6.
9. "Parametric amplification and frequency conversion in optical fibers," Stolen, R. H., Bjorkholm, J. E. *IEEE J. Quantum Electron.*; 1982; vol. QE-18, pp. 1062–72.
10. "Confluent hypergeometric solutions for parametric interactions in optical fibers," Marhic, M. E., Yang, F. S., Kazovsky, L. G. In *Proc. Nonlinear Optical Materials, Fundamentals and Applications Topical Meeting*, Princeville, Kauai HI, August 1998; paper PD005.
11. http://www.wolfram.com.
12. http://www.mathworks.com.
13. "Widely tunable spectrum translation and wavelength exchange by four-wave mixing in optical fibers," Marhic, M. E., Park, Y., Yang, F. S., Kazovsky, L. G. *Opt. Lett.*; 1996; vol. 21, pp. 1906–8.
14. "Unified analysis of modulational instability induced by cross-phase modulation in optical fibers," Tanemura, T., Kikuchi, K. *J. Opt. Soc. Amer. B*; 2003; vol. 20, pp. 2502–14.
15. "Quantum noise properties of parametric amplifiers driven by two pump waves," McKinstrie, C. J., Radic, S., Raymer, M.G.; *Optics Express*; 2004; vol. 12, pp. 5037–66.
16. "Parametric amplifiers driven by two pump waves with dissimilar frequencies," McKinstrie, C. J., Radic, S. *Opt. Lett.*; 2002; vol. 27, pp. 1138–40.
17. "40-Gb/s optical switching and wavelength multicasting in a two-pump parametric device," Lin, Q., Jiang, R., Marki, C. F., McKinstrie, C. J., Jopson, R., Ford, J.; Agrawal, G. P., Radic S. *IEEE Photon. Technol. Lett.*; 2005; vol. 17, pp. 2376–8.
18. "Bessel function solution for the gain of a one-pump fiber optical parametric amplifier," Marhic, M. E., Curri, V., Kazovsky, L. G. In *Proc Nonlinear Optical Materials, Fundamentals and Applications Topical Meeting*, Princeville, Kauai HI, August 1998; paper TuC21.
19. "Raman response function for silica fibers," Lin, Q., Agrawal, G. P. *Opt. Lett*; 2006; vol. 31, pp. 3086–8.
20. "Combined processes of stimulated Raman scattering and four-wave mixing in optical fibers," Chen, Y. *J. Opt. Soc. Amer. B*; 1990; vol. 7, pp. 43–52.
21. "Unified analysis of four-photon mixing, modulational instability, and simulated Raman scattering under various polarization conditions in fibers," Golovchenko, E. A., Pilipetskii, A. N. *J. Opt. Soc. Amer. B*; 1994; vol. 11, pp. 92–101.
22. "Raman-assisted parametric frequency conversion in a normally dispersive single-mode fiber," Sylvestre, T., Maillotte, H., Lantz, E., Tchofo Dinda, P. *Opt. Lett.*; 1999; vol. 24, pp. 1561–3.
23. "Raman-noise-induced noise-figure limit for $\chi^{(3)}$ parametric amplifiers," Voss, P. L., Kumar, P. *Opt. Lett.* 2004; vol. 29, pp. 445–7.

24. "200-nm-bandwidth fiber optical amplifier combining parametric and Raman gain," Ho, M. C., Uesaka, K., Marhic, M. E., Akasaka, Y., Kazovsky, L. G. *J. Lightwave Technol.*; 2001; vol. 19, pp. 977–81.
25. "Complete experimental characterization of the influence of parametric four-wave mixing on stimulated Raman gain," Vanholsbeeck, F., Emplit, P., Coen, S. *Opt. Lett.*; 2003; vol. 28, pp. 1960–2.
26. "Experimental demonstration of a squeezing-enhanced power-recycled Michelson interferometer for gravitational wave detection," McKenzie, K., Shaddock, D. A., McClelland, D. E., Buchler, B. C., Lam, P. K. *Phys. Rev. Letters*; 2002; vol. 88, pp. 231102/1–4.
27. "Optical amplification in a nonlinear fiber interferometer," Marhic, M. E., Hsia, C. H., Jeong, J. M. *Electron. Lett.*; 1991; vol. 27, pp. 210–1.

4 Vector OPA theory

4.1 Introduction

In Chapter 3 we derived optical parametric amplifier (OPA) equations under the assumption that all four waves involved were in the same state of linear polarization along the fiber length. This led to a relatively simple set of four scalar equations. By solving these equations we then obtained a set of solutions that can provide a good preliminary understanding of the performance of fiber OPAs.

In this chapter we enlarge the class of OPAs by considering more general possibilities for the SOPs of the waves. We begin by developing a vector formalism for light propagating in arbitrary SOPs in isotropic fibers. We apply it to circularly polarized (CP) and linearly polarized (LP) waves; since some of the waves may have identical or orthogonal SOPs, this leads to a class of 12 basic types of OPA. We then turn to fibers with constant birefringence, such as polarization-maintaining fibers (PMFs), and explore the impact of the birefringence on the gain spectrum. Finally we consider the typical case of fibers that have weak birefringence with random longitudinal variations, and we investigate the effect of these variations.

4.2 Isotropic fibers

The treatment here follows the general approach introduced in [1].

4.2.1 Classification of basic OPA types

We make the following basic assumptions. (i) The fiber is lossless. (ii) The fiber does not exhibit birefringence, either linear or circular. (iii) The fiber operates in the fundamental mode at all frequencies. (iv) The fiber material is isotropic, hence the nonlinear polarization can be calculated from one basic scalar [2]. (v) The third-order nonlinear interactions considered are self-phase modulation (SPM), cross-phase modulation (XPM), and four-wave mixing (FWM). They are described by means of the fiber nonlinearity coefficient γ, which is assumed to be independent of frequency. (vi) Other nonlinear effects, namely stimulated Brillouin scattering (SBS) and stimulated Raman scattering (SRS), are not considered, for reasons we now explain.

Regarding SBS, in practice it will always be efficiently suppressed by pump frequency dithering, or by other means, so that its effect on OPA performance can be neglected; SRS gain, however, cannot be suppressed and plays a role in OPA operation. Nevertheless, in a well-phase-matched OPA the parametric gain exceeds the SRS gain, so that the latter can be viewed as a relatively minor perturbation [3]. Hence to simplify the analysis we do not include SRS gain here. Even so we obtain a good first-order approximation. In principle, SRS gain could be included at a later stage in the analysis to obtain more accurate results (as in subsection 3.6.4) [4, 5].

We consider only situations where all four waves maintain their original SOPs along the isotropic fiber. If we first apply this requirement to the propagation of the two pumps, the need to avoid ellipse rotation dictates that they should be either LP or CP. Also, they should be either in the same SOP or in orthogonal SOPs. Altogether, there are four possible choices for the pump SOPs of two-pump OPAs: (i) parallel LP; (ii) perpendicular LP; (iii) identical CP; (iv) orthogonal CP. For one-pump OPA there are two possibilities: the pump should be either LP or CP.

Knowing the pump SOPs, we now turn to the signal and idler. For LP pumps, the signal will be subject to the optical Kerr effect due to the pumps (i.e. induced linear birefringence) and so the signal SOP will be affected unless the signal is itself in an LP state, parallel to either pump. A similar reasoning holds for CP pumps. So, we restrict ourselves to OPAs with all four waves either LP or CP.

An additional condition is imposed by the symmetry requirements of the SOPs of the four waves. It can be traced to the properties of the third-order susceptibility tensor $\chi^{(3)}$ and is stated as follows. If the pumps have identical (orthogonal) SOPs then so do the signal and the idler. For a one-pump OPA, the signal and idler must have the same SOP.

We can now make a list of all basic OPA types that satisfy these constraints. We let X, Y, R, and L denote LP states parallel to the x- or y-axis and CP states with right or left handedness, respectively. We use the notation ABCD for the state of an OPA, where A, B, C, and D stand for the SOPs of the signal, first pump, second pump, and idler, respectively. For one-pump OPAs we have these possibilities: XXX, YXY, RRR, LRL. For two-pump OPAs we have XXXX, YXXY, XYXY, YYXX, RRRR, LRRL, RRLL, LRLR. For non-birefringent fibers, other SOP combinations would yield OPAs that are physically equivalent to the ones listed: we can go from one to the other by exchanging X and Y, R and L, or the labels of the signal and idler and the labels of the pumps. For birefringent fibers, however, this degeneracy can be broken because the fiber dispersion does not remain invariant under these transformations; then the list of physically distinguishable OPAs is considerably longer. Where the degeneracy remains we will frame our discussion using the most convenient OPA type.

4.2.2 Derivation of OPA equations

This derivation parallels that of Chapter 3, and so we will concentrate mainly on those aspects that are different. We begin with a plane-wave analysis, considering spatial variations only along the z-axis. The transverse mode profile will eventually be taken into account by means of A_{eff}.

Vector OPA theory

The wave equation for the electric field vector $\vec{E}(z,t)$ is

$$\frac{\partial^2 \vec{E}}{\partial z^2} - \frac{n^2}{c^2}\frac{\partial^2 \vec{E}}{\partial t^2} = -\mu_0 \frac{\partial^2 \vec{P}_{NL}}{\partial t^2}. \tag{4.1}$$

The nonlinear polarization density $\vec{P}_{NL}(z,t)$ is given by

$$\vec{P}_{NL}(z,t) = \varepsilon_0 \chi^{(3)} \vdots \vec{E}\vec{E}\vec{E}, \tag{4.2}$$

where $\chi^{(3)}$ is the third-order susceptibility tensor and is generally given for fields with components parallel to the x-, y-, and z- axes of a Cartesian coordinate system. This can be inconvenient when using fields with SOPs that are not linear, such as CP states. In such situations it is advantageous in calculating $\vec{P}_{NL}(z,t)$ not to implicitly assume Cartesian coordinates. In isotropic materials, this can be achieved by using an alternate form for $\vec{P}_{NL}(z,t)$, i.e. [2]

$$\vec{P}_{NL}(z,t) = \varepsilon_0 \chi^{(3)}_{xxxx}(\vec{E}\cdot\vec{E})\vec{E}. \tag{4.3}$$

We now use Eq. (4.3) to calculate SPM, XPM, and FWM terms that arise when the four waves interact. We let the total electric field be of the form

$$\vec{E}(z,t) = \frac{\rho}{2}\sum_{k=1}^{4}\left\{\vec{A}_k(z)\exp[i(\beta_k z - \omega_k t)] + \text{c.c}\right\}, \tag{4.4}$$

where $\vec{A}_k(z)$ is the complex vector phasor (a slowly varying envelope, or SVE) of the field at ω_k and $\beta_k = \beta(\omega_k)$ is the propagation constant (wavevector) of the kth wave; ρ is a constant that will be determined later.

We substitute Eq. (4.4) into Eq. (4.2) and then substitute the latter into Eq. (4.1). We consider only terms at ω_ℓ and equate them on both sides. This gives

$$\frac{\rho}{2}\left(\frac{d^2\vec{A}_\ell}{dz^2} + 2i\beta_\ell\frac{d\vec{A}_\ell}{dz}\right) \approx i\rho\beta_\ell\frac{d\vec{A}_\ell}{dz}$$

$$= -\frac{\chi^{(3)}_{xxxx}\omega_\ell^2}{c^2}\left\{\{(\vec{E}\cdot\vec{E})\vec{E}\}_\ell \exp[-i(\beta_\ell z - \omega_\ell t)]\right\} \tag{4.5}$$

or

$$\frac{d\vec{A}_\ell}{dz} = i\frac{\chi^{(3)}_{xxxx}\omega_\ell}{\rho n c}\left\{(\vec{E}\cdot\vec{E})\vec{E}\right\}_\ell \exp[-i(\beta_\ell z - \omega_\ell t)]. \tag{4.6}$$

The notation $\{\ \}_\ell$ indicates that we keep only the terms in $e^{-i\omega_\ell t}$ within $\{\ \}_\ell$. To simplify Eq. (4.6), we have neglected the second-order derivative of $\vec{A}_\ell$, as it is slowly varying, used $\beta_\ell = n\omega_\ell/c$, and also used the fact that $\{\ \}_\ell$ only contains terms at ω_ℓ.

When $(\vec{E}\cdot\vec{E})\vec{E}$ is fully expanded, it contains $8^3 = 512$ terms. It is thus desirable to find a means of simplifying this expression. To this effect we introduce the following vector operation:

$$[\vec{a},\vec{b},\vec{c}] = \frac{1}{3}[(\vec{a}\cdot\vec{b})\vec{c} + (\vec{b}\cdot\vec{c})\vec{a} + (\vec{c}\cdot\vec{a})\vec{b}], \tag{4.7}$$

where $\vec{a}$, $\vec{b}$, and $\vec{c}$ are three vectors. It can be shown that if $\vec{a}$, $\vec{b}$, and $\vec{c}$ correspond to any combination of LP (CP) SOPs then $[\vec{a}, \vec{b}, \vec{c}]$ is also LP (CP).

We now introduce the operation

$$[\vec{a}, \vec{b}, \vec{c}, \vec{d}] = [\vec{a}, \vec{b}, \vec{c}] \cdot \vec{d}$$
$$= \frac{1}{3}[(\vec{a} \cdot \vec{b})(\vec{c} \cdot \vec{d}) + (\vec{b} \cdot \vec{c})(\vec{a} \cdot \vec{d}) + (\vec{c} \cdot \vec{a})(\vec{b} \cdot \vec{d})]. \quad (4.8)$$

Both these operations are independent of the order of the vectors in the square brackets. Each operation is linear with respect to any of the vectors, i.e.

$$[\alpha_1 \vec{a}_1 + \alpha_2 \vec{a}_2, \vec{b}, \vec{c}] = \alpha_1 [\vec{a}_1, \vec{b}, \vec{c}] + \alpha_2 [\vec{a}_2, \vec{b}, \vec{c}],$$
$$[\alpha_1 \vec{a}_1 + \alpha_2 \vec{a}_2, \vec{b}, \vec{c}, \vec{d}] = \alpha_1 [\vec{a}_1, \vec{b}, \vec{c}, \vec{d}] + \alpha_2 [\vec{a}_2, \vec{b}, \vec{c}, \vec{d}],$$

where α_1 and α_2 are scalars.

We then have $\{(\vec{E} \cdot \vec{E})\vec{E}\}_\ell \exp[-i(\beta_\ell z - \omega_\ell t)] = 3\rho^3 \vec{R}_\ell/8$ where, for a two-pump OPA,

$$\vec{R}_\ell = [\vec{A}_\ell, \vec{A}_\ell^*, \vec{A}_\ell] + 2 \sum_{j \neq l=1}^{4} [\vec{A}_j, \vec{A}_j^*, \vec{A}_\ell] + 2[\vec{A}_m, \vec{A}_n, \vec{A}_k^*] e^{i \Delta \beta_{klmn} z}. \quad (4.9)$$

The three terms on the right-hand side correspond respectively to SPM, XPM, and non-degenerate FWM. The integers k, l, m, and n are such that if $l = 1$ or 2 then $k = 3 - l$, $m = 3$, $n = 4$ and if $l = 3$ or 4 then $k = 7 - l$, $m = 1$, $n = 2$; $\Delta \beta_{klmn} = \beta_m + \beta_n - \beta_k - \beta_\ell$.

For a one-pump OPA, we remove the second pump ($\vec{A}_2 = 0$) and consider interactions between only three distinct waves. Equation (4.9) becomes

$$\vec{R}_1 = [\vec{A}_1, \vec{A}_1^*, \vec{A}_1] + 2 \sum_{j \neq l=1}^{4} [\vec{A}_j, \vec{A}_j^*, \vec{A}_1] + 2[\vec{A}_3, \vec{A}_4, \vec{A}_1^*] e^{i \Delta \beta z}, \quad (4.10)$$

$$\vec{R}_\ell = [\vec{A}_\ell, \vec{A}_\ell^*, \vec{A}_\ell] + 2 \sum_{j \neq l=1}^{4} [\vec{A}_j, \vec{A}_j^*, \vec{A}_\ell] + [\vec{A}_1, \vec{A}_1, \vec{A}_k^*] e^{-i \Delta \beta z},$$
$$l = 3, 4, \quad k = 7 - l. \quad (4.11)$$

where $\Delta \beta = \beta_3 + \beta_4 - 2\beta_1$.

We impose the condition $\vec{A}_k \cdot \vec{A}_k^* = P_k$, the power of the kth wave. If we normalize the electric field so that its square is equal to the irradiance, we see that ρ^2 should be the reciprocal of an area. We thus choose $\rho^{-2} = A_{\text{eff}}$, the mode effective area. This can also be derived from first principles, as in Chapter 3.

The propagation equation for $\vec{A}_\ell$ can then be rewritten as

$$\frac{d\vec{A}_\ell}{dz} = i\gamma \vec{R}_\ell, \quad (4.12)$$

where $\gamma = 3\omega_c \chi_{xxxx}^{(3)}/(8cn A_{\text{eff}})$ is the same fiber nonlinearity coefficient as in Chapter 3. Having established the basic formalism, we now specialize it to write down the equations governing two-pump and one-pump fiber OPAs.

4.2.3 Small-signal gain

We assume that the propagation of the pumps is not affected by the presence of the signal and the idler. The first step is thus to calculate the evolution of the pump fields along the fiber. With these, we can then set up the equations for the signal and idler fields.

(a) Pump equations

We begin with the two-pump case. To set up the pump equations, we neglect the effect of the signal and idler on the pump. The pumps interact only through the first two terms of Eq. (4.9). From the latter, we then obtain

$$\frac{d\vec{A}_k}{dz} = i\gamma([\vec{A}_k, \vec{A}_k, \vec{A}_k^*] + 2[\vec{A}_k, \vec{A}_\ell, \vec{A}_\ell^*]), \qquad k = 1, 2, \quad l = 3 - k. \qquad (4.13)$$

For LP or CP states, the two square brackets in Eq. (4.13) are proportional to $\vec{A}_k$. The proportionality constants are obtained by dot-multiplying them by $\hat{e}_k^*$, where $\hat{e}_k$ is the normalized Jones vector of $\vec{A}_k$; $\hat{e}_k$ is such that $\vec{A}_k = A_k \hat{e}_k$, where A_k is the scalar complex amplitude of $\vec{A}_k$. Then

$$\frac{dA_k}{dz} = i\gamma([\hat{e}_k, \hat{e}_k^*, \hat{e}_k, \hat{e}_k^*]P_k + 2[\hat{e}_k, \hat{e}_k^*, \hat{e}_\ell, \hat{e}_\ell^*]P_\ell)A_k$$
$$= i\gamma(a_{kk}P_k + 2a_{kl}P_\ell)A_k = ip_k A_k, \qquad k = 1, 2, \quad l = 3 - k, \qquad (4.14)$$

where

$$p_k = \gamma(a_{kk}P_k + 2a_{kl}P_\ell), \qquad k = 1, 2, \quad l = 3 - k \qquad (4.15)$$

and

$$a_{kl} = [\hat{e}_k, \hat{e}_k^*, \hat{e}_\ell, \hat{e}_\ell^*] = \frac{1}{3}(1 + |\hat{e}_k \cdot \hat{e}_\ell|^2 + |\hat{e}_k \cdot \hat{e}_\ell^*|^2). \qquad (4.16)$$

The coefficient a_{kl} has the following properties: it is always real, $1/3 \leq a_{kl} \leq 1$, $a_{kl} = a_{lk}$, and $a_{kk} = (2 + |\hat{e}_k \cdot \hat{e}_k|^2)/3$. If the two pumps are both LP or CP then $a_{11} = a_{22}$. For all the two-pump OPAs considered, we have $a_{12} = a_{34}$, $a_{13} = a_{24}$, and $a_{14} = a_{23}$.

For one-pump OPAs, we let $P_2 = 0$. Then Eq. (4.10) yields the pump equation, which contains only the first term (SPM). Hence $p_1 = \gamma a_{11} P_1$ and $p_2 = 0$.

The solution of Eq. (4.14), valid for one- or two-pump OPAs, is

$$A_k(z) = A_k(0)e^{ip_k z}, \qquad k = 1, 2. \qquad (4.17)$$

(b) Signal and idler equations

Here we will neglect the signal and idler SPM and XPM interactions between the signal and idler. Then the equations begin with two XPM terms, due to the two pumps, which are similar to the second term on the right-hand side of Eq. (4.14). In addition, there is a FWM term coupling the signal and idler. For LP or CP pumps the coupling from signal to idler and then back always leads back to the original signal SOP. Hence the SOP choices made in our list of OPAs are self-consistent. This implies that in the signal equation, for example, both the XPM term and the FWM term have the same SOP as the signal; this allows us to arrive at a scalar equation.

For two pumps, the equations for A_k take the form

$$\frac{dA_k}{dz} = i\gamma(2a_{k1}P_1 + 2a_{k2}P_2)A_k + 2i\gamma[\vec{A}_1, \vec{A}_2, \vec{A}_\ell^*] \cdot \hat{e}_k^* \exp(-i\Delta\beta z)$$
$$= 2i\gamma(a_{k1}P_1 + a_{k2}P_2)A_k + 2i\gamma A_1 A_2[\hat{e}_1, \hat{e}_2, \hat{e}_k^*, \hat{e}_\ell^*]A_\ell^* \exp(-i\Delta\beta z)$$
$$= 2i\gamma(a_{k1}P_1 + a_{k2}P_2)A_k + ir_k A_\ell^* \exp[i(p_1 + p_2 - \Delta\beta)z)]$$
$$= ip_k A_k + ir_k A_\ell^* \exp[i(p_1 + p_2 - \Delta\beta)z)], \tag{4.18}$$

where

$$\Delta\beta = \beta_3 + \beta_4 - \beta_1 - \beta_2,$$
$$p_k = 2\gamma(a_{k1}P_1 + a_{k2}P_2), \qquad r_k = 2\gamma A_1(0)A_2(0)[\hat{e}_1, \hat{e}_2, \hat{e}_k^*, \hat{e}_\ell^*]$$
$$k = 3, 4, \quad l = 7 - k. \tag{4.19}$$

Here r_k is the amplitude of the FWM coupling coefficient for the four waves. Note that $r_3 = r_4$, so we can use a single notation, r, for both the signal and idler equations. We define $P_0 = P_1 + P_2$ as the total pump power.

For a single pump, Eqs. (4.18) and (4.19) become

$$\frac{dA_k}{dz} = ip_k A_k + ir_k A_\ell^* \exp[i(2p_1 - \Delta\beta)z)], \quad k = 3, 4, \quad l = 7 - k, \tag{4.20}$$

$$p_k = 2\gamma a_{k1}P_1, \quad r_k = \gamma[A_1(0)]^2[\hat{e}_1, \hat{e}_1, \hat{e}_k^*, \hat{e}_\ell^*], \quad k = 3, 4, \quad l = 7 - k, \tag{4.21}$$

where $\Delta\beta = \beta_3 + \beta_4 - 2\beta_1$. Again we have $r_3 = r_4 = r$.

For all one-pump OPAs considered, $a_{13} = a_{14}$.

(c) Solution of the signal and idler propagation equations

The solution proceeds as in subsection 3.5.3. We introduce the total wavevector mismatch

$$\kappa = \Delta\beta + \Delta\beta_{NL}, \tag{4.22}$$

where

$$\Delta\beta_{NL} = p_3 + p_4 - p_1 - p_2 \qquad \text{for two pumps,} \tag{4.23}$$
$$\Delta\beta_{NL} = p_3 + p_4 - 2p_1 \qquad \text{for one pump.}$$

Equations (3.73)–(3.81) can now be used again. Note that these equations have the same form for either one- or two-pump OPAs and for all possible SOP combinations. We thus have a general solution that covers all 12 types of OPA identified above. Each type of OPA will have its own characteristics, determined by the SOPs of its waves.

For $\Delta\beta_{NL}$, symmetry and other considerations allow a reduction to the simple form $\Delta\beta_{NL} = \gamma P_0 u$, where u has the following forms:

$$u = 2(a_{13} + a_{14} - a_{12}) - a_{11} \qquad \text{for a two-pump OPA,} \tag{4.24}$$
$$u = 2(2a_{13} - a_{11}) \qquad \text{for a one-pump OPA.} \tag{4.25}$$

For CP OPAs $a_{kl} = 2/3$ for all k and l, so that u has one value for all two-pump OPAs and another for all for one-pump OPAs.

Table 4.1. Parameters $a_{k\ell}$, b, and u for one-pump OPAs

OPA type spi	a_{11}	a_{13}	$b = \dfrac{g_{max}}{\gamma P_0}$	$u = \dfrac{\Delta\beta_{NL}}{\gamma P_0}$
XXX	1	1	1	2
YXY	1	1/3	1/3	−2/3
RRR	2/3	2/3	2/3	4/3
LRL	2/3	2/3	0	4/3

Table 4.2. Parameters a_{kl}, b, and u for two-pump OPAs

OPA type sppi	a_{11}	a_{12}	a_{13}	a_{14}		$b = \dfrac{g_{max}}{2\gamma\sqrt{P_1 P_2}}$	$u = \dfrac{\Delta\beta_{NL}}{\gamma P_0}$
XXXX	1	1	1	1		1	1
YXXY	1	1	1/3	1/3		1/3	−5/3
XXYY	1	1/3	1	1/3		1/3	1
YXYX	1	1/3	1/3	1		1/3	1
RRRR	2/3	2/3	2/3	2/3	2/3		2/3
LRRL	2/3	2/3	2/3	2/3	0		2/3
RRLL	2/3	2/3	2/3	2/3	2/3		2/3
LRLR	2/3	2/3	2/3	2/3	2/3		2/3

Since they play such important roles in determining the gain spectra, we have calculated b and u for the 12 types of fiber OPA; they are listed in Tables 4.1 and 4.2, along with the quantities necessary for their calculation. We notice some interesting particular cases: YXXY and YXY are the only OPAs with $\Delta\beta_{NL} < 0$. This condition might be useful in fibers with particular dispersion properties, for it would be desirable for optimizing the shape of the gain spectrum in ways not feasible with $\Delta\beta_{NL} > 0$ [6, 7]. The cases LRRL and LRL have $r = 0$, which implies that they exhibit little gain at any wavelength since g^2 is generally negative.

We can now also examine the cases leading to polarization-independent OPAs. The cases XXYY and YXYX (Table 4.2) are degenerate in the sense that they have exactly the same r and $\Delta\beta_{NL}$, and this leads to the ability to amplify any input SOP with the same gain. The pumps, however, introduce linear birefringence into the signal through the Kerr effect, and so the signal SOP evolves along the fiber, if $P_1 \neq P_2$. Nevertheless, this does not affect the gain and should not be a problem in practice. Similar considerations apply to the cases RRLL and LRLR, indicating that the two orthogonal CP pumps also lead to a polarization-independent OPA [7].

4.2.4 Pump depletion

In Chapter 3 we studied in detail the possibility of pump depletion for the scalar case, where all the waves are in the same linear state of polarization. We did this for both one-pump and two-pump OPAs as well as for wavelength exchange. We showed that exact solutions exist, which are expressed in terms of Jacobian elliptic functions.

The investigation of pump depletion can also be carried out for all the more general combinations of SOPs that we have investigated in this chapter. The development is a generalization of that in Chapter 3 and follows it closely. For this reason, we place the detailed derivations for the general case in Appendix 3. The general expressions of Appendix 3 reduce to those of Chapter 3 upon substitution of the parameters for the XXXX configuration into the equations.

Here we merely set up the basic coupled equations necessary for studying pump depletion, by keeping all the terms in Eqs. (4.9)–(4.11). The rest of the derivations and discussion can be found in Appendix 3.

(a) Four-wave case (non-degenerate)
The required four equations are obtained from Eqs. (4.9) and (4.12). They are

$$-i\frac{d\vec{A}_\ell}{dZ} = [\vec{A}_\ell, \vec{A}_\ell^*, \vec{A}_\ell] + 2 \sum_{j \neq l = 1}^{4} [\vec{A}_j, \vec{A}_j^*, \vec{A}_\ell]$$

$$+ 2[\vec{A}_m, \vec{A}_n, \vec{A}_k^*]e^{i\Delta\beta_{klmn}Z/\gamma}, \qquad l = 1-4. \qquad (4.26)$$

(b) Three-wave case (degenerate)
We obtain from Eqs. (4.10)–(4.12)

$$-i\frac{d\vec{A}_1}{dZ} = [\vec{A}_1, \vec{A}_1^*, \vec{A}_1] + 2 \sum_{j \neq l = 1}^{4} [\vec{A}_j, \vec{A}_j^*, \vec{A}_1] + 2[\vec{A}_3, \vec{A}_4, \vec{A}_1^*]e^{i\Delta\beta Z/\gamma}, \qquad (4.27)$$

$$-i\frac{d\vec{A}_\ell}{dZ} = [\vec{A}_\ell, \vec{A}_\ell^*, \vec{A}_\ell] + 2 \sum_{j \neq l = 1}^{4} [\vec{A}_j, \vec{A}_j^*, \vec{A}_\ell] + [\vec{A}_1, \vec{A}_1, \vec{A}_k^*]e^{-i\Delta\beta Z/\gamma},$$

$$l = 3, 4, \quad k = 7 - l. \qquad (4.28)$$

4.2.5 Six-wave model

In subsection 3.5.6 we investigated the case of a two-pump OPA with the signal close to one of the pumps, and we saw that it was necessary to consider four waves in addition to the pumps in order to obtain an accurate description. We did this in the case where all the waves are in the same state of linear polarization. In this chapter we extend the six-wave-model analysis to all possible types of OPAs with either linearly polarized (LP) or circularly polarized (CP) waves. In particular, these include OPAs with orthogonal pumps, which have received a great deal of attention because they yield polarization-insensitive OPAs. By using analytical calculations as much as possible, we are able to provide physical insights about the origin of these phenomena. We obtain closed-form solutions when the signal is very close to one of the pumps and find interesting differences from the case where all waves have the same linear SOP; for orthogonal linear SOPs one can have the gain near the pumps higher than at the center wavelength. For circular

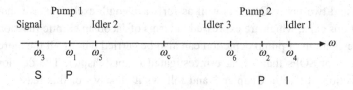

Fig. 4.1 Frequency assignments for the six-wave model: S, P, P, I refer respectively to the signal, first pump, second pump, and idler.

SOPs the gain spectrum can be the same as that expected from the four-wave model, exhibiting a smooth behavior near the pumps.

(a) Notation

We will use the notation introduced in subsection 4.2.1 to refer to the OPAs of interest. A member of this family of OPAs is designated by SPPI, where S, P, P, I respectively stand for the SOPs of the waves of the four-wave model, listed by increasing frequency, as shown in Fig. 4.1: the signal, at frequency ω_3; the first pump, at ω_1; the second pump, at ω_2; the idler, at ω_4.

In Fig. 4.1 are also shown the two additional waves considered in the six-wave model. They are the wave symmetric to the signal with respect to the first pump, at $\omega_5 = 2\omega_1 - \omega_3$ and the wave symmetric to the original idler with respect to the second pump, at $\omega_6 = 2\omega_2 - \omega_4$. The center frequency ω_c is such that $2\omega_c = \omega_1 + \omega_2 = \omega_3 + \omega_4 = \omega_5 + \omega_6$.

The three waves with even (odd) subscripts are grouped together. The waves at ω_3 and ω_5 (ω_4 and ω_6) can be viewed as sidebands associated with the pump at ω_1 (ω_2).

We can use the notation introduced for the four-wave model because for all possible cases the SOP of the new wave symmetric to the signal (idler) with respect to the pump adjacent to it is the same as the SOP to the signal (idler) itself. Therefore, by using this simple rule we know immediately the SOPs of the two new waves simply by inspection of the four-letter name. For example, for XYXY the SOP of the wave symmetric to the signal with respect to the first pump is X (the first letter), while that of the wave symmetric of the idler with respect to the second pump is Y (the last letter).

(b) The coupled equations and their solution

We consider first the pump equations, neglecting pump depletion, i.e. any terms due to the signal and the idlers. The equations are therefore the same as for the four-wave model, i.e. Eq. (4.14). The solution is

$$A_k = A_{k0} e^{i\varphi_k} \qquad k = 1, 2,$$

where $\varphi_k = \gamma(a_{kk} P_k + 2a_{kl} P_l)z$ is the nonlinear phase shift of the kth pump at the distance z.

The coupled equations for the remaining four waves are then (see Eq. (4.7) for the bracket notation)

$$
\begin{aligned}
-i\frac{d\vec{A}_3}{dZ} &= 2[\vec{A}_1, \vec{A}_1^*, \vec{A}_3] + 2[\vec{A}_2, \vec{A}_2^*, \vec{A}_3] + 2[\vec{A}_1, \vec{A}_2, \vec{A}_4^*]e^{i\Delta\beta_{1234}Z} \\
&\quad + [\vec{A}_1, \vec{A}_1, \vec{A}_5^*]e^{i\Delta\beta_{1135}Z} + 2[\vec{A}_1, \vec{A}_2^*, \vec{A}_6]e^{i\Delta\beta_{1623}Z}, \\
-i\frac{d\vec{A}_4}{dZ} &= 2[\vec{A}_1, \vec{A}_1^*, \vec{A}_4] + 2[\vec{A}_2, \vec{A}_2^*, \vec{A}_4] + 2[\vec{A}_1, \vec{A}_2, \vec{A}_3^*]e^{i\Delta\beta_{1234}Z} \\
&\quad + [\vec{A}_2, \vec{A}_2, \vec{A}_6^*]e^{i\Delta\beta_{2246}Z} + 2[\vec{A}_1, \vec{A}_2^*, \vec{A}_5]e^{i\Delta\beta_{1524}Z}, \\
-i\frac{d\vec{A}_5}{dZ} &= 2[\vec{A}_1, \vec{A}_1^*, \vec{A}_5] + 2[\vec{A}_2, \vec{A}_2^*, \vec{A}_5] + 2[\vec{A}_1, \vec{A}_2, \vec{A}_6^*]e^{i\Delta\beta_{1256}Z} \\
&\quad + [\vec{A}_1, \vec{A}_1, \vec{A}_3^*]e^{i\Delta\beta_{1135}Z} + 2[\vec{A}_1, \vec{A}_2^*, \vec{A}_4]e^{i\Delta\beta_{1425}Z}, \\
-i\frac{d\vec{A}_6}{dZ} &= 2[\vec{A}_1, \vec{A}_1^*, \vec{A}_6] + 2[\vec{A}_2, \vec{A}_2^*, \vec{A}_6] + 2[\vec{A}_1, \vec{A}_2, \vec{A}_5^*]e^{i\Delta\beta_{1256}Z} \\
&\quad + [\vec{A}_2, \vec{A}_2, \vec{A}_4^*]e^{i\Delta\beta_{2246}Z} + 2[\vec{A}_1, \vec{A}_2^*, \vec{A}_3]e^{i\Delta\beta_{1326}Z}
\end{aligned}
\quad (4.29)
$$

where $\Delta\beta_{klmn} = \beta_k + \beta_\ell - \beta_m - \beta_n$. Dot-multiplying Eqs. (4.29) by $\hat{e}_k^*$, $k = 3\text{–}6$, leads to

$$
\begin{aligned}
-i\frac{dA_3}{dZ} &= 2(a_{13}P_1 + a_{23}P_2)A_3 + 2c_{1234}(z)A_{10}A_{20}A_4^*e^{i(\varphi_1+\varphi_2)} \\
&\quad + 2c_{1135}(z)A_{10}^2A_5^*e^{2i\varphi_1} + 2c_{1623}(z)A_{10}A_{20}^*A_6e^{i(\varphi_1-\varphi_2)}, \\
-i\frac{dA_4}{dZ} &= 2(a_{14}P_1 + a_{24}P_2)A_4 + 2c_{1234}(z)A_{10}A_{20}A_3^*e^{i(\varphi_1+\varphi_2)} \\
&\quad + c_{2246}(z)A_{20}^2A_6^*e^{2i\varphi_2} + 2c_{2514}(z)A_{10}^*A_{20}A_5e^{i(\varphi_2-\varphi_1)}, \\
-i\frac{dA_5}{dZ} &= 2(a_{15}P_1 + a_{25}P_2)A_5 + 2c_{1256}(z)A_{10}A_{20}A_6^*e^{i(\varphi_1+\varphi_2)} \\
&\quad + c_{1135}(z)A_{10}^2A_3^*e^{2i\varphi_1} + 2c_{1425}(z)A_{10}A_{20}^*A_4e^{i(\varphi_1-\varphi_2)}, \\
-i\frac{dA_6}{dZ} &= 2(a_{16}P_1 + a_{26}P_2)A_6 + 2c_{1256}(z)A_{10}A_{20}A_5^*e^{i(\varphi_1+\varphi_2)} \\
&\quad + c_{2246}(z)A_{20}^2A_4^*e^{2i\varphi_2} + 2c_{2316}(z)A_{10}^*A_{20}A_3e^{i(\varphi_2-\varphi_1)},
\end{aligned}
\quad (4.30)
$$

where we have introduced the notation

$$c_{klmn}(z) = a_{klmn}e^{i\Delta\beta_{klmn}z} \quad (4.31)$$

and

$$a_{klmn} = [\hat{e}_k, \hat{e}_\ell, \hat{e}_m^*, \hat{e}_n^*]; \quad (4.32)$$

note that $a_{kl} = a_{klkl} = [\hat{e}_k, \hat{e}_\ell, \hat{e}_k^*, \hat{e}_\ell^*]$.

Six distinct values of $\Delta\beta_{klmn}$ enter Eq. (4.30). They correspond to three different types of FWM, as follows.

Type 1 $\Delta\beta_{1135}$ and $\Delta\beta_{2246}$ correspond to degenerate FWM about the pumps. This type of term is encountered when studying one-pump parametric amplification about a single pump. Because each pump in Fig 4.1 is relatively far from the ZDW, the wavevector mismatches due to these terms grow rapidly as the signal wavelength moves away from a pump. This variation leads to the appearance of sharp dips or peaks in the gain spectrum, surrounding each pump. As a first approximation, one can think of these features as being associated with the one-pump OPA gain spectrum due to each pump acting independently of the other. This provides a correct estimate for the spectral region where these features are found.

Type 2 $\Delta\beta_{1623}$ and $\Delta\beta_{2514}$ correspond to non-degenerate FWM, in which the pumps are involved in an asymmetric manner. By itself, this type of FWM leads to WE. As for type 1, these terms grow quickly away from the pumps and so their presence contributes to the rapid spectral variations around the pumps. They provide a coupling mechanism for the one-pump gain spectra of type 1, and therefore prevent the latter from being observed independently.

Type 3 $\Delta\beta_{1234}$ and $\Delta\beta_{1256}$ correspond to non-degenerate FWM, in which the pumps are involved in a symmetric manner. They correspond to the usual two-pump parametric amplification arising from the simple four-wave model. For a two-pump OPA optimized to have a large bandwidth, these terms vary much more slowly with wavelength than those of types 1 and 2 and so do not contribute to the rapid spectral variations around the pumps.

From the preceding discussion it should be clear that the description of the gain spectra near the pumps is complex, because there are several types of phase-matching terms, varying at different rates with wavelength. For this reason, we will not attempt to obtain a full analytic description of the spectrum valid for all wavelengths. Instead, we will restrict our attention to a narrow wavelength range near each pump, where $\Delta\beta'_{klmn}$ remains essentially constant. (In the following examples, this range is of the order of 1 nm or less and does not include the sharp variations extending over a few nanometers around the pumps.) In that case it can be shown that the various wavevector mismatches $\Delta\beta'_{klmn}$ in Eq. (4.30) approach zero, so that $c_{klmn}(z)$ can be replaced by a_{klmn}. Under these circumstances the only remaining exponentials are those containing φ_1 and φ_2. These can be eliminated by making the transformations $A_3 = B_3 e^{i\varphi_1}$, $A_4 = B_4 e^{i\varphi_2}$, $A_5 = B_5 e^{i\varphi_1}$, and $A_6 = B_6 e^{i\varphi_2}$. Upon complex-conjugating the second and third equations in Eq. (4.30), we then obtain the set of equations

$$-i\frac{dB_3}{dZ} = [(2a_{13} - a_{11})P_1 + (2a_{23} - 2a_{21})P_2]B_3 + 2a_{1234}A_{10}A_{20}B_4^*$$
$$+ a_{1135}A_{10}^2 B_5^* + 2a_{1623}A_{10}A_{20}^* B_6,$$

$$i\frac{dB_4^*}{dZ} = [(2a_{14} - a_{22})P_1 + (2a_{24} - 2a_{21})P_2]B_4^* + 2a_{1234}^* A_{10}^* A_{20}^* B_3$$
$$+ a_{2246}^*(A_{20}^*)^2 B_6 + 2a_{2514}^* A_{10} A_{20}^* B_5^*,$$

$$i\frac{dB_5^*}{dZ} = [(2a_{15} - a_{11})P_1 + (2a_{25} - 2a_{21})P_2]B_5^* + 2a_{1256}^* A_{10}^* A_{20}^* B_6$$
$$+ a_{1135}^*(A_{10}^*)^2 B_3 + 2a_{1425}^* A_{10}^* A_{20} B_4^*,$$

$$-i\frac{dB_6}{dZ} = [(2a_{16} - a_{22})P_1 + (2a_{26} - 2a_{21})P_2]B_6 + 2a_{1256}A_{10}A_{20}B_5^*$$
$$+ a_{2246}A_{20}^2 B_4^* + 2a_{2316}A_{10}^* A_{20}B_3,$$

(4.33)

which can be rewritten as

$$\frac{d}{dz}\begin{bmatrix} B_3 \\ B_4^* \\ B_5^* \\ B_6 \end{bmatrix} = i\gamma \begin{bmatrix} b_{33} & 2a_{1234}A_{10}A_{20} & a_{1135}A_{10}^2 & 2a_{1623}A_{10}A_{20}^* \\ -2a_{1234}^*A_{10}^*A_{20}^* & b_{44} & -2a_{2514}^*A_{10}A_{20}^* & -a_{2246}^*(A_{20}^*)^2 \\ -a_{1135}^*(A_{10}^*)^2 & -2a_{1425}^*A_{10}^*A_{20} & b_{55} & -2a_{1256}^*A_{10}^*A_{20}^* \\ 2a_{2316}A_{10}^*A_{20} & a_{2246}A_{20}^2 & 2a_{1256}A_{10}A_{20} & b_{66} \end{bmatrix} \begin{bmatrix} B_3 \\ B_4^* \\ B_5^* \\ B_6 \end{bmatrix},$$
(4.34)

where

$$\begin{aligned} b_{33} &= [(2a_{13} - a_{11})P_1 + (2a_{23} - 2a_{21})P_2], \\ b_{44} &= -[(2a_{14} - a_{22})P_1 + (2a_{24} - 2a_{21})P_2], \\ b_{55} &= -[(2a_{15} - a_{11})P_1 + (2a_{25} - 2a_{21})P_2], \\ b_{66} &= [(2a_{16} - a_{22})P_1 + (2a_{26} - 2a_{21})P_2]. \end{aligned}$$
(4.35)

Equation (4.34) has the form $d\tilde{B}/dz = N\tilde{B}$, where $\tilde{B}$ is the column vector in Eq. (4.34). The solution can be expressed as

$$\tilde{B}(z) = \exp(Nz)\,\tilde{B}(0) = \sum_{m=0}^{\infty} \frac{(Nz)^m}{m!} \tilde{B}(0). \quad (4.36)$$

The nature of the solution depends on the particular forms of the matrix coefficients. In general it is not possible to see at this stage whether the OPA gain near the pumps is different from that predicted by the four-wave model. To proceed further, we must calculate the matrix coefficients for particular cases. In order to simplify the calculations, we concentrate on the cases where the two pumps have equal power. We also assume without loss of generality that both pump phasors are real at $t = 0$. Then $A_{10} = A_{20} = \sqrt{P_0/2}$, where P_0 is the total pump power, and N takes the form $N = (i\gamma P_0/2)M$, where

$$M = \begin{bmatrix} c_{33} & 2a_{1234} & a_{1135} & 2a_{1623} \\ -2a_{1234}^* & -c_{44} & -2a_{2514}^* & -a_{2246}^* \\ -a_{1135}^* & -2a_{1425}^* & -c_{55} & -2a_{1256}^* \\ 2a_{2316} & a_{2246} & 2a_{1256} & c_{66} \end{bmatrix} \quad (4.37)$$

and

$$\begin{aligned} c_{33} &= 2(a_{13} + a_{23} - a_{12}) - a_{11}, \\ c_{44} &= 2(a_{14} + a_{24} - a_{12}) - a_{22}, \\ c_{55} &= 2(a_{15} + a_{25} - a_{12}) - a_{11}, \\ c_{66} &= 2(a_{16} + a_{26} - a_{12}) - a_{22}. \end{aligned}$$
(4.38)

The a_{klmn} and a_{kl} are real. As mentioned earlier, A_3 and A_5 always have the same SOP, and so do A_4 and A_6. As a result, $a_{13} = a_{15}$, $a_{23} = a_{25}$, $a_{14} = a_{16}$, $a_{24} = a_{26}$, and so $c_{33} = c_{55}$, $c_{44} = c_{66}$. The last two relations imply that the trace of M vanishes.

In addition $a_{13} = a_{24}$ and $a_{23} = a_{14}$, and therefore the c_{kk} all have the same value, which we denote by a.

We also have $a_{1623} = a_{1425} = a_{2514}^* = a_{1423}$, $a_{1234} = a_{1256}$, and $a_{1135} = a_{2246} = a_{1133} = a_{2244} = a_{13} = a_{24}$.

With these simplifications, M can be rewritten as

$$M = \begin{bmatrix} M_a & M_b \\ -M_b & -M_a \end{bmatrix}, \qquad (4.39)$$

where M_a and M_b are 2×2 matrices given by

$$M_a = \begin{bmatrix} a & b \\ -b & -a \end{bmatrix}, \quad M_b = \begin{bmatrix} c & d \\ -d & -c \end{bmatrix}, \qquad (4.40)$$

with $b = 2a_{1234}$, $c = a_{13}$, and $d = 2a_{1423}$. Note that the matrix M_a that couples B_3 and B_4^* is similar to that obtained in Appendix 1 for the four-wave model. Its eigenvalues (EVs) are $\pm(b^2 - a^2)^{1/2}$.

The matrix M depends only on a, b, c, and d:

$$M = \begin{bmatrix} a & b & c & d \\ -b & -a & -d & -c \\ -c & -d & -a & -b \\ d & c & b & a \end{bmatrix}. \qquad (4.41)$$

The determinant of M is

$$\begin{aligned} \mathrm{Det}(M) &= (a+b+c+d)(a+b-c-d)(a-b+c-d)(a-b-c+d) \\ &= [(a+b)^2 - (c+d)^2][(a-b)^2 - (c-d)^2]. \end{aligned} \qquad (4.42)$$

The characteristic equation for the EVs is

$$\mathrm{Det}(M - \lambda I) = \lambda^4 - 2(a^2 - b^2 - c^2 + d^2)\lambda^2 + \mathrm{Det}(M) = 0. \qquad (4.43)$$

Since Eq. (4.43) is of second order in λ^2 it can always be solved algebraically. The EVs are given by

$$\lambda = \pm[(a \pm d)^2 - (b \pm c)^2]^{1/2}, \qquad (4.44)$$

where the two $\pm$ signs inside the square bracket are the same, but the $\pm$ sign outside is independent of the two inside. We denote the EVs by

$$\begin{aligned} \lambda_1 &= [(a+d)^2 - (b+c)^2]^{1/2}, & \lambda_2 &= -\lambda_1, \\ \lambda_3 &= [(a-d)^2 - (b-c)^2]^{1/2}, & \lambda_4 &= -\lambda_3. \end{aligned} \qquad (4.45)$$

It can be shown that the four eigenvectors V_k of M, $k = 1$–4, are given by

$$V_{1,2} = \begin{bmatrix} b+c \\ -a-d+\lambda_{1,2} \\ -a-d+\lambda_{1,2} \\ b+c \end{bmatrix}, \quad V_{3,4} = \begin{bmatrix} -b+c \\ a-d-\lambda_{3,4} \\ -a+d+\lambda_{3,4} \\ b-c \end{bmatrix}. \qquad (4.46)$$

These eigenvectors have interesting symmetry properties. We call $V_{1,2}$ the symmetric eigenvectors, because B_3 and B_6 have the same amplitude and phase, and so do B_4 and B_5. Similarly, we call $V_{3,4}$ the antisymmetric eigenvectors, because the fields in each pair have the same amplitudes but opposite signs.

4.2 Isotropic fibers

Table 4.3. Parameters a, b, c, d and also Det(M), M^2, and the eigenvalues of M for two-pump OPAs

	OPA type sppi							
	XXXX	XXYY	YXYX	YXXY	RRRR	RRLL	LRLR	LRRL
a	1	1	1	−5/3	2/3	2/3	2/3	−2
b	2	2/3	2/3	2/3	4/3	4/3	4/3	0
c	1	1	1/3	1/3	2/3	2/3	0	0
d	2	2/3	2/3	2/3	4/3	4/3	0	4/3
Det(M)	0	0	0	0	0	0	16/9	400/81
Eigenvalues	0	0	0	0	0	0	$2i/\sqrt{3}$	10/3
	0	0	0	$4/\sqrt{3}$	0	0	$-2i/\sqrt{3}$	2/3
	0	0	4/3	$-4/\sqrt{3}$	0	0	$2i/\sqrt{3}$	−2/3
	0	0	−4/3	0	0	0	$-2i/\sqrt{3}$	−10/3
$M^2 = 0$?	Yes	Yes	No	No	Yes	Yes	No	No

It can be verified that the symmetric and asymmetric eigenvectors are orthogonal, in the sense that $V_{1,2}^t V_{3,4} = 0$. These symmetry properties can be used in some cases to rapidly find the decomposition of an input wave on the eigenvectors; this can then be used to express the evolution of all the waves as a linear superposition of eigenvectors, each propagating as determined by its EV. In particular, we can then find a good approximation for the gain by considering the dominant eigenvector(s).

For example, it can easily be verified that

$$(\lambda_4 - \lambda_3)(V_1 - V_2) + (\lambda_1 - \lambda_2)(V_3 - V_4) = 2(\lambda_4 - \lambda_3)(\lambda_1 - \lambda_2)[0 \ \ 1 \ \ 0 \ \ 0]^t, \quad (4.47)$$

which shows that if only A_4 is present at the input, it excites the four eigenvectors with weights proportional to $\lambda_4 - \lambda_3$ and $\lambda_1 - \lambda_2$.

If two or four EVs vanish, these considerations about eigenvectors are no longer applicable, because the eigenvectors associated with $\lambda = 0$ are not uniquely defined. The eigenvalues and EVs can also be used to calculate the exponentiated matrix of Eq. (4.36). However, as this direct calculation is relatively tedious for a 4 × 4 matrix we provide in Appendix 4 an alternate method for calculating the output fields in the general case.

(c) Discussion of individual cases

In Table 4.3 are listed the parameters a, b, c, d for all eight types of two-pump OPAs identified in [1], along with corresponding values of Det(M) and EVs.

Also indicated is whether M is nilpotent ($M^2 = 0$). If $M^2 = 0$, the series of Eq. (4.36) terminates after the first two terms, and therefore the fields grow in a linear manner with z instead of exponentially as in the case of a well-phased-matched OPA. In such situations, it is therefore likely that the gain will be low around the pump adjacent to the signal compared with its value at the center wavelength.

However, if $M^2 \neq 0$ then the series in Eq. (4.36) does not terminate and the field components vary exponentially or periodically with distance. In such cases, one needs to inspect the EVs to see whether this growth matches or perhaps exceeds that predicted by the four-wave model for a well-phase-matched OPA, which is valid at the center wavelength.

Before examining the spectra of the OPAs we note that, when the pumps have the same SOP, the relative orientation of the signal with respect to that of the adjacent pump does not change as one scans the signal wavelength is tuned across the pumps with a fixed SOP.

This is not true when the pumps are orthogonal: if the signal SOP is kept constant, its orientation with respect to the adjacent pump changes as the signal wavelength is tuned across the pumps. In such cases, the gain spectra around the two pumps should therefore be labeled appropriately. As an example, if the OPA is XXYY for $\lambda_3 < \lambda_c$ then it becomes YXYX for $\lambda_3 > \lambda_c$. As a result, the gain spectra may be different around the two pumps.

Inspection of Table 4.3 shows that XXXX, XXYY, RRRR, and RRLL satisfy $M^2 = 0$. Therefore, their gains should be low around the pump adjacent to the signal, as is the case for XXXX [8, 9].

Cases YXYX, YXXY, and LRRL fall into another class, since $M^2 \neq 0$. Thus we expect exponential variation of the fields. However, since the EVs are real, upon multiplication by $i\gamma P_0/2$ the argument of the exponentials becomes purely imaginary and the spatial variation of the fields is then periodic, rather than exponential. This leads to bounded fields, which would be exceeded by any real positive exponential after a long distance (provided that g does not vanish). Thus in this case there should also be a dip around the pump, provided that the gain is sufficiently large.

The LRLR case has $M^2 \neq 0$, and therefore we expect exponential growth. All four EVs are imaginary; they have the same magnitude and opposite signs. Upon multiplication by $i\gamma P_0/2$ the arguments of the exponentials become real, and so the spatial variations of the fields are real exponentials. We have $g = \gamma P_0/\sqrt{3} = 0.577\gamma P_0$, to be compared with the gain predicted by the four-wave model. The latter is obtained from the eigenvalues of the matrix M_a, which are the same as for M. Thus we expect that there will be no gain dip or peak around the pump.

A simple way to prove this is as follows. Since $M_b = 0$, $\vec{A}_5$ and $\vec{A}_6$ are coupled only to each other. Therefore, if only the signal is present at the input, only the signal and the main idler can grow. The two other idlers do not grow. Thus for this particular choice of SOPs the two extra idlers need not be included as we try to obtain a suitable set of equations. Therefore the usual four-wave model should provide an accurate approximation for the gain spectrum, even if the LCP signal is very close to the RCP pump. We conclude that the gain spectrum near that pump should be the same for either the four-wave model or the six-wave model.

Because the spectrum is the same as that predicted by the four-wave model and has a slow variation near the pump, one might think of using it to obtain a wide gain spectrum, free of dips or peaks near the pump. In practice, however, this may be hard to implement because maintaining circular SOPs in long fibers is very difficult [1].

4.2.6 Discussion

The theory presented here predicts the existence of 12 separate types of fiber OPAs, when LP or CP waves are used, and it would be interesting to verify their existence experimentally. This, however, may be difficult in most cases, because actual fibers always have some random birefringence, not included in our model.

This effect has prevented verification of the performance of LRLR to date, because a CP state launched into a fiber with random linear birefringence cannot be maintained over more than a few centimeters. The use of spun fibers [10, 11] may permit the use of OPAs with CP states, since the principal states of polarization (PSPs) of such fibers are CP and such SOPs can thus propagate up to the birefringence coherence length, which could be hundreds of meters long. Another possibility is to use pulsed pumps, with high peak powers, and fibers just a few centimeters long, in which case CP states could be well preserved. Circularly polarized states have been used successfully in the context of modulation instability (MI) in a spun fiber with very low residual birefringence [12]. Experiments with twisted dispersion-shifted fiber (DSF) several hundred meters long have been performed, allowing the demonstration of ellipse rotation [13], as well as polarization-independent four-wave mixing and wavelength exchange [14].

The situation is not as difficult for LP states, since typical fibers exhibit mostly linear birefringence and have nearly linear PSPs. Thus, by aligning LP input states with the PSPs, one can in principle propagate such states over hundreds of meters.

In OPA work it is common to use fibers that are hundreds or thousands of meters long. In such situations, one can expect that the SOPs will undergo many changes and will eventually be uniformly distributed over the Poincaré sphere. A consequence is that the resulting SOPs have a higher probability of being nearly LP than being nearly CP, since the CP states occur for just the two poles of the sphere, whereas the LP states occur over the entire equator [15]. Therefore we intuitively expect that, regardless of whether the input SOPs are LP or CP, the OPA will behave much like an OPA using LP states in a non-birefringent fiber. This explains why most OPA experiments to date, performed with randomly birefringent fibers, have been well explained on the basis of the simple theory developed for LP states in non-birefringent fibers; it also explains why experiments with CP states are more difficult than those with LP states.

With a uniform SOP distribution over the Poincaré sphere, one can invoke probability arguments and derive an averaged version of the basic propagation equations. This reasoning has been used successfully to investigate the propagation of solitons in fibers with random birefringence [16]. Among the well-known results in this area is that for SPM and XPM the γ value for an isotropic fiber should be multiplied by 8/9; this has been verified experimentally [17]. We will extend this to fiber OPAs in Section 4.4.

4.3 Fibers with constant birefringence

We now consider fibers with constant birefringence, such as polarization-maintaining fibers (PMFs). Such fibers are most easily fabricated to exhibit linear birefringence. The presence of a fast and a slow axis of linear polarization, which are orthogonal, provides a natural system of coordinates for decomposing other SOPs. We denote these axes by x and y; they are perpendicular to the direction of propagation, which is the z-axis. A wave with angular frequency ω propagating in the fiber can then be represented by a Jones vector of the form

$$\vec{A}(z) = [A_x(z) \quad A_y(z)]^t = [A_x(0)e^{i\beta^x z} \quad A_y(0)e^{i\beta^y z}]^t,$$

where $\beta^x = n_x \omega/c$ and $\beta^y = n_y \omega/c$ are the wavevectors for the x- and y-polarized components, respectively, and n_x and n_y are the corresponding refractive indices. The birefringence is defined as $\Delta n = n_x - n_y$.

In general β^x and β^y have different dispersion properties. More specifically, all the coefficients in their power series expansions with respect to ω are different. As a first approximation, however, it is often assumed that Δn is independent of frequency. Then $\beta^x - \beta^y = \Delta n\, \omega/c$, i.e. β^x and β^y differ by a term proportional to ω. As a result all their derivatives beyond the first ones are equal; they thus have the same zero-dispersion wavelength (ZDW). However, one should bear in mind that real fibers may not necessarily have this simple property.

4.3.1 Waves linearly polarized along the axes

Since birefringent fibers preserve linear SOPs parallel to the birefringence axes, a simple one-pump OPA configuration is obtained by using an x-polarized signal and a y-polarized pump; then the idler is x-polarized. The propagation equations governing such an OPA are very similar to those for an XYX in the isotropic case, Section 4.2. The coefficients of the SPM, XPM, and FWM terms are the same. A difference appears only when $\Delta\beta$ is written explicitly in terms of the individual wavevectors. Specifically, we have

$$\Delta\beta^{xyx} = \beta_3^x + \beta_4^x - 2\beta_1^y = \beta_3^x + \beta_4^x - 2\beta_1^x + 2\Delta n\, \omega_1/c$$
$$= \Delta\beta^{xxx} + 2\Delta n\, \omega_1/c, \tag{4.48}$$

where $\Delta\beta^{xxx} = \beta_3^x + \beta_4^x - 2\beta_1^x$ is the wavevector mismatch that would apply if all waves were x-polarized. Thus we see that, compared with XYX in an isotropic fiber, the only effect of birefringence is to add the extra term $2\Delta n\, \omega_1/c$ to $\Delta\beta$. Since Δn can be quite large for some fibers, this extra term can strongly modify the shape of the gain spectrum. We will investigate the resulting gain shapes in Chapter 5 and show that birefringence can be used to create narrow gain regions far from the pump [18].

Similarly, one can also implement several other schemes alluded to in section 4.2 for an isotropic fiber, namely XYYX, XXYY, and XYXY. For these schemes we have

$$\Delta\beta^{xyyx} = \Delta\beta^{xxxx} + \Delta n\, (\omega_1 + \omega_2)/c = \Delta\beta^{xxxx} + 2\Delta n\omega_c/c \tag{4.49}$$

and

$$\Delta\beta^{xxyy} = \Delta\beta^{xyxy} = \Delta\beta^{xxxx} + \Delta n\, (\omega_3 - \omega_1)/c$$
$$= \Delta\beta^{xxxx} + \Delta n\, (\Delta\omega_s - \Delta\omega_p)/c. \tag{4.50}$$

We see that the shift for XYYX is similar to that for XYX and could thus also generate narrow gain regions away from the pumps.

The shift for XXYY and XYXY is much smaller. However, its linear dependence on ω_4 indicates that the corresponding gain spectrum will not be symmetric with respect to ω_c, a fact of practical importance. This feature has been observed experimentally [19].

It should be mentioned for completeness that PMFs can be used to implement XXX and XXXX. This may seem like a trivial application for such fibers, but one should remember that currently non-PMFs are not fully isotropic and thus cannot really be used for implementing XXX and XXXX. The weak random birefringence in such fibers can cause severe problems (see Section 4.4 and Chapter 6), which can be completely avoided by using PMFs.

The preceding discussion is valid regardless of the relative strengths of the waves. Therefore the expressions for $\Delta\beta$ modified by birefringence can be used whether or not there is significant pump depletion. Thus all the expressions obtained in Appendix 3 can be extended to the case of constant birefringence, with suitably modified $\Delta\beta$'s. (If applicable, the $\Delta\beta$'s can be modified simply by using Δn; if not, the variations in β^x and β^y with respect to ω must be taken accurately into account.

4.3.2 Waves with arbitrary SOPs

We now consider the more complicated situation where the input waves have arbitrary SOPs, i.e. each has an x and a y component. Because the x and y components now have different propagation constants, the derivation of the SPM, XPM, and FWM terms that enter the OPA equations must be reexamined. We now let

$$A_k^s(z) = A_k^s(0) \exp\left[i\left(\beta_k^s z - \omega_k t\right)\right], \quad s = x, y, \quad (4.51)$$

and substitute in Eq. (4.9). Looking for the terms at the various frequencies for a two-pump OPA we obtain

$$-i\frac{dA_1^x}{dZ} = \{[A_1^x \hat{x} e^{i\beta_1^x z} + A_1^y \hat{y} e^{i\beta_1^y z}, A_1^x \hat{x} e^{i\beta_1^x z} + A_1^y \hat{y} e^{i\beta_1^y z}, (A_1^x \hat{x} e^{i\beta_1^x z} + A_1^y \hat{y} e^{i\beta_1^y z})^*]$$

$$+ 2[A_1^x \hat{x} e^{i\beta_1^x z} + A_1^y \hat{y} e^{i\beta_1^y z}, A_2^x \hat{x} e^{i\beta_2^x z} + A_2^y \hat{y} e^{i\beta_2^y z},$$

$$(A_2^x \hat{x} e^{i\beta_2^x z} + A_2^y \hat{y} e^{i\beta_2^y z})^*]\} \cdot \hat{x} e^{-i\beta_1^x z},$$

$$-i\frac{dA_1^y}{dZ} = \{[A_1^x \hat{x} e^{i\beta_1^x z} + A_1^y \hat{y} e^{i\beta_1^y z}, A_1^x \hat{x} e^{i\beta_1^x z} + A_1^y \hat{y} e^{i\beta_1^y z}, (A_1^x \hat{x} e^{i\beta_1^x z} + A_1^y \hat{y} e^{i\beta_1^y z})^*]$$

$$+ 2[A_1^x \hat{x} e^{i\beta_1^x z} + A_1^y \hat{y} e^{i\beta_1^y z}, A_2^x \hat{x} e^{i\beta_2^x z} + A_2^y \hat{y} e^{i\beta_2^y z},$$

$$(A_2^x \hat{x} e^{i\beta_2^x z} + A_2^y \hat{y} e^{i\beta_2^y z})^*]\} \cdot \hat{y} e^{-i\beta_1^y z},$$

$$-i\frac{dA_2^x}{dZ} = \{[A_2^x \hat{x} e^{i\beta_2^x z} + A_2^y \hat{y} e^{i\beta_2^y z}, A_2^x \hat{x} e^{i\beta_2^x z} + A_2^y \hat{y} e^{i\beta_2^y z}, (A_2^x \hat{x} e^{i\beta_2^x z} + A_2^y \hat{y} e^{i\beta_2^y z})^*]$$

$$+ 2[A_2^x \hat{x} e^{i\beta_2^x z} + A_2^y \hat{y} e^{i\beta_2^y z}, A_1^x \hat{x} e^{i\beta_1^x z} + A_1^y \hat{y} e^{i\beta_1^y z},$$

$$(A_1^x \hat{x} e^{i\beta_1^x z} + A_1^y \hat{y} e^{i\beta_1^y z})^*]\} \cdot \hat{x} e^{-i\beta_2^x z},$$

$$-i\frac{dA_2^y}{dZ} = \{[A_2^x\hat{x}e^{i\beta_2^x z} + A_2^y\hat{y}e^{i\beta_2^y z}, A_2^x\hat{x}e^{i\beta_2^x z} + A_2^y\hat{y}e^{i\beta_2^y z}, (A_2^x\hat{x}e^{i\beta_2^x z} + A_2^y\hat{y}e^{i\beta_2^y z})^*]$$
$$+ 2[A_2^x\hat{x}e^{i\beta_2^x z} + A_2^y\hat{y}e^{i\beta_2^y z}, A_1^x\hat{x}e^{i\beta_1^x z} + A_1^y\hat{y}e^{i\beta_1^y z},$$
$$(A_1^x\hat{x}e^{i\beta_1^x z} + A_1^y\hat{y}e^{i\beta_1^y z})^*]\} \cdot \hat{y}e^{-i\beta_2^y z},$$

$$-i\frac{dA_3^x}{dZ} = 2\{[A_1^x\hat{x}e^{i\beta_1^x z} + A_1^y\hat{y}e^{i\beta_1^y z}, (A_1^x\hat{x}e^{i\beta_1^x z} + A_1^y\hat{y}e^{i\beta_1^y z})^*, A_3^x\hat{x}e^{i\beta_3^x z} + A_3^y\hat{y}e^{i\beta_3^y z}]$$
$$+ [A_2^x\hat{x}e^{i\beta_2^x z} + A_2^y\hat{y}e^{i\beta_2^y z}, (A_2^x\hat{x}e^{i\beta_2^x z} + A_2^y\hat{y}e^{i\beta_2^y z})^*, A_3^x\hat{x}e^{i\beta_3^x z} + A_3^y\hat{y}e^{i\beta_3^y z}]$$
$$+ [A_1^x\hat{x}e^{i\beta_1^x z} + A_1^y\hat{y}e^{i\beta_1^y z}, A_2^x\hat{x}e^{i\beta_2^x z} + A_2^y\hat{y}e^{i\beta_2^y z},$$
$$(A_4^x\hat{x}e^{i\beta_4^x z} + A_4^y\hat{y}e^{i\beta_4^y z})^*]\} \cdot \hat{x}e^{-i\beta_3^x z},$$

$$-i\frac{dA_3^y}{dZ} = 2\{[A_1^x\hat{x}e^{i\beta_1^x z} + A_1^y\hat{y}e^{i\beta_1^y z}, (A_1^x\hat{x}e^{i\beta_1^x z} + A_1^y\hat{y}e^{i\beta_1^y z})^*, A_3^x\hat{x}e^{i\beta_3^x z} + A_3^y\hat{y}e^{i\beta_3^y z}]$$
$$+ [A_2^x\hat{x}e^{i\beta_2^x z} + A_2^y\hat{y}e^{i\beta_2^y z}, (A_2^x\hat{x}e^{i\beta_2^x z} + A_2^y\hat{y}e^{i\beta_2^y z})^*, A_3^x\hat{x}e^{i\beta_3^x z} + A_3^y\hat{y}e^{i\beta_3^y z}]$$
$$+ [A_1^x\hat{x}e^{i\beta_1^x z} + A_1^y\hat{y}e^{i\beta_1^y z}, A_2^x\hat{x}e^{i\beta_2^x z} + A_2^y\hat{y}e^{i\beta_2^y z},$$
$$(A_4^x\hat{x}e^{i\beta_4^x z} + A_4^y\hat{y}e^{i\beta_4^y z})^*]\} \cdot \hat{y}e^{-i\beta_3^y z},$$

$$-i\frac{dA_4^x}{dZ} = 2\{[A_1^x\hat{x}e^{i\beta_1^x z} + A_1^y\hat{y}e^{i\beta_1^y z}, (A_1^x\hat{x}e^{i\beta_1^x z} + A_1^y\hat{y}e^{i\beta_1^y z})^*, A_4^x\hat{x}e^{i\beta_4^x z} + A_4^y\hat{y}e^{i\beta_4^y z}]$$
$$+ [A_2^x\hat{x}e^{i\beta_2^x z} + A_2^y\hat{y}e^{i\beta_2^y z}, (A_2^x\hat{x}e^{i\beta_2^x z} + A_2^y\hat{y}e^{i\beta_2^y z})^*, A_4^x\hat{x}e^{i\beta_4^x z} + A_4^y\hat{y}e^{i\beta_4^y z}]$$
$$+ [A_1^x\hat{x}e^{i\beta_1^x z} + A_1^y\hat{y}e^{i\beta_1^y z}, A_2^x\hat{x}e^{i\beta_2^x z} + A_2^y\hat{y}e^{i\beta_2^y z},$$
$$(A_3^x\hat{x}e^{i\beta_3^x z} + A_3^y\hat{y}e^{i\beta_3^y z})^*]\} \cdot \hat{x}e^{-i\beta_4^x z}.$$

$$-i\frac{dA_4^y}{dZ} = 2\{[A_1^x\hat{x}e^{i\beta_1^x z} + A_1^y\hat{y}e^{i\beta_1^y z}, (A_1^x\hat{x}e^{i\beta_1^x z} + A_1^y\hat{y}e^{i\beta_1^y z})^*, A_4^x\hat{x}e^{i\beta_4^x z} + A_4^y\hat{y}e^{i\beta_4^y z}]$$
$$+ [A_2^x\hat{x}e^{i\beta_2^x z} + A_2^y\hat{y}e^{i\beta_2^y z}, (A_2^x\hat{x}e^{i\beta_2^x z} + A_2^y\hat{y}e^{i\beta_2^y z})^*, A_4^x\hat{x}e^{i\beta_4^x z} + A_4^y\hat{y}e^{i\beta_4^y z}]$$
$$+ [A_1^x\hat{x}e^{i\beta_1^x z} + A_1^y\hat{y}e^{i\beta_1^y z}, A_2^x\hat{x}e^{i\beta_2^x z} + A_2^y\hat{y}e^{i\beta_2^y z},$$
$$(A_3^x\hat{x}e^{i\beta_3^x z} + A_3^y\hat{y}e^{i\beta_3^y z})^*]\} \cdot \hat{y}e^{-i\beta_4^y z}.$$

(4.52)

We note that on the right-hand sides there are a number of terms multiplied by exponentials of the form

$$\exp[i(\beta_1^y + \beta_2^y - \beta_1^x - \beta_2^x)z], \quad \exp[i(\beta_1^y + \beta_2^y - \beta_3^x - \beta_4^x)z],$$

etc. The average of such terms is zero over a beat length. As a result, it is often assumed that all these terms can be neglected when studying nonlinear propagation in long fibers. (However, it is important to note that this argument fails for isotropic fibers; in this case all the terms should be kept, as in Section 4.2.)

With this assumption, Eq. (4.52) can be expressed in terms of the pump powers relating to the x- and y- polarizations:

$$-i\frac{dA_1^x}{dZ} = \left(P_1^x + \frac{2}{3}P_1^y + 2P_2^x + \frac{2}{3}P_2^y\right) A_1^x, \tag{4.53}$$

$$-i\frac{dA_1^y}{dZ} = \left(P_1^y + \frac{2}{3}P_1^x + 2P_2^y + \frac{2}{3}P_2^x\right) A_1^y, \tag{4.54}$$

$$-i\frac{dA_2^x}{dZ} = \left(P_2^x + \frac{2}{3}P_2^y + 2P_1^x + \frac{2}{3}P_1^y\right) A_2^x, \tag{4.55}$$

$$-i\frac{dA_2^y}{dZ} = \left(P_2^y + \frac{2}{3}P_2^x + 2P_1^y + \frac{2}{3}P_1^x\right) A_2^y, \tag{4.56}$$

$$-i\frac{dA_3^x}{dZ} = 2\left(P_1^x + \frac{1}{3}P_1^y + P_2^x + \frac{1}{3}P_2^y\right) A_3^x$$
$$+ 2A_1^x A_2^x A_4^{x*} \exp\left[i\left(\beta_1^x + \beta_2^x - \beta_3^x - \beta_4^x\right)z\right] \tag{4.57}$$

$$i\frac{dA_3^y}{dZ} = 2\left(P_1^y + \frac{1}{3}P_1^x + P_2^y + \frac{1}{3}P_2^x\right) A_3^y$$
$$+ 2A_1^y A_2^y A_4^{y*} \exp\left[i\left(\beta_1^y + \beta_2^y - \beta_3^y - \beta_4^y\right)z\right] \tag{4.58}$$

$$-i\frac{dA_4^x}{dZ} = 2\left(P_1^x + \frac{1}{3}P_1^y + P_2^x + \frac{1}{3}P_2^y\right) A_4^x$$
$$+ 2A_1^x A_2^x A_3^{x*} \exp\left[i\left(\beta_1^x + \beta_2^x - \beta_3^x - \beta_4^x\right)z\right] \tag{4.59}$$

$$-i\frac{dA_4^y}{dZ} = 2\left(P_1^y + \frac{1}{3}P_1^x + P_2^y + \frac{1}{3}P_2^x\right) A_4^y$$
$$+ 2A_1^y A_2^y A_3^{y*} \exp\left[i\left(\beta_1^y + \beta_2^y - \beta_3^y - \beta_4^y\right)z\right], \tag{4.60}$$

which are considerably simpler.

In fibers with constant birefringence the pump powers in the x- and y- polarizations remain constant and Eqs. (4.53)–(4.56) can be integrated, yielding exponential solutions for the pump fields. Substituting these into Eqs. (4.53)–(4.56) yields a system of four linear equations for the field components of the signal and the idler.

Inspection of Eqs. (4.57)–(4.60) shows that the x-components of the signal and idler are coupled only to each other, and similarly for the y-components. The resulting systems of equations are of the general form studied in Appendix 1, and their solutions can therefore be written in terms of exponential functions.

Looking at the FWM terms in the x- and the y-equations, we see that their maximum parametric gain coefficients are respectively equal to $\gamma\sqrt{P_1^x P_2^x}$ and $\gamma\sqrt{P_1^y P_2^y}$. Thus the maximum gain for the signal light polarized along one direction is not affected by the pump power components that are orthogonally polarized. The maximum gains for the two polarizations are generally different.

However, examination of the SPM and XPM terms shows a mixture of x- and y-terms. Since these terms affect the phase matching, the wavelengths of the gain maxima along one axis will generally be affected by the presence of orthogonally polarized pump components. This could possibly be used for tuning the gain spectrum of OPAs realized with PMFs.

4.4 Fibers with random birefringence

For much experimental OPA work to date, performance has been consistent with the simple theory developed for ideal fibers without birefringence, in which SOPs remain the same (linear) over the entire fiber length and for which one uses the nonlinearity coefficient γ, generally obtained by measuring the SPM or XPM. This general agreement is actually remarkable considering that the SOPs in typical fibers, which are not PMFs, vary randomly over the fiber length. This fiber characteristic is responsible for the phenomenon of polarization-mode dispersion (PMD), which has been the object of many studies in recent years, as it limits the performance of high-speed optical communication systems. In fibers with typical PMD of the order of 0.1 ps km$^{-1/2}$ the beat length is of the order of a few meters, which indicates that the SOPs can be substantially altered over such a distance. Since experiments with continuous-wave fiber OPAs may require hundreds or even thousands of meters of fiber, SOPs in such fibers are thus likely to cover the entire Poincaré sphere, with a high density of states in the vicinity of any particular state.

The effect of random fiber birefringence on nonlinear effects has been studied in particular contexts. It has been shown that the SPM that occurs in long fibers is accounted for by using a value of γ that is 8/9 times that for very short fibers. This result was established by assuming that in long fibers the SOPs have a random distribution, uniform over the Poincaré sphere [17]. Similar reasoning has been used for establishing the basic equations for soliton propagation in very long fiber links [16, 20, 21].

Carena *et al.* investigated the role of fiber birefringence in parametric amplification [22]. However, they were primarily interested in the phenomenon as it occurs in communication systems, using long fibers (50 km) and relatively low pump powers (40 mW), a regime different from that required in the production of compact fiber OPAs. While their numerical studies provided interesting examples of the effect of fiber birefringence on parametric gain, Carena *et al.* did not provide analytical expressions for gains that could be used for the design of discrete fiber OPAs.

In this section we are going to use a simplified model for the propagation equations, assuming that all SOPs evolve in the same manner. We will first perform an analytical averaging of the propagation equations, assuming that the SOPs uniformly cover the Poincaré sphere. This will lead to a new set of OPA equations, similar to those obtained in isotropic fibers but with modified parameters. We will then validate this analytical averaging by modeling the actual propagation of waves in a series of waveplates with random parameters, by means of a Runge–Kutta technique.

4.4.1 Basic OPA equations for arbitrary states of polarization

In Section 4.2 we investigated the various types of fiber OPA that can be obtained by using waves that are all either linearly polarized (LP) or circularly polarized (CP), in a fiber without birefringence. For that purpose a set of basic OPA differential equations was introduced that is valid regardless of the SOPs of the four waves involved (i.e. they can all be in different states of *elliptical* polarization). Here we modify these equations to include the effects of fiber birefringence. We assume that the pumps are not depleted; hence P_1 and P_2 are constants of motion. The birefringence is accounted for by the Hermitian matrix

$$M = \begin{bmatrix} \varepsilon_\ell & -i\varepsilon_c \\ i\varepsilon_c & -\varepsilon_\ell \end{bmatrix} \tag{4.61}$$

where ε_ℓ and ε_c are real coefficients, respectively corresponding to linear and circular birefringence. We assume that M is independent of wavelength. This is a good approximation if the birefringence is weak and the wavelength spacings are not very large.

For two-pump OPAs the pump equations are then (again using the notation of Eq. (4.7))

$$\frac{d\vec{A}_k}{dz} = iM\vec{A}_k + i\gamma\left([\hat{e}_k, \hat{e}_k^*, \hat{e}_k]P_k + 2[\hat{e}_\ell, \hat{e}_\ell^*, \hat{e}_k]P_\ell\right)A_k, \quad k=1,2,\ l=3-k. \tag{4.62}$$

The signal and idler equations are

$$\frac{d\vec{A}_k}{dz} = iM\vec{A}_k + 2i\gamma\left([\hat{e}_1, \hat{e}_1^*, \hat{e}_k]P_1 + [\hat{e}_2, \hat{e}_2^*, \hat{e}_k]P_2\right)A_k$$
$$+ 2i\gamma[\hat{e}_1, \hat{e}_2, \hat{e}_\ell^*]A_1 A_2 A_\ell^* e^{-i\Delta\beta z}, \quad k=3,4,\ l=7-k. \tag{4.63}$$

For each of these equations, the left-hand side can be rewritten as

$$\frac{d\vec{A}_k}{dz} = \frac{dA_k}{dz}\hat{e}_k + \frac{d\hat{e}_k}{dz}A_k. \tag{4.64}$$

Since M actually describes the evolution of $\hat{e}_k$, we also have

$$\frac{d\hat{e}_k}{dz} = iM\hat{e}_k. \tag{4.65}$$

Hence

$$\frac{d\vec{A}_k}{dz} = \frac{dA_k}{dz}\hat{e}_k + iMA_k\hat{e}_k. \tag{4.66}$$

We see that making this substitution in Eqs. (4.62) and (4.63) eliminates the terms containing M. Finally, in order to obtain scalar equations for the complex amplitudes A_k, we dot-multiply the modified Eqs. (4.62) and (4.63) by the respective complex-conjugated unit vectors $\hat{e}_k^*$. This leads to

$$\frac{dA_k}{dz} = i\gamma(a_{kk}P_k + 2a_{kl}P_\ell)A_k, \quad k=1,2,\ l=3-k, \tag{4.67}$$

$$\frac{dA_k}{dz} = 2i\gamma(a_{k1}P_1 + a_{k2}P_2)A_k + 2i\gamma\sqrt{P_1 P_2}[\hat{e}_1, \hat{e}_2, \hat{e}_\ell^*, \hat{e}_k^*]A_\ell^* e^{-i\Delta\beta z},$$
$$k=3,4,\ l=7-k, \tag{4.68}$$

where

$$a_{kl} = [\hat{e}_k, \hat{e}_k^*, \hat{e}_l, \hat{e}_l^*] = \frac{1}{3}(1 + |\hat{e}_k \cdot \hat{e}_l|^2 + |\hat{e}_k \cdot \hat{e}_l^*|^2). \tag{4.69}$$

For a one-pump OPA the pump equation is

$$\frac{dA_1}{dz} = i\gamma a_{11} P_1 A_1 \tag{4.70}$$

and the signal and idler equations are

$$\frac{dA_k}{dz} = 2i\gamma a_{k1} P_1 A_k + i\gamma a_{11kl} P_1 A_l^* e^{-i\Delta\beta z}, \qquad k = 3, 4, \ l = 7 - k, \tag{4.71}$$

where $a_{11kl} = [\hat{e}_1, \hat{e}_1, \hat{e}_l^*, \hat{e}_k^*]$. We note that Eqs. (4.67), (4.68), (4.70), and (4.71) have the same form as in the absence of birefringence. The difference from the previous derivation, however, is that now the SOP of each wave evolves in a random manner, governed by the variations in the fiber birefringence. As a result the coefficients of the equations, which depend on SOPs, vary in a random manner. Clearly it is generally not possible to obtain a closed-form solution for equations with such varying coefficients. However, we will see that, by making some reasonable assumptions about the evolution of SOPs in birefringent fibers, we will be able to average the coefficients of the equations and obtain closed-form expressions for the gain.

4.4.2 Preliminary relations

To render the analysis tractable, we make the assumption that if we launch N waves into a long fiber with randomly varying birefringence, their Jones vectors at any distance, $\hat{e}_k(z)$, $k = 1 - N$, can be obtained from their input Jones vectors by means of a common unitary matrix $U(z) = \exp[i \int_0^z M(\xi) d\xi]$, which represents only the fiber's linear birefringence. Hence we have

$$\hat{e}_k(z) = U(z)\hat{e}_k(0) \quad \text{and} \quad \hat{e}_l(z) = U(z)\hat{e}_l(0). \tag{4.72}$$

There are two assumptions behind this.

1. First, it is assumed that the nonlinear interactions do not modify the SOPs. This should be true in fibers with a large nonlinear length L_{NL}. In this case effects due to ellipse rotation and nonlinear birefringence should average to zero, owing to the frequent changes of sign related to the rapid SOP variation induced by the fiber's linear birefringence.
2. Second, it is assumed that waves at different frequencies experience identical phase shifts. This is a good approximation if the linear birefringence is weak, if the wavelength spacings are not very large, and if the fiber is short. However, it is not uncommon to run into experimental situations where one or more of these assumptions is or are violated; in such situations, this analysis will become invalid. (This type of situation will be considered in detail in Chapter 6.)

4.4 Fibers with random birefringence

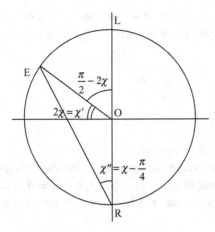

Fig. 4.2 Definition of the angles χ, χ', and χ'' used for the averaging procedure. The circle is the cross-section of the Poincaré sphere containing the poles (L and R, representing left and right circular polarization) and an arbitrary SOP represented by the point E. The angle $2\chi = \chi'$ is the latitude of E.

An important consequence of Eq. (4.72) is that if we consider the kth and the lth waves then their Hermitian product is conserved, because

$$\hat{e}_k^*(z) \cdot \hat{e}_l(z) = [U(z)\hat{e}_k(0)]^* \cdot U(z)\hat{e}_l(0)$$
$$= \hat{e}_k^*(0) U^{*t}(z) U(z) \cdot \hat{e}_l(0) = \hat{e}_k^*(0) \cdot \hat{e}_l(0). \quad (4.73)$$

(We have used the fact that $U^{*t}(z)U(z) = I$, the identity matrix, because $U(z)$ is unitary.) Equation (4.73) in particular implies that if two waves are initially in identical or orthogonal SOPs then they maintain this relationship everywhere along the fiber. These are the two cases that we will be considering below, as they correspond to the most interesting choices for the input SOPs of pumps and signals.

4.4.3 Averaging over the Poincaré sphere

We now make the assumption that a great deal of randomization occurs along the fiber, so that we can divide it into sections in which the SOP can be assumed to cover the Poincaré sphere uniformly. For each segment we then calculate the average value of the equation coefficients, using the appropriate probability density function for each coordinate. Figure 4.2 shows the angular coordinates that we use for locating a point on the sphere [23].

To calculate the various dot products in the basic equations, we use the basis of CP unit vectors $\{\hat{r}, \hat{l}\}$, where $\hat{r}$ is right-CP and $\hat{l}$ is left-CP. Then any SOP is represented by the unit vector $\hat{e} = u\hat{r} + v\hat{l}$. With this notation the dot product of any two vectors is $\hat{e}_1 \cdot \hat{e}_2 = u_1 v_2 + u_2 v_1$.

We may express u and v in terms of the standard angles χ and ψ locating SOPs on the Poincare sphere [23]: χ is half the latitude and ψ is half the longitude. It can be shown that

$$u = \frac{1}{\sqrt{2}} e^{i\psi}(\cos\chi + \sin\chi) = e^{i\psi}\cos\chi'', \quad (4.74)$$

$$v = \frac{1}{\sqrt{2}} e^{-i\psi}(\cos\chi - \sin\chi) = e^{-i\psi}\sin\chi'' \quad (4.75)$$

where $\chi'' = \chi - \pi/4$ is measured as shown in Fig. 4.2.

With any unit vector $\hat{e}_1$ we can associate another unit vector, $\hat{e}_5$, orthogonal to it; we can choose $\hat{e}_5 = u_1\hat{r} - v_1\hat{l}$. We have $\hat{e}_1 \cdot \hat{e}_1^* = \hat{e}_5 \cdot \hat{e}_5^* = 1$, $\hat{e}_1 \cdot \hat{e}_5^* = 0$. It can also be shown that $\hat{e}_1 \cdot \hat{e}_1 = -\hat{e}_5 \cdot \hat{e}_5 = \cos\chi' = \sin(2\chi'')$ and that $\hat{e}_1 \cdot \hat{e}_5 = i\cos\chi'\sin(2\psi)$, where $\chi' = 2\chi$ (Fig. 4.2).

We now proceed with averaging Eqs. (4.67), (4.68), (4.70), and (4.71) over the Poincaré sphere, assuming a uniform density of SOPs on the sphere. We use $\langle \cdot \rangle$ to represent the average of a term. In terms of the angles χ and ψ, the average of a function $f(\psi, \chi)$ is

$$\langle f(\psi, \chi) \rangle = \frac{1}{2\pi} \int_{\psi=0}^{\pi} \int_{2\chi=-\pi/2}^{\pi/2} f(\psi, \chi) \cos(2\chi) d(2\chi) d\psi$$

$$= \frac{1}{2\pi} \int_{\psi=0}^{\pi} \int_{\chi'=-\pi/2}^{\pi/2} f(\psi, \chi) \cos(\chi') d(\chi') d\psi. \quad (4.76)$$

The right-hand sides of Eqs. (4.67), (4.68), (4.70), and (4.71) contain SPM and XPM terms, expressed in terms of the coefficients a_{kl}. Hence we need to calculate $\langle a_{kl} \rangle$; it is given by

$$\langle a_{kl} \rangle = \frac{1}{3}\langle 1 + |\hat{e}_k \cdot \hat{e}_\ell|^2 + |\hat{e}_k \cdot \hat{e}_\ell^*|^2 \rangle \quad (4.77)$$

In particular,

$$3\langle a_{kk} \rangle = \langle 2 + |\hat{e}_k \cdot \hat{e}_k|^2 \rangle = 2 + \langle \cos^2\chi' \rangle = 2 + \frac{1}{2}\int_{\chi'=-\pi/2}^{\pi/2} \cos^3(\chi') d(\chi')$$

$$= 2 + \frac{1}{2}\int_{\chi'=-\pi/2}^{\pi/2} (1 - \sin^2\chi') d(\sin\chi')$$

$$= 2 + \int_{x=0}^{1} (1-x^2) dx = \frac{8}{3}. \quad (4.78)$$

Hence $\langle a_{kk} \rangle = 8/9$. This is the correction factor for SPM, which has been arrived at by others.

We will also need $\langle a_{kl} \rangle$, where k and l correspond to orthogonal states. Since orthogonality is preserved by the unitary transformations describing the random birefringence, we average over all pairs of orthogonal states. An example would be

$$3\langle a_{15} \rangle = \langle 1 + |\hat{e}_1 \cdot \hat{e}_5|^2 \rangle = 1 + \langle \cos^2\chi' \sin^2(2\psi) \rangle$$

$$= 1 + \frac{1}{4}\int_{\chi'=-\pi/2}^{\pi/2} \cos^3(\chi') d(\chi') = \frac{4}{3}. \quad (4.79)$$

Hence $\langle a_{15} \rangle = 4/9$.

Table 4.4. Parameters for one-pump OPAs with (b) and without (nb) birefringence

OPA type spi	$\langle a_{11}\rangle$	$\langle a_{13}\rangle$	$b = \dfrac{g_{\max}}{\gamma P_0}$	$u = \dfrac{\Delta\beta_{\rm NL}}{\gamma P_0}$
XXX (nb)	1	1	1	2
XXX (b)	8/9	8/9	8/9	16/9
YXY (nb)	1	1/3	1/3	−2/3
YXY (b)	8/9	4/9	2/9	0

With $\langle a_{kk}\rangle = 8/9$ and $\langle a_{15}\rangle = 4/9$, we can now calculate the average coefficients for the SPM and XPM terms in Eq. (4.69), whether the initial SOPs are identical or orthogonal.

Finally, we need to calculate terms of the type $\langle[\hat{e}_1, \hat{e}_2, \hat{e}_\ell^*, \hat{e}_k^*]\rangle = \langle[\hat{e}_1, \hat{e}_2, \hat{e}_k^*, \hat{e}_\ell^*]\rangle$, which come from the FMW terms coupling signal and idler. (Note that the two terms in a pair of such coupled equations are always identical.) These will take on a variety of forms, depending on the number of pumps, on whether their SOPs are identical or orthogonal, and on the initial signal SOP (and the resulting idler SOP). Assuming for reasons of symmetry that if the pumps have identical (orthogonal) SOPs then so do the signal and idler, there are only three possible forms for these terms, namely

$$\langle[\hat{e}_1, \hat{e}_1, \hat{e}_1^*, \hat{e}_1^*]\rangle = \langle a_{11}\rangle = \frac{8}{9}, \tag{4.80}$$

$$\langle[\hat{e}_1, \hat{e}_1, \hat{e}_5^*, \hat{e}_5^*]\rangle = \langle a_{1155}\rangle = \frac{1}{3}\langle(\hat{e}_1 \cdot \hat{e}_1)(\hat{e}_5^* \cdot \hat{e}_5^*)\rangle$$

$$= \frac{1}{6}\int_{\chi'=-\pi/2}^{\pi/2}\cos^3(\chi')d(\chi') = \frac{2}{9}, \tag{4.81}$$

$$\langle[\hat{e}_1, \hat{e}_5, \hat{e}_1^*, \hat{e}_5^*]\rangle = \langle a_{15}\rangle = \frac{1}{3}\left(1 + \langle|\hat{e}_1 \cdot \hat{e}_5|^2\rangle\right)/3$$

$$= \frac{1}{3} + \frac{1}{6\pi}\int_{\psi=0}^{\pi/2}\int_{\chi'=-\pi/2}^{\pi/2}\sin^2(2\psi)\cos^3(\chi')d\chi'd\psi = \frac{4}{9}. \tag{4.82}$$

With these averaged coefficients, we can now examine the gain characteristics of all the possible basic OPA configurations. We note that the averaged equations are all of the general form previously studied in the absence of random birefringence, so that we can immediately infer the important parameters $g_{\max}$ and $\Delta\beta_{\rm NL}$ from the averaged coefficients. Table 4.4 shows the results for one-pump OPAs and Table 4.5 those for two-pump OPAs. We have included for reference the results for OPAs with no birefringence, denoted by (nb); (b) indicates that the OPA does have random birefringence. Notation such as XXX was defined in Section 4.2 for non-birefringent fibers; we keep the same notation for birefringent fibers simply as a description of the input SOPs.

We see that for XXX and XXXX the effect of birefringence is to multiply all the a_{kl} by 8/9. As a result, $g_{\max}$ and $\Delta\beta_{\rm NL}$ are reduced by the same factor. In other words, for birefringent fibers the gain curves for the cases where all input waves are linearly polarized are the same as for non-birefringent fibers, except that γ is multiplied by 8/9. Therefore, from a practical standpoint it is actually difficult to determine whether the

Table 4.5. Parameters for two-pump OPAs with (b) and without (nb) birefringence

OPA type sppi	$\langle a_{11} \rangle$	$\langle a_{12} \rangle$	$\langle a_{13} \rangle$	$\langle a_{14} \rangle$	$b = \dfrac{g_{max}}{2\gamma \sqrt{P_1 P_2}}$	$u = \dfrac{\Delta\beta_{NL}}{\gamma P_0}$
XXXX (nb)	1	1	1	1	1	1
XXXX (b)	8/9	8/9	8/9	8/9	8/9	8/9
YXXY (nb)	1	1	1/3	1/3	1/3	−5/3
YXXY (b)	8/9	8/9	4/9	4/9	2/9	−8/9
XXYY (nb)	1	1/3	1	1/3	1/3	1
XXYY (b)	8/9	4/9	8/9	4/9	4/9	8/9
YXYX (nb)	1	1/3	1/3	1	1/3	1
YXYX (b)	8/9	4/9	4/9	8/9	4/9	8/9

fiber has birefringence or not, since the gain curves in both cases would be very similar, and the value of γ is generally not known with sufficient accuracy to distinguish between the two cases on the basis of which theoretical spectrum gives a better match for the experimental results.

Since many experiments to date have been performed with XXX or XXXX, this explains why theories developed for non-birefringent fibers have been sufficient to explain results obtained with fibers that actually have a great deal of random birefringence.

From this conclusion about XXX and XXXX one might be tempted to generalize to all types of OPA, and simply to scale γ by 8/9 to calculate g_{max} and $\Delta\beta_{NL}$ in all cases. However, examination of Tables 4.4 and 4.5 shows that this would be incorrect; in fact the only other items to which the 8/9 scaling applies are the $\Delta\beta_{NL}$ values for XXYY and YXYX; for all other items, the correction factors differ from 8/9, sometimes considerably.

An interesting observation is that g_{max} for XXYY and YXYX is actually *increased* by 1/3 (i.e. multiplied by 4/3) for the birefringent OPAs as compared with the non-birefringent OPAs. The existence of these two OPAs already leads to the ability to make OPAs whose gain is independent of the signal SOP, i.e. which are polarization-insensitive. We see that the presence of birefringence does not modify this conclusion but that birefringence actually leads to a higher gain. This is clearly a benefit, as we see that realistic fibers, exhibiting considerable birefringence, are quite suitable for making polarization-insensitive OPAs, with a gain actually larger than if we used birefringence-free fibers.

The ratio of the parametric gain coefficient g for XXYY and XXXX is 1/2 in the presence of random birefringence but would be only 1/3 in an ideally isotropic fiber. The difference is large enough that one should be able to distinguish between these two cases by experimental means. This has been done [15], and it was found that in commercially available fibers, which do exhibit significant random birefringence, the measured ratio was close to 1/2, in agreement with the theory. This provides a confirmation of the predictions made on the basis of averaging for a uniform distribution of SOPs on the Poincaré sphere.

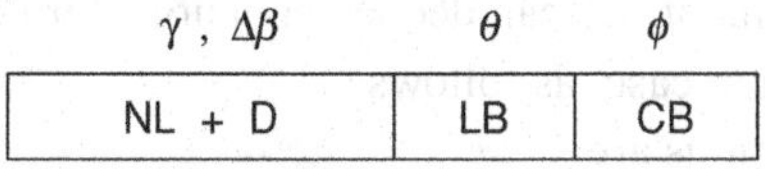

γ, $\Delta\beta$	θ	ϕ
NL + D	LB	CB

Fig. 4.3 Technique for calculating the output fields of a fiber segment.

4.4.4 Numerical integration of the basic equations

The analytical averaging technique used in the preceding section is based on the mathematical abstraction that the SOPs uniformly cover the Poincaré sphere during propagation over each short fiber segment, and one might argue that this is valid only in the limit of infinite randomization, that is physically unrealistic. So it is desirable to check the validity of the results by comparing with a model that uses a finite amount of randomization. To do so, we now divide the fiber into small segments and assume a new SOP for each segment; this corresponds to a randomization that is much closer to reality than assuming uniform sphere coverage in each small segment.

In each elementary segment we separate the various contributions and treat them one at a time; this is similar to what is done in the split-step Fourier method (SSFM), which will be described in detail in Chapter 6. We also combine some of these elementary contributions to speed up the calculations.

Figure 4.3 shows in schematic form how the input–output calculations are performed. We first consider an isotropic fiber with dispersion (D) and nonlinearity (NL); we studied propagation in such fibers in Section 4.2 and so we can make use of the equations in that section. This is then followed by a fiber exhibiting linear birefringence (LB), and another one with circular birefringence (CB); the combined role of these two sections is to provide a random rotation on the Poincaré sphere for the SOPs. Dispersion is included in the first isotropic stage but not in the LB and CB sections. So the LB stage provides a fixed phase shift θ_{xy} between the x- and y-axes, regardless of wavelength. Similarly, the CP stage provides rotation by the same angle ϕ for all wavelengths; this ensures that all SOPs evolve in the same way, as in the preceding section. (This model is good primarily for small wavelength spacings; it becomes unreliable for large spacings, for which waveplate dispersion cannot be neglected. In Chapter 6 we will make use of the nonlinear Schrödinger equation (NLSE) to investigate cases for which it cannot be assumed that the SOPs of all waves evolve in the same manner.)

The fiber is assumed to consist of a number N_z of the segments shown in Fig. 4.3. In each isotropic fiber segment all nonlinear effects such as ellipse rotation and the optical Kerr effect are included. We then use random unitary matrices to introduce SOP changes between segments. By repeating this procedure N_z times, we thus obtain an approximation to the propagation in an actual fiber with continuously varying random birefringence.

Specifically, we use a Runge–Kutta-type method for integrating. We denote the length of a fiber segment by $dz = L/N_z$. Assuming that $N_z \gg 1$, the fields do not change much over a segment and thus the changes can be well approximated by terms proportional to dz. Denoting by $\vec{A}_k(ndz)$ and $\vec{A}'_k(ndz)$ the slowly varying envelope of the kth wave at the beginning and at the end of the nth segment, we can deduce equations for the pump(s), signal and idler in the two- and one-pump case, as follows.

For two-pump OPAs, the pump equations are

$$\vec{A}'_k = \vec{A}_k + i\gamma dz([\vec{A}_k, \vec{A}_k, \vec{A}^*_k] + 2[\vec{A}_k, \vec{A}_\ell, \vec{A}^*_\ell]), \quad k=1,2, \quad l=3-k, \quad (4.83)$$

and the signal and idler equations are

$$\vec{A}'_k = \vec{A}_k + 2i\gamma dz([\vec{A}_1, \vec{A}^*_1, \vec{A}_k] + [\vec{A}_2, \vec{A}^*_2, \vec{A}_k] + [\vec{A}_1, \vec{A}_2, \vec{A}^*_\ell]e^{-i\Delta\beta z},$$
$$k=3,4, \quad l=7-k. \quad (4.84)$$

For one-pump OPAs, we have the pump equation

$$\vec{A}'_1 = \vec{A}_1 + i\gamma dz[\vec{A}_1, \vec{A}_1, \vec{A}^*_1], \quad (4.85)$$

and the signal and idler equations are

$$\vec{A}'_k = \vec{A}_k + i\gamma dz(2[\vec{A}_1, \vec{A}^*_1, \vec{A}_k] + [\vec{A}_1, \vec{A}_1, \vec{A}^*_\ell]e^{-i\Delta\beta z}),$$
$$k=3,4, \quad l=7-k. \quad (4.86)$$

The effect of the nth waveplate is represented by the unitary matrix

$$U_n = \begin{bmatrix} \cos\theta_n \, e^{i\varphi_n} & \sin\theta_n \\ -\sin\theta_n & \cos\theta_n \, e^{-i\varphi_n} \end{bmatrix}, \quad (4.87)$$

where θ_n and φ_n are random angles; θ_n corresponds to linear birefringence, and φ_n to circular birefringence. These angles are uniformly distributed over $[0, \pi]$. We then have

$$\vec{A}_k[(n+1)dz] = U_n \vec{A}'_k(ndz). \quad (4.88)$$

The OPA equations have been integrated in this manner for different values of N_z. For typical gains, of the order of 20 dB, good convergence is obtained for $N_z \geq 1000$. This procedure gives the values for $\Delta\beta_{NL}$ and g_{max} listed in Tables 4.4 and 4.5, except for the cases YXXY and YXY; convergence is also more difficult to achieve in these cases. The reason for this is that in these two cases the gain calculated in subsection 4.4.3 is small compared with the cases where all SOPs are the same (i.e. XXXX and XXX). Hence, if there is some leakage from the original signal and idler SOPs into the orthogonal SOPs, owing to ellipse rotation and the nonlinear Kerr effect, spurious polarization components can grow and their power may exceed that of the initial components (this has been verified by numerical simulations). Since YXXY and YXY have low gain they do not appear to be of much practical interest; hence it is not essential to improve the modeling of these cases at the present time.

For all the other types of SOP this situation is reversed, i.e. the gain for the undesired SOPs is low and hence little leakage into the wrong polarization occurs; the numerical results are in good agreement with the analytical ones.

4.4.5 Discussion

In the preceding subsections we have explained why, in many experiments performed with randomly birefringent fibers, it has been possible to use the simple theory developed for non-birefringent fibers. This shows that random birefringence may not be a problem for the design and operation of either high-gain or polarization-independent OPAs as long as they are not very wideband. Random birefringence can also be potentially beneficial under some circumstances (i.e. it leads to increased gain for polarization-independent OPAs).

The theory, however, requires that there be sufficient random birefringence to be able to perform the averaging many times over the fiber length. The theory may thus apply well in long fibers (1 km, or more) but less so in short fibers (say 20 m long). If it is desired to be in the regime where averaging applies, but with shorter fibers, it may be possible to induce strong *random* fiber birefringence *intentionally*, by bending, twisting, or squeezing the fiber in many locations; winding on an appropriate former could be a way to achieve this.

It is also important to note that the general form of the averaged equations is similar to those in non-birefringent fibers. It follows that the six types of OPA in birefringent fibers studied here and the 12 types in non-birefringent fibers studied in Section 4.2 form an equivalence class of 18 OPAs, in the sense that any two-pump OPA can be made to have exactly the same gain spectrum as any other OPA using the same type of fiber, by adjusting pump powers and pump frequencies (provided that a certain algebraic equation has real roots). This equivalence can provide a basis for the comparison and optimization of fiber OPAs. This will be discussed in more detail in Chapter 5.

4.5 Conclusion

In this chapter we have extended the mathematical formalism developed in Chapter 3 to include OPAs in which the waves do not remain in the same linear SOP along the entire fiber. We first investigated the ideal case of isotropic fibers, identified a total of 12 one- or two-pump OPAs for which each wave retained its initial LP or CP state, and showed how to calculate the gain of each. We also considered the case of fibers with constant birefringence and showed that they could be used for shaping gain spectra. We then developed a model including random fiber birefringence in the case where all SOPs can be assumed to evolve in the same manner; this model is applicable for weak birefringence and/or small wavelength spacings. In this case we were able to average the OPA expressions and so arrive at a solvable set of equations. The conclusion was that OPAs in such fibers can exhibit spectra similar to those in isotropic fibers, with minor or sometimes significant differences. The study of the more general case, where it cannot be assumed that all SOPs evolve in the same manner, will be undertaken in Chapter 6 with the help of the nonlinear Schrödinger equation and the split-step Fourier method.

References

1. "Fiber optical parametric amplifiers with linearly- or circularly-polarized waves," Marhic, M. E., Wong, K. K. Y., Kazovsky, L. G. *J. Opt. Soc. Amer. B.*; 2003; vol. 20, pp. 2425–33.
2. "Study of optical effects due to an induced polarization third order in the electric field strength," Maker, P. D., Terhune, R. W. *Phys. Rev.*; 1965; vol. 137, pp. A801–8.
3. "200-nm bandwidth fiber optical amplifier combining parametric and Raman gain," Ho, M. C., Marhic, M. E., Akasaka, Y., Kazovsky, L. G. *J. Lightwave Technol.*; 2001; vol. 19, pp. 977–81.
4. "Parametric and Raman amplification in birefringent fibers," Trillo, S., Wabnitz, S. *J. Opt. Soc. Amer. B*; 1992; vol. 9, pp. 1061–82.
5. "Unified analysis of four-photon mixing, modulational instability, and simulated Raman scattering under various polarization conditions in fibers," Golovchenko, E. A., Pilipetskii, A. N. *J. Opt. Soc. Amer. B*; 1994; vol. 11, pp. 92–101.
6. "Broadband fiber-optical parametric amplifiers," Marhic, M. E., Kagi, N., Chiang, T. K., Kazovsky, L. G. *Opt. Lett.*; 1996; vol. 21, pp. 573–5.
7. "Broadband fiber-optical parametric amplifiers and wavelength converters with low-ripple Chebyshev gain spectra," Marhic, M. E., Park, Y., Yang, F. S., Kazovsky, L. G. *Opt. Lett.*; 1996; vol. 21, pp. 1354–6.
8. "Unified analysis of modulational instability induced by cross-phase modulation in optical fibers," Tanemura, T., Kikuchi, K. *J. Opt. Soc. Amer. B*; 2003; vol. 20, pp. 2502–14.
9. "Quantum noise properties of parametric amplifiers driven by two pump waves," McKinstrie, C. J., Radic, S., Raymer, M. G. *Optics Express*; 2004; vol. 12, pp. 5037–66.
10. "Fiber optical parametric amplifier with circularly-polarized pumps," Marhic, M. E., Wong, K. K. Y., Kazovsky, L. G. *Electron. Lett.*; 2003; vol. 39, pp. 350–1.
11. "Polarization optics of twisted single mode fibers," Ulrich, R., Simon, A. *Appl. Opt.*; 1975; vol. 18, pp. 2241–51.
12. "Strong reduction of optimum pump power for efficient wave conversion in optical fibers with dual-frequency circularly polarized pump waves," Tchofo Dinda, P., Millot, G. *Opt. Lett.*; 2002; vol. 27, pp. 225–7.
13. "Observation of elliptical polarization rotation in a long twisted fiber," Tanemura, T., Kikuchi, K. *Opt. Lett.*; 2006; vol. 31, pp. 882–4.
14. "Polarization-insensitive asymmetric four-wave mixing using circularly polarized pumps in a twisted fiber," Tanemura, T., Katoh, K., Kikuchi, K. *Optics Express*; 2005; vol. 13, pp. 7497–505.
15. "Parametric amplification in fibers with random birefringence," Marhic, M. E., Wong, K. K. Y., Kazovsky, L. G. In *Proc. Optical Fiber Communication Conf.*, Los Angeles CA, February 2004; paper TuC2.
16. "Stability of solitons in randomly varying birefringent fibers," Wai, P. K. A., Menyuk, C. R., Chen, H. H. *Opt. Lett.*; 1991; vol. 16, pp. 1231–3.
17. "Measurement of normalization factor of n_2 for random polarization in optical fibers," Chernikov, S. V., Taylor, J. R. *Opt. Lett.*; 1996; vol. 21, 1559–61.
18. "Phase matching in birefringent fibers," Stolen, R. H, Bosch, M. A., Lin, C. *Opt. Lett.*; 1981; vol. 6, pp. 213–5.
19. "Polarization dependent parametric gain in amplifiers with orthogonally multiplexed optical pumps," Radic, S., Mckinstrie, C., Jopson, R. In *Proc. Optical Fiber Communication Conf.*, Atlanta GA, March 2003; *Technical Digest*, vol. 2, pp. 508–10.

20. "Polarization multiplexing with solitons," Evangelides, S. G., Mollenauer, L. F., Gordon, J. P., Bergano, N. S. *J. Lightwave Technol.*; 1992; vol. 10, pp. 28–35.
21. "Path average measurements of optical fiber nonlinearity using solitons," Andersen, J. K., Lou, J. W., Nowak, G. A., Xia, T., Islam, M. N., Fortenberry, R. M., Newton, S. A. *Lightwave Technol.*; 1998; vol. 16, pp. 2328–34.
22. "On the joint effects of fiber parametric gain and birefringence and their influence on ASE noise," Carena, A., Curri, V., Gaudino, R., Poggiolini, P., Benedetto, S. *J. Lightwave Technol.*; 1998; vol. 16, pp. 1149–57.
23. *Principles of Optics*, Born, M., Wolf, E. Pergamon, New York; 1970; p. 31.

5 The optical gain spectrum

5.1 Introduction

In Chapter 3 we derived expressions for the gain of fiber OPAs in terms of several parameters, including $\Delta\beta$, the wavevector mismatch. We now turn our attention to the frequency dependence of $\Delta\beta$, which we need to obtain the shape of the gain spectrum versus signal frequency; we will also study how the gain spectrum is influenced by the choice of pump frequencies and by the fiber's dispersion properties.

For the OPAs that we investigated in Chapter 4, for undepleted pumps the idler gain is of the form

$$G_4(z) = \left|\frac{r}{g}\sinh(gz)\right|^2.$$

The parametric gain coefficient g is obtained from $g^2 = r^2 - (\kappa/2)^2$; r is proportional to the pump power. The gain coefficient reaches its maximum value $g_{\max} = r$ when $\kappa = \Delta\beta + \Delta\beta_{NL} = 0$, where $\Delta\beta = \beta(\omega_3) + \beta(\omega_4) - \beta(\omega_1) - \beta(\omega_2)$; $\Delta\beta_{NL}$ is the nonlinear contribution to the wavevector mismatch, and it is of the order of γP_0. Hence G_4 depends on the wave frequencies and the fiber dispersion via $\Delta\beta$, and on the pump power via r and $\Delta\beta_{NL}$.

5.2 The effect of pump power on gain bandwidth

Before studying in detail the shape of the gain spectrum, let us establish the important fact that the spectrum is broadened by increasing pump power. To see this, let us first look at the maximum gain, obtained when $\kappa = 0$; it is $G_{4,\max}(z) = |\sinh(rz)|^2$. As the signal frequency is changed the gain is reduced, by an amount that depends only on g. For example, for a 3 dB drop in gain, g will need to be reduced to some value $g' = \delta r < r$. This in turn implies that $\Delta\beta$ must be changed to $\Delta\beta' = 2r\sqrt{1-\delta^2} - \Delta\beta_{NL}$. Since r and $\Delta\beta_{NL}$ are proportional to the pump power, we see that $\Delta\beta$ needs to scale as the pump power. As a result, the 3 dB bandwidth of the OPA, expressed in terms of $\Delta\beta$, scales as the pump power. We will see below that when the dependence of $\Delta\beta$ on ω_3 is taken into account, one finds that this 3 dB gain bandwidth increases with pump power, albeit in a sublinear fashion.

This dependence of OPA gain bandwidth on pump power is what makes it possible to achieve very wideband fiber OPAs. This effect is not present in other types of optical amplifier such as Raman amplifiers (RAs), erbiuim-doped fiber amplifiers (EDFAs), semiconductor optical amplifiers (SOAs), or Brillouin amplifiers.

5.3 The effect of fiber dispersion on the gain spectrum

5.3.1 One-pump OPA

To investigate possible shapes for the gain spectrum, we now need to express $\Delta\beta$ in terms of the optical frequency ω_3 of the signal and to study the influence of the available parameters on the gain spectrum. It is convenient to refer the frequencies to the pump frequency by letting $\Delta\omega_s = \omega_3 - \omega_1$. We can then expand $\Delta\beta$ as a power series in terms of $\Delta\omega_s$ [1, 2]:

$$\Delta\beta = \beta(\omega_3) + \beta(\omega_4) - 2\beta(\omega_1)$$
$$- \beta(\omega_1 + \Delta\omega_s) + \beta(\omega_1 - \Delta\omega_s) - 2\beta(\omega_1)$$
$$= 2\sum_{m=1}^{\infty} \frac{\beta^{(2m)}}{(2m)!}(\Delta\omega_s)^{2m} = 2\beta_e(\Delta\omega_s) - 2\beta(\omega_1), \quad (5.1)$$

where $\beta^{(l)} = d^l\beta/d\omega^l\big|_{\omega=\omega_1}$ and $\beta_e(\Delta\omega_s)$ is the even part of $\beta(\omega_1 + \Delta\omega_s)$ about ω_1, i.e.

$$\beta_e(\Delta\omega_s) = \frac{1}{2}[\beta(\omega_1 + \Delta\omega_s) + \beta(\omega_1 - \Delta\omega_s)] = \sum_{m=0}^{\infty} \frac{\beta^{(2m)}}{(2m)!}(\Delta\omega_s)^{2m}. \quad (5.2)$$

We note that $\Delta\beta$ depends only on the even derivatives of $\beta(\omega)$ at ω_1, namely $\beta^{(2)}(\omega_1)$, $\beta^{(4)}(\omega_1)$, $\beta^{(6)}(\omega_1)$, etc. and not directly on the odd derivatives $\beta^{(1)}(\omega_1)$, $\beta^{(3)}(\omega_1)$, etc. In particular, $\Delta\beta$ does not depend directly on the dispersion slope, which is proportional to $\beta^{(3)}(\omega_1)$.

Also, once the pump frequency is fixed, Eq. (5.2) depends only on $(\Delta\omega_s)^2$. This implies that $\Delta\beta$, g, $G_3(z)$, and $G_4(z)$ are all even functions of $\Delta\omega_s$. *Hence the signal and idler gain spectra are symmetric with respect to the center frequency.* This allows us to limit discussions and graphs to one side of ω_1, since everything is the same on the other side. (Note that this symmetry can be broken if the four-wave model of Chapter 3 is not sufficient to describe the nonlinear interactions; this occurs in the case of fibers with linear birefringence as well as in the case where six waves are required to describe the situation properly, as detailed in subsection 3.5.6.)

While the use of the angular frequency ω is convenient for these theoretical calculations, in experimental work it is generally more convenient to deal with the wavelength $\lambda = 2\pi c/\omega$. It is important to note that the gain symmetry is generally lost if one plots gain spectrum versus wavelength difference $\Delta\lambda_s = \lambda_3 - \lambda_1$, because then

$$(\Delta\omega_s)^2 = \left[\frac{2\pi c \Delta\lambda_s}{\lambda_1(\lambda_1 + \Delta\lambda_s)}\right]^2 \quad (5.3)$$

is not an even function of $\Delta\lambda_s$. This asymmetry is quite noticeable for very wideband OPAs, with a bandwidth of several hundred nanometers. For relatively narrowband OPAs, however, the effect is hardly noticeable: the gain spectrum is nearly symmetric when plotted versus $\Delta\lambda_s$.

The power series of Eq. (5.2) is most useful when it can be truncated after the first few terms, which is the case when the frequency excursions are not very large. Then $\Delta\beta$ is given by a simple low-order polynomial, which can be used for studying and optimizing the gain spectrum.

The simplest form for $\Delta\beta$ is $\Delta\beta = \beta^{(2)}(\Delta\omega_s)^2$, which holds when (i) the pump wavelength is not too close to λ_0 (where by definition $\beta^{(2)} = 0$) and (ii) the frequency excursions are not too large. For example, this would be the case for standard single-mode fiber (SSMF) used near 1550 nm, which exhibits a large chromatic dispersion $D = 17\,\text{ps}\,\text{nm}^{-1}\,\text{km}^{-1}$ and hence a large $\beta^{(2)} = -(\lambda^2/2\pi c)D$. Such a large value of $\beta^{(2)}$ leads to a large $\Delta\beta$ even for small frequency excursions and thus to a small gain bandwidth. In practice, with a pump power of a few hundred milliwatts the OPA bandwidth in such a case would be just a few nanometers.

Clearly then, in order to obtain much larger bandwidths one must reduce $\beta^{(2)}$. To do this, it is necessary to arrange for λ_1 to be very close to λ_0; in practice "very close" means within a few nanometers. One might in fact think that λ_1 and λ_0 should coincide, so that $\beta^{(2)} = 0$; in this case $\Delta\beta$ would be well approximated by the next surviving term, i.e. $\Delta\beta = \beta^{(4)}(\Delta\omega_s)^4/12$. While this choice would certainly lead to greatly improved bandwidth, it turns out that keeping a small but finite value for $\beta^{(2)}$ may actually be helpful in optimizing the spectrum shape and, in particular, in maximizing the bandwidth.

We are thus led to choose for $\Delta\beta$ a model that contains the first two terms, i.e.

$$\Delta\beta = \beta^{(2)}(\Delta\omega_s)^2 + \frac{\beta^{(4)}}{12}(\Delta\omega_s)^4. \tag{5.4}$$

In practice $\beta^{(2)}$ can be adjusted almost at will, in both magnitude and sign, by adjusting λ_1 to be near λ_0, since in that region we have $\beta^{(2)} \approx \beta^{(3)}(\omega_1 - \omega_0)$, where

$$\beta^{(3)} = \left(\frac{\lambda^2}{2\pi c}\right)^2 \frac{dD}{d\lambda}$$

when $\beta^{(2)} = 0$; $dD/d\lambda$ is the dispersion slope, generally provided by manufacturers, in $\text{ps}\,\text{nm}^{-2}\,\text{km}^{-1}$.

We note that when plotted versus $(\Delta\omega_s)^2$ the graph of $\Delta\beta$ is a parabola that always passes through the origin. Depending upon the values of $\beta^{(2)}$ and $\beta^{(4)}$, a variety of shapes can result for $g(\Delta\lambda_s)$. Two examples are shown in Figs. 5.1 and 5.2. In Fig. 5.1 we have assumed that $\beta^{(4)} > 0$ and have chosen $\beta^{(2)} < 0$ with a magnitude such that the ΔB parabola reaches an extremum vertically aligned with the left of the semicircle (see below). This ensures that g^2 remains positive over as wide a range of $\Delta\lambda_s$ as possible, i.e. this choice provides the largest gain bandwidth achievable with the given $\beta^{(4)}$.

In Fig. 5.1 we also show how to construct the graph of g versus $\Delta\lambda_s$ by graphical means. This is achieved by means of three curves, in three quadrants. The first curve, C_1, is the graph of $(\Delta\omega_s)^2$ versus $\Delta\lambda_s$; according to our preceding discussion, it is nearly

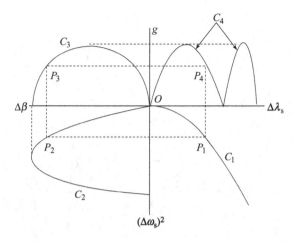

Fig. 5.1 Graphical construction of OPA gain coefficient curve, for $\beta^{(4)} > 0$.

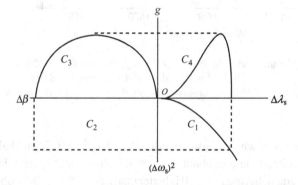

Fig. 5.2 Graphical construction of OPA gain coefficient curve, for $\beta^{(4)} < 0$.

parabolic. The second curve, C_2, is the graph of $\Delta\beta$ versus $(\Delta\omega_s)^2$, which is parabolic. The third curve, C_3, is the graph of g versus $\Delta\beta$ (note that $\Delta\beta$ becomes increasingly negative to the left of O); it is a half-ellipse that has been reduced to a semicircle by an appropriate choice of scales along the axes. With these three curves the graph of g versus $\Delta\lambda_s$, C_4, is constructed as follows. Choose a value of $\Delta\lambda_s$ on the horizontal axis and draw the vertical line passing through it; it intersects C_1 at P_1. Draw the horizontal line through P_1; it intersects C_2 at P_2. Draw the vertical line through P_2; it intersects C_3 at P_3. Draw the horizontal line through P_3; it intersects the vertical line through P_1 at P_4, which is the desired point on the double-humped curve C_4. This type of graphical construction is very helpful for visualizing how the shapes of the various curves combine to determine the shape of the gain spectrum. In particular, it is helpful for maximizing the gain bandwidth, for obtaining a flatter gain spectrum, etc.

Figure 5.2. is an example where $\beta^{(4)} < 0$ but it has the same magnitude as for Fig. 5.1. Here we have also chosen $\beta^{(2)}$ to maximize the gain bandwidth; graphical considerations indicate that this occurs for $\beta^{(2)} = 0$, resulting in the shape of C_2 shown. We see that the

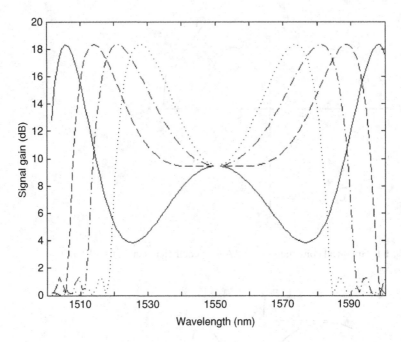

Fig. 5.3 Theoretical signal gain spectra for $\gamma = 2\ \text{W}^{-1}\ \text{km}^{-1}$, $P_0 = 7\ \text{W}$, $L = 200\ \text{m}$, $\beta^{(3)} = 1.2 \times 10^{-40}\ \text{s}^3\ \text{m}^{-1}$, $\beta^{(4)} = -5 \times 10^{-55}\ \text{s}^4\ \text{m}^{-1}$, and $\lambda_0 = 1550.0\ \text{nm}$. Solid line, $\lambda_1 = 1549.8\ \text{nm}$; broken line, $\lambda_1 = 1550.0\ \text{nm}$; broken-and-dotted line, $\lambda_1 = 1550.2\ \text{nm}$; dotted line, $\lambda_1 = 1550.5\ \text{nm}$.

gain bandwidth in this case, when measured along the $(\Delta\omega_s)^2$-axis, is just half that in Fig. 5.1; this translates into a ratio of about $1/\sqrt{2} \approx 0.7$ along the $\Delta\lambda_s$-axis. Hence we see that, all else being equal, having $\beta^{(4)} > 0$ is better than having $\beta^{(4)} < 0$ for obtaining a large gain bandwidth. However, we see that the gain curve is now single-humped instead of being double-humped as in Fig. 5.1; this may be viewed as a positive aspect.

The idler conversion gain G_4 is related to g by Eq. (3.79). While the gain bandwidth can be defined as the region where $g^2 > 0$, other features of the conversion gain cannot be easily understood by looking only at the graph of g. For example, G_4 is finite when $g = 0$, because it is obtained as the limit of a ratio of the form $0/0$; specifically, one finds that $G_4 = (rL)^2$. Hence the graph of G_4 in fact will not touch the $\Delta\lambda_s$-axis in the way that the graph of g does in Figs. 5.1 and 5.2.

Realistic examples of $G_3(\Delta\lambda_s) = 1 + G_4(\Delta\lambda_s)$ are shown in Fig. 5.3. The parameters correspond to a common type of DSF.

It can be seen that a variety of shapes can be obtained by tuning λ_1 near λ_0. However, these shapes are very sensitive to $\lambda_1 - \lambda_0$: a change of 0.1 nm will cause a noticeable distortion of the gain spectrum.

Figure 5.4 shows the kind of gain spectra that could be obtained for a fiber with the same parameters as in Fig. 5.3 but with a positive $\beta^{(4)}$ value. We note in particular that for $\lambda_1 = 1550.5\ \text{nm}$ we could obtain a gain spectrum with a high gain over a fairly wide region. Such a spectrum is not available for negative $\beta^{(4)}$, and this illustrates the

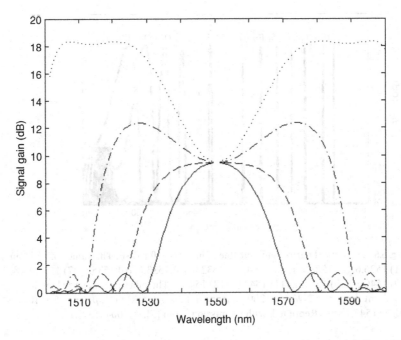

Fig. 5.4 All parameters are the same as in Fig. 5.3, except that the sign of $\beta^{(4)}$ has been changed, i.e. $\beta^{(4)} = +5 \times 10^{-55}$ s^4 m^{-1}.

desirability of being able to choose fiber parameters in such a way as to obtain interesting gain spectra.

The flat gain spectrum, shown by the broken line in Fig. 5.4, is interesting, and as such it has been studied in great detail [3]. However, as can be seen from Fig. 5.4, the shape of this flat gain spectrum is greatly affected by even very slight variations in pump wavelength (or alternatively in zero-dispersion wavelength (ZDW)). This indicates that actually obtaining such idealized spectra will be very difficult. Indeed, even with the recent availability of highly nonlinear fibers (HNLFs) with $\beta^{(4)} > 0$ [4], it has not been possible to observe spectra even remotely similar to the flat-top spectrum seen in Fig. 5.4 [5, 6]. It thus appears that the utilization of this spectral feature for making broadband gain spectra is probably not as promising as the use of two-pump fiber OPAs, to be described further on in this section.

Narrowband tunable gain spectra

So far, we have concentrated on obtaining a gain region in the vicinity of $\Delta\lambda_s = 0$ and increasing its width, for the purpose of making wideband optical amplifiers. However, the complex dependence of the gain on dispersion also makes it possible to obtain gain spectra with very different characteristics, which might be useful for other applications. One possibility is to achieve a single narrow gain region, far removed from λ_1. This might be useful for wavelength conversion, filtering with gain, tunable narrowband amplification, etc.

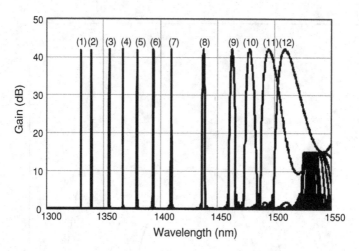

Fig. 5.5 Theoretical narrowband tunable gain spectra for DSF. The values of λ_1 in nm are (1) 1526.6, (2) 1528.1, (3) 1530.4, (4) 1532.0, (5) 1533.5, (6) 1535.1, (7) 1536.6, (8) 1538.9, (9) 1540.5, (10) 1541.3, (11) 1542.0, (12) 1542.8. The other parameter values are $\gamma = 2.3$ W^{-1} km^{-1}, $P_0 = 12$ W, $L = 200$ m, $\beta_3 = 1.14 \times 10^{-40}$ s^3 m^{-1}, $\beta_4 = -5 \times 10^{-55}$ s^4 m^{-1}, and $\lambda_0 = 1542.3$ nm. (Reprinted, with permission, from [7]. © 2004 IEEE.)

How this can be accomplished can be understood with the help of Fig. 5.1, for which $\beta^{(4)} < 0$. By tuning the pump further and further away from λ_0, we can make $\beta^{(2)}$ positive and of increasing in magnitude. This causes the parabola $\Delta\beta((\Delta\omega_s)^2)$ to intersect the $\Delta\beta = 0$ axis further from the origin and with an increasing slope. The OPA gain will be obtained in that region, as long as $-4\gamma P_0 < \Delta\beta < 0$; this will correspond to a decreasing range of $\Delta\lambda_s$ as the gain region moves further from the origin. This regime was investigated in [7], where it was found that, for the same parameters as in Fig. 5.3, the gain bandwidth $\delta\lambda$ at a distance $\Delta\lambda$ from the pump was

$$\delta\lambda = \frac{3.56 \times 10^6}{\Delta\lambda^3}, \qquad (5.5)$$

where the wavelengths are in nanometers. This yields $\delta\lambda = 1.36\,(0.54)$ nm for $\Delta\lambda = 100\,(200)$ nm.

By this approach, it is in principle possible to create a narrow gain region several hundred nanometers away from λ_1. The bandwidth would be less than 1 nm wide if the fiber had a uniform λ_0. However, longitudinal λ_0 variations will cause $\beta^{(2)}$, and hence the location of the gain region, to fluctuate. The overall effect will be a broadening of the gain region and a reduction in the peak gain; nevertheless, a relatively narrow gain region will still be produced [7–9].

Figure 5.5 shows plots of theoretical narrowband tunable gain spectra obtained by tuning the pump near λ_0. The fiber is simple DSF, and it is assumed to be uniform along its length. It can be seen that gain bandwidths of the order of 1 nm are obtained far from λ_0, as predicted by the above equations.

5.3.2 Two-pump OPA

Two-pump OPAs have gain characteristics fairly similar to those of one-pump OPAs. However, they have an additional independent parameter, that can be used the better to optimize the gain spectrum. Here we will concentrate on this point.

We let $\omega_c = (\omega_1 + \omega_2)/2$, $\Delta\omega_s = \omega_3 - \omega_c$, $\Delta\omega_p = \omega_1 - \omega_c = \omega_c - \omega_2$. Then as before we can expand $\Delta\beta$ as a double power series in terms of $\Delta\omega_s$ and $\Delta\omega_p$ [10]:

$$\Delta\beta = \beta(\omega_s) + \beta(\omega_i) - \beta(\omega_1) - \beta(\omega_2)$$
$$= \beta(\omega_c + \Delta\omega_s) + \beta(\omega_c - \Delta\omega_s) - \beta(\omega_c + \Delta\omega_p) - \beta(\omega_c - \Delta\omega_p)$$
$$= 2\sum_{m=1}^{\infty} \frac{\beta^{(2m)}}{(2m)!}\left[(\Delta\omega_s)^{2m} - (\Delta\omega_p)^{2m}\right]$$
$$= 2[\beta_e(\Delta\omega_s) - \beta_e(\Delta\omega_p)], \quad (5.6)$$

where $\beta^{(l)} = d^l\beta/d\omega^l|_{\omega=\omega_c}$; $\beta_e(\Delta\omega_s)$ is the even part of $\beta(\omega_c + \Delta\omega_s)$, i.e.

$$\beta_e(\Delta\omega_s) = \frac{1}{2}[\beta(\omega_c + \Delta\omega_s) + \beta(\omega_c - \Delta\omega_s)]$$
$$= \sum_{m=0}^{\infty} \frac{\beta^{(2m)}}{(2m)!}(\Delta\omega_s)^{2m}. \quad (5.7)$$

Again we find that the signal and idler gain spectra are symmetric with respect to the center frequency.

We will find it convenient to use ω_c and $\Delta\omega_p$ as the two independent parameters, instead of ω_1 and ω_2. Equation (5.6) shows that if we change $\Delta\omega_p$, while keeping ω_c fixed, the graph of $\Delta\beta(\Delta\omega_s)$ is simply translated along the $\Delta\beta$-axis, by an amount that depends only on $\Delta\omega_p$; $\Delta\beta(\Delta\omega_s)$ no longer has to pass through the origin, as was the case for a one-pump OPA. This important property in principle allows us to shift the graph of g along the $\Delta\beta$-axis by an (almost) arbitrary amount. This can then be used to ensure that the dispersion properties are optimized to yield maximum gain at a particular frequency, as well as a desirable gain shape in the vicinity of that frequency. The range of the shift may not be infinite, however, as it depends on the behavior of the function $\beta_e(\Delta\omega_p)$, which may be bounded (this is indeed the case when the series in Eq. (5.7) is limited to two terms, as studied in [10]).

The two-term model for $\Delta\beta$ is as follows:

$$\Delta\beta = \beta^{(2)}\left[(\Delta\omega_s)^2 - (\Delta\omega_p)^2\right] + \frac{\beta^{(4)}}{12}\left[(\Delta\omega_s)^4 - (\Delta\omega_p)^4\right]. \quad (5.8)$$

Here, $\beta^{(2)}$ can be adjusted almost at will, in both magnitude and sign, by adjusting λ_c near λ_0, since in that region we have $\beta_2 \approx \beta_3(\omega_c - \omega_0)$. For a given $\beta^{(2)}$ we can translate the graph of $\Delta\beta(\Delta\omega_s)$ by adjusting $\Delta\omega_p$. The amount of translation can be conveniently obtained as

$$\Delta\beta(\Delta\omega_s = 0) = -\left[\beta^{(2)}(\Delta\omega_p)^2 + \frac{\beta^{(4)}}{12}(\Delta\omega_p)^4\right]. \quad (5.9)$$

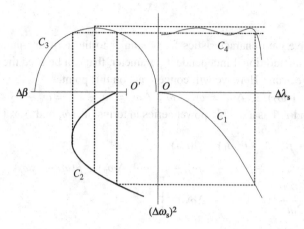

Fig. 5.6 Graphical construction of the gain coefficient graph for a two-pump OPA, showing that a flattened gain spectrum C_4 can be obtained.

Before proceeding with an algebraic optimization of the shape of the gain spectrum, let us consider in Fig. 5.6 a graphical construction similar to Figs. 5.1 and 5.2, in order to show that two-pump OPAs can provide flatter gain spectra than their one-pump counterparts. The basic idea is that since the parabola C_2 no longer has to pass through the origin, we can have $\Delta\beta(\Delta\omega_s = 0)$ close to the optimum value for high gain, and also have a suitable slope at that point (i.e. a suitable $\beta^{(2)}$). Then we can arrange for the top of the parabola to remain near the optimum $\Delta\beta$ over a large range of $\Delta\omega_s$, which will lead to a relatively flat-top gain spectrum. This is illustrated in Fig. 5.6. Note that here we have separated the origin of the $\Delta\beta$ axis, O', from that of the other three axes, O, so that C_3 and C_4 are separated.

The shape of C_4 in particular can be optimized so that it has two equal maxima, as well as two equal minima, for $\Delta\lambda_s > 0$. The first condition is satisfied simply by ensuring that the maximum of C_2 is to the left of $\Delta\beta = -\gamma P_0$; then C_2 will cross this value twice, providing two equal maxima. The second condition is somewhat more difficult to satisfy. It requires that the minimum value of $\Delta\beta$ be symmetric to $\Delta\beta(\Delta\omega_s = 0)$ with respect to $\Delta\beta = -\gamma P_0$; this leads to an algebraic relation between the fiber parameters. When these conditions are satisfied, $\Delta\beta(\Delta\omega_s)$ is an even fourth-order polynomial, with a minimum at $\Delta\omega_s = 0$. As such, it can be written in terms of a fourth-order Chebyshev polynomial:

$$\frac{\Delta\beta}{\gamma P_0} = -1 + \rho T_4 \left(\frac{\Delta\omega_s}{\Delta\omega_T}\right), \tag{5.10}$$

where $T_4(x) = 8x^4 - 8x^2 + 1$, ρ is the relative ripple in $\Delta\beta$, and $\Delta\omega_T$ represents the range (to one side of zero) over which the ripple remains between the Chebyshev extrema; we use it as a measure of the OPA bandwidth. From Eq. (5.10) we can calculate g^2, which is such that

$$\left(\frac{g}{\gamma P_0}\right)^2 = 1 - \frac{\rho^2}{4}\left[T_4\left(\frac{\Delta\omega_s}{\Delta\omega_T}\right)\right]^2 = 1 - \frac{\rho^2}{8}\left[1 + T_8\left(\frac{\Delta\omega_s}{\Delta\omega_T}\right)\right]. \tag{5.11}$$

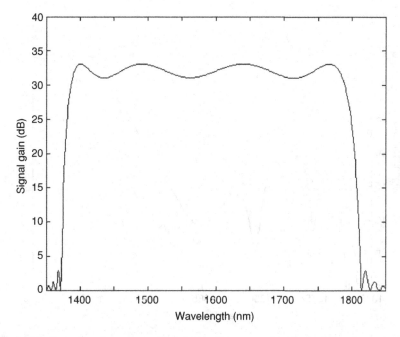

Fig. 5.7 Example of two-pump fiber OPA with a very wide Chebyshev gain spectrum, calculated with the four-wave model. The parameter values are $\gamma = 25$ W^{-1} km^{-1}, $\lambda_0 = 1561.00$ nm, $\beta^{(4)} = 2 \times 10^{-56}$ s^4 m^{-1}, $D_\lambda = 0.024$ ps nm^{-2} km^{-1}, $L = 30$ m, $\lambda_1 = 1385.00$ nm, $\lambda_2 = 1791.45$ nm, and $P_{10} = P_{20} = 3$ W.

This shows that the relative ripple in g^2 is $\rho^2/8$, which is quite small even for moderate values of ρ. This is therefore an effective method for obtaining flat gain spectra.

We let $\Delta\omega'_4$ denote $\Delta\omega_T$ for $\rho = 0$. It can be shown that $\Delta\omega'_4 = (12\gamma P_0/\beta^{(4)})^{1/4}$ and that

$$\rho = \frac{1}{8}\left(\frac{\Delta\omega_T}{\Delta\omega'_4}\right)^4. \tag{5.12}$$

Equation (5.12) shows that there is a trade-off between ρ and $\Delta\omega_T$; hence very flat gain spectra can be obtained but only over a restricted bandwidth.

Numerical examples show that one could in principle obtain very flat gain spectra over tens of nanometers by this method. Experiments have borne out the prediction that two-pump OPAs can exhibit flatter gain spectra than one-pump OPAs [11].

Figure 5.7 shows a calculated gain spectrum for a two-pump OPA with the following parameters: $\gamma = 25$ W^{-1} km^{-1}; $\lambda_0 = 1561.00$ nm; $\beta^{(4)} = +2 \times 10^{-56}$ s^4 m^{-1}; $D_\lambda = 0.024$ ps nm^{-2} km^{-1}; $L = 30$ m; $\lambda_1 = 1385.00$ nm; $\lambda_2 = 1791.45$ nm; $P_{10} = P_{20} = 3$ W. The fiber was a recently developed HNLF, with $\beta^{(4)} > 0$ [4], which is a requirement for obtaining the very flat Chebyshev gain spectra described above. The pump parameters were chosen to yield a very broad gain spectrum. The high-gain region (say with gain values over 30 dB) is over 400 nm wide.

The four-wave model was used, and therefore the gain-dips around the pump wavelengths (investigated in subsection 3.5.6) do not appear. To see them, we also performed

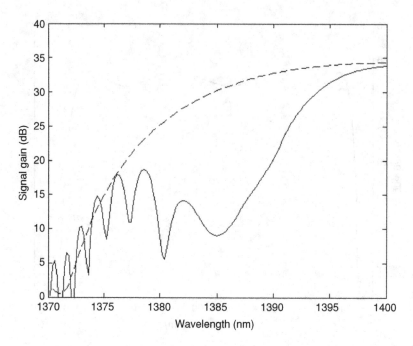

Fig. 5.8 Expanded view of the gain spectrum for the same OPA as in Fig. 5.7. Broken line, four-wave model; solid line, six-wave model.

calculations with the six-wave model in the vicinity of the pumps. Figure 5.8 shows an expanded view of the gain spectrum for the same OPA, centered about $\lambda_1 = 1385.00$ nm. As expected from Eq. (3.120) the gain in decibels at λ_1 predicted by the six-wave model is close to $10 \log[1 + (\gamma P_{10} L)^2] = 7.8$ dB. It can be seen that the gain above 1400 nm is virtually unaffected by the presence of the extra idlers. A similar situation exists around $\lambda_2 = 1791.45$ nm. Altogether, even the more accurate six-wave model still predicts a high-gain region extending over 375 nm.

Figure 5.9 shows an example of gain spectra calculated for a two-pump OPA that has $\beta^{(4)} < 0$ and therefore cannot exhibit a Chebyshev gain spectrum. It shows results obtained by the four-wave model and results obtained by the six-wave model. The parameters were chosen to give a reasonably flat spectrum; it can be seen that the spectrum calculated from the four-wave model is flatter than for a typical one-pump OPA.

In this example the gain dips have a non-negligible width and they are at inconvenient locations, causing a significant loss of usable gain bandwidth.

Overall, we see that the availability of fibers with $\beta^{(4)} > 0$ allows the design of two-pump fiber OPAs with superior gain spectrum characteristics.

5.4   OPAs with similar gain spectra

In Section 3.4 we showed that different OPAs can have the same longitudinal gain profile and overall gain, provided that Eqs. (3.41)–(3.43) are satisfied. These relations can tell

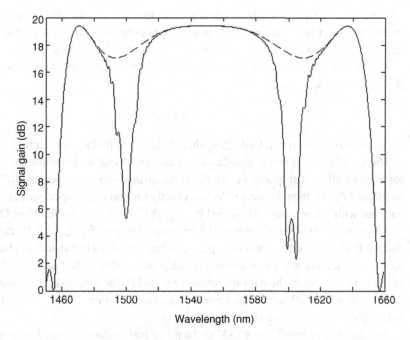

Fig. 5.9 Two-pump OPA gain spectra obtained by the four-wave model (broken line) and the six-wave model (solid line). The parameter values are $\gamma = 18$ W^{-1} km^{-1}, $\lambda_0 = 1550.00$ nm, $\beta^{(4)} = -5 \times 10^{-56}$ s^4 m^{-1}, $D_\lambda = 0.02$ ps nm^{-2} km^{-1}, $L = 160$ m, $\lambda_1 = 1500.00$ nm, $\lambda_2 = 1602.00$ nm, and $P_{10} = P_{20} = 0.5$ W.

us about the gain equivalence for particular wavevector mismatches but cannot provide information about gain spectra, since they do not contain explicit information about the dispersion of the fibers.

To obtain information about gain spectra, it is thus necessary to make additional assumptions. We now assume that we are comparing two OPAs that are made from the same type of fiber. Then Eqs. (3.41)–(3.43) lead to

$$\frac{\Delta\beta(\Delta\omega_s)}{\Delta\beta(\Delta\omega'_s)} = \frac{P_0}{P'_0} = \frac{L'}{L}. \tag{5.13}$$

To have two different OPAs, we need $P_0/P'_0 \neq 1$. Since the two fibers now have the same dispersion we cannot have $\Delta\omega_s = \Delta\omega'_s$, because then the leftmost ratio in Eq. (5.13) would be equal to 1 and would differ from the middle ratio.

To find possible relations between $\Delta\omega_s$ and $\Delta\omega'_s$ such that Eq. (5.13) can be satisfied over a wide range of frequencies, we need to make some assumptions about the form of the function $\Delta\beta(\Delta\omega_s)$. Specifically, we are going to consider cases where $\Delta\beta(\Delta\omega_s)$ can be approximated by a low-order power series in $\Delta\omega_s$.

(a) Second-order approximation

Here we assume a one-pump OPA and that $\Delta\beta(\Delta\omega_s)$ can be well approximated as

$$\Delta\beta(\Delta\omega_s) = \beta^{(2)}(\Delta\omega_s)^2. \tag{5.14}$$

This model is suitable when the pump is not too close to λ_0, so that $\beta^{(2)}$ is fairly large and the second-order term dominates the power series expansion of $\Delta\beta(\Delta\omega_s)$ (i.e. provided that $\Delta\omega_s$ is not too large).

To satisfy Eq. (5.13) we need to scale $\Delta\beta(\Delta\omega_s)$ as P_0. One way to do this is to keep $\beta^{(2)}$ constant, and to have

$$|\Delta\omega_s| \propto (P_0)^{1/2}. \tag{5.15}$$

(In the case of a two-pump OPA, $\Delta\omega_p$ should also scale in the same manner.)

The significance of this is as follows. Let us assume that we know the gain spectrum for a given OPA, with pump power P_0. If we change the pump power to P_0' and the length to LP_0/P_0' then the new OPA has exactly the same gain spectrum as the initial one, but with a frequency axis scaled by $(P_0'/P_0)^{1/2}$. As an example, if an OPA has a gain bandwidth of 100 GHz (about 0.8 nm near 1550 nm) for $P_0 = 1$ W then raising P_0 to 100 W will result in the same spectral shape but with a bandwidth of about 1 THz (about 8 nm). Note that this reasoning is independent of how the bandwidth is defined, because all features of the gain spectrum are scaled by the same magnification factor.

Since these two OPAs have the same gain spectrum, within a scaling of the frequency axis, we say that they have *similar* gain spectra.

Another way to satisfy Eq. (5.13) as the pump power changes would be to keep $\Delta\omega_s$ unchanged and to arrange to have

$$\beta^{(2)} \propto P_0. \tag{5.16}$$

This is possible because $\beta^{(2)} = \beta^{(3)}(\omega_1 - \omega_0)$, so that $\beta^{(2)}$ can be varied simply by tuning the pump wavelength. In this case, since there is no scaling of the frequency axis the two gain spectra are actually exactly *identical*.

More generally, it is in principle also possible to combine the scaling of $\beta^{(2)}$ with that of $\Delta\omega_s$. The most general case would correspond to

$$\beta^{(2)} \propto f(P_0) \quad \text{and} \quad |\Delta\omega_s| \propto [P_0/f(P_0)]^{1/2}, \tag{5.17}$$

where $f(P_0)$ is an arbitrary positive function of P_0. In all such cases the gain spectra of the various OPAs will be similar, the scaling of the frequency axis being given by Eq. (5.17).

(b) Fourth-order approximation

In this case we assume that $\Delta\beta(\Delta\omega_s)$ has the form of an even fourth-order polynomial in $\omega_s - \omega_1$, which is often a good model for broadband fiber OPAs when the pump is close to λ_0, so that $\beta^{(2)}$ is fairly small. Specifically, we let

$$\Delta\beta(\Delta\omega_s) = \beta^{(2)}(\Delta\omega_s)^2 + \frac{\beta^{(4)}}{12}(\Delta\omega_s)^4. \tag{5.18}$$

As ω_p is tuned about ω_0, we can assume that $\beta^{(4)}$ remains fixed but $\beta^{(2)}$ varies linearly, i.e. $\beta^{(2)} = \beta^{(3)}(\omega_1 - \omega_0)$.

To satisfy Eq. (5.13), again we need to scale $\Delta\beta(\Delta\omega_s)$ as P_0. Since $\beta^{(4)}$ is fixed but $\beta^{(2)}$ can be varied, we see that we will have the desired scaling of $\Delta\beta(\Delta\omega_s)$ if and only if we have the following scaling rules:

$$|\Delta\omega_s| \propto (P_0)^{1/4} \quad \text{and} \quad \beta^{(2)} \propto (P_0)^{1/2}. \tag{5.19}$$

In contrast with the previous case, which had several possible solutions, this solution is unique.

All OPAs obtained using of Eq. (5.19) will have similar gain spectra. If we change the pump power to P_0', the length to LP_0/P_0', and $\beta^{(2)}$ to $\beta^{(2)}(P_0'/P_0)^{1/2}$, by tuning the pump wavelength, then the new OPA will have exactly the same gain spectrum as the initial one but with the frequency axis scaled by $(P_0'/P_0)^{1/4}$. As an example, if an OPA has a gain bandwidth of 10 THz (about 80 nm near 1550 nm) for $P_0 = 1$ W then raising P_0 to 100 W will result in a bandwidth of over 30 THz (about 300 nm).

(c) Sixth-order approximation

Such an approximation may be necessary for an OPA with a very large bandwidth, e.g. covering hundreds of nanometers. Then

$$\Delta\beta(\Delta\omega_s) = \beta^{(2)}(\Delta\omega_s)^2 + \frac{\beta^{(4)}}{12}(\Delta\omega_s)^4 + \frac{\beta^{(6)}}{360}(\Delta\omega_s)^6. \tag{5.20}$$

The second-order mismatch $\beta^{(2)}$ can be adjusted at will by tuning the pump near λ_0, but $\beta^{(4)}$ and $\beta^{(6)}$ remain essentially fixed. To have the sixth-order term scale as P_0 we would need $|\Delta\omega_s| \propto (P_0)^{1/6}$, but to have the fourth-order term do the same requires $|\Delta\omega_s| \propto (P_0)^{1/4}$. Since these two requirements are in contradiction, it is therefore not possible to find OPAs with gain spectra depending on the sixth order of dispersion that will be similar.

5.5 Equivalent gain spectra for OPAs using pumps with different SOPs

In the preceding section we studied the conditions under which OPAs of the same type, and made from the same type of fiber, would exhibit similar gain spectra, i.e. gain spectra that are the same except for a scaling of the frequency axis. Here we examine another type of relationship between gain spectra, namely the possibility that different types of fiber OPA (i.e. fiber OPAs that use different SOPs, as seen in Chapter 4), but still made from the same type of fiber, may be designed to exhibit the *same* gain spectra.

For any of the 12 types of OPA studied in subsection 4.3.2, g^2 can be rewritten as

$$g^2 = -\frac{1}{4}(\kappa - 2r)(\kappa + 2r)$$

$$= -\frac{1}{4}(\Delta\beta + \Delta\beta_{NL} - 2r)(\Delta\beta + \Delta\beta_{NL} + 2r). \tag{5.21}$$

Viewed as an equation for $\Delta\beta$, the equation $g^2 = 0$ has the two roots $\Delta\beta_\pm = -\Delta\beta_{NL} \pm 2r$. The OPA gain region is such that $g^2 > 0$, i.e. it corresponds to $\Delta\beta_- < \Delta\beta < \Delta\beta_+$. The distance between the two roots is $\delta\beta = \Delta\beta_+ - \Delta\beta_- = 4r$. For g real and positive, the graph of g versus $\Delta\beta$ is a half-ellipse, symmetric about $\Delta\beta_{NL} = -(\Delta\beta_+ + \Delta\beta_-)/2$. Maximum gain is reached for $\Delta\beta = -\Delta\beta_{NL}$ and is $g_{max} = r$. We thus see that all these half-ellipses are similar in the sense that the ratio of width and maximum parametric gain is the same for all these OPAs, i.e.

$$\frac{\delta\beta}{g_{max}} = 4. \qquad (5.22)$$

This fact can be used to simplify further the shape of the graphs. If we use a unit along the g-axis equal to twice that along the $\Delta\beta$-axis then the graphs become semicircles, centered at $g = 0$ and $\Delta\beta = -\Delta\beta_{NL}$.

To show the gain equivalence of the various fiber OPAs, we first choose the pump powers in such a way that they have the same g_{max}. Since the maximum signal power gain for an OPA of length L, with no idler at the input, is $G_{3,max}(L) = |\cosh(rL)|^2$, all the OPAs will have exactly the same maximum signal power gain. According to Eq. (5.22), their graphs of g versus $\Delta\beta$ will also all have the same width along the $\Delta\beta$-axis, $\delta\beta = 4g_{max}$, and they will then differ only by the position of their maximum along the $\Delta\beta$-axis, i.e. $-\Delta\beta_{NL}$.

To investigate the shapes of the gain spectra, we express $\Delta\beta$ in terms of the optical frequency ω_s of the signal, as in Section 5.3. The following discussion concerns primarily two-pump OPAs for which the pump spacing can be used as a variable parameter, but it can also be applied to single-pump OPAs by letting the pump spacing vanish.

Equation (5.8) shows that as $\Delta\omega_p$ changes the graph of $\Delta\beta(\Delta\omega_s)$ is simply translated along the $\Delta\beta$-axis, by an amount that depends only on $\Delta\omega_p$. This important property in principle allows us to shift the graph of g^2 along the $\Delta\beta$-axis by an (almost) arbitrary amount. This can then be used to ensure that, for a particular type of OPA, the dispersion properties are optimized to yield maximum gain at a particular frequency as well as a desirable gain shape in the vicinity of that frequency. Again, however, the range of the shift may not be infinite as it depends on the behavior of the function $\beta_e(\Delta\omega_p)$, which may be bounded (this is indeed the case when the series in Eq. (5.8) is limited to two terms, as studied in [10]).

In summary, because (i) all OPAs with the same g_{max} have the same $g(\Delta\beta)$ within a translation along the $\Delta\beta$-axis, and (ii) the graph of $\Delta\beta(\Delta\omega_s)$ can be translated arbitrarily along the $\Delta\beta$-axis by adjusting $\Delta\omega_p$, *several types of OPA using the same sort of fiber can exhibit exactly the same gain spectrum* $G_3(\Delta\omega_s)$, centered about the same ω_c. The only differences between them will be that they will generally require different $\Delta\omega_p$ values and pump powers.

We can prove this more formally, as follows. Consider a first OPA, OPA1, with

$$g^2(\Delta\omega_s) = r^2 - \frac{1}{4}[2\Delta\beta_e(\Delta\omega_s) - 2\Delta\beta_e(\Delta\omega_p) + \Delta\beta_{NL}]^2, \qquad (5.23)$$

and a second OPA, OPA2, with primed notation for those quantities that are not the same as for OPA1. We assume that the two OPAs use exactly the same sort of fiber and have the same ω_c; also, by our equalization of maximum gains, $r' = r$. Then the parametric gain of OPA2 is given by

$$g'^2(\Delta\omega_s) = r^2 - \frac{1}{4}[2\Delta\beta_e(\Delta\omega_s) - 2\Delta\beta_e(\Delta\omega'_p) + \Delta\beta'_{NL}]^2. \tag{5.24}$$

Thus, in order for the two OPAs to have the same $g^2(\Delta\omega_s)$ and hence the same signal gain spectrum $G_3(\Delta\omega_s)$, we must have

$$2\Delta\beta_e(\Delta\omega_p) + \Delta\beta_{NL} = 2\Delta\beta_e(\Delta\omega'_p) - \Delta\beta'_{NL}. \tag{5.25}$$

Equation (5.25) can be viewed as an equation for $\Delta\omega'_p$ when all the other parameters are given. When Eq. (5.25) has one or more roots, the resulting OPA2 will have exactly the same gain spectrum as OPA1. The two OPAs will in general require different pump spacings $\Delta\omega_p$ and $\Delta\omega'_p$. They will also in general require different total pump powers, because r depends differently on pump power for different OPAs. The existence of suitable roots must be examined for each function $\Delta\beta_e(\Delta\omega'_p)$.

An example is treated in detail in [12]. The problem was to find whether two-pump OPAs could provide the same gain spectrum as XXX, for a fiber with $\beta^{(4)} < 0$. This could be useful in practice, to achieve a gain independent of the signal SOP and/or to avoid idler-broadening problems due to pump dithering. It was found that we have the choice between XXXX and RLRL to replace XXX. In both two-pump cases the pumps fall at the same location, unfortunately within the desired gain bandwidth. The RLRL case has the additional advantage of being polarization independent, but it requires 50% more pump power than XXX or XXXX.

This typical example illustrates the design of OPAs with identical gain spectra and also some of the trade-offs that one faces when choosing between various types of OPA to achieve particular objectives.

5.6 Saturated gain spectra

In the preceding sections we concentrated on the gain spectra of OPAs operating in the small-signal gain regime, i.e. without pump depletion. That regime is particularly important for optical communication, where linear amplification is generally desired for preserving pulse shapes.

For other applications, however, such as high-power wavelength conversion (WC) in non-communication applications, it may be desirable to achieve a high conversion efficiency, which requires strong pump depletion. In this case the small-signal analysis of the gain and conversion spectra is no longer valid and one must redo the analysis, using the mathematical tools appropriate for this regime. Specifically, we must use the results obtained in subsections 3.5.1. and 3.6.1. In the present section we are going to present a simplified approach, which has the merit of showing (i) under what circumstances complete pump depletion can be achieved, and (ii) how nearly complete pump depletion can be obtained over large bandwidths, leading to flat broadband conversion spectra. We treat in detail the one-pump case; the two-pump case can be handled in a similar manner.

5.6.1 Conditions for complete pump depletion

The frequency relation $2\omega_1 = \omega_3 + \omega_4$ shows that when the pump releases two photons both signal and idler gain one photon. Therefore, if the pump is completely depleted then both signal and idler powers must have increased by $P_{10}/2$. For the idler power, this must be its maximum, η_3; hence we must have $\eta_3 = P_{10}/2$. As shown in subsection 3.6.1, we must then have $\Delta\beta = \Delta\beta_0$, where

$$\Delta\beta_0 = \gamma\left(P_{30} - \frac{P_{10}}{2}\right), \tag{5.26}$$

which can be interpreted as a nonlinear wavevector mismatch. (This condition is more general than in [13], where it was assumed that $P_{30} \ll P_{10}$.) We also see that $P_{10}/2$ is a double root, i.e. that $\eta_3 = \eta_4 = P_{10}/2$. This leads to $k = 1$, which implies that the sn function reduces to tanh, a non-periodic function. Hence one needs an infinitely long fiber to obtain complete pump depletion.

Equation (5.26) yields an expression for the optimum wavevector mismatch which is quite different from that for the small-signal regime, $\Delta\beta_0 = -2\gamma P_{10}$. This indicates that spectrum optimization in these two regimes leads to very different results.

5.6.2 Obtaining nearly complete pump depletion over a large bandwidth

Obtaining complete pump depletion at a particular frequency can be done, provided that Eq. (5.26) can be satisfied. In practice, however, it is important to investigate under what circumstances nearly complete pump depletion can be obtained over a wide wavelength range [13, 14]. The advantages of being able to do so are clear: one can in principle then tune the signal wavelength over this range while keeping the pump wavelength fixed, and thus obtain an idler wavelength that will be tuned over a similar range, with high conversion efficiency. This is more desirable than if high conversion efficiency is obtained only over a narrow range. Designing fiber OPAs with such wide conversion spectra is therefore the topic of this section.

We will begin with a qualitative discussion of the conditions necessary to obtain large pump depletion over a large bandwidth. We will then continue with an investigation of the gain spectra by means of analytic solutions.

(a) Qualitative description of the conversion gain spectra
We have as usual

$$\Delta\beta = \beta^{(2)}(\Delta\omega_s)^2 + \frac{\beta^{(4)}}{12}(\Delta\omega_s)^4. \tag{5.27}$$

According to Eq. (5.26), complete pump depletion can be obtained when $\Delta\beta = \Delta\beta_0$. Equation (5.27) can then be viewed as a second-order polynomial equation for $u = (\Delta\omega_s)^2$. This equation can have zero, one, or two real roots for u, which are positive or equal to 0, depending on the parameters entering the equation. We denote these roots by u_1 and u_2 ($u_1 < u_2$) when there are two of them and by u_1 when there is only one.

5.6 Saturated gain spectra

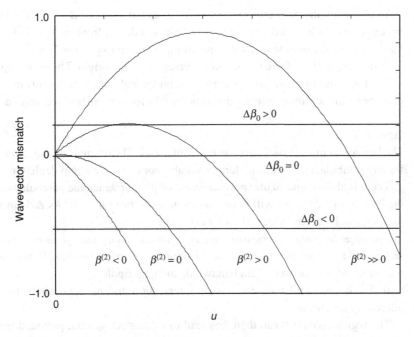

Fig. 5.10 Graphical representation of the wavevector mismatch $\Delta\beta$ in inverse atomic units versus u for various values of $\beta^{(2)}$. The horizontal lines correspond to different values of $\Delta\beta_0$.

Intuitively, to obtain near-complete pump depletion over a wide bandwidth, $\Delta\beta$ should remain close to $\Delta\beta_0$ over a wide frequency region. For this it is helpful if $\Delta\beta$ crosses $\Delta\beta_0$ once or twice within that region. This is similar to the analysis performed in [10] to study the shape of $\Delta\beta$ in relation to $-2\gamma P_{10}$ (the optimum for achieving maximum small-signal gain bandwidth).

We assume that the fiber is given, so that γ and $\beta^{(4)}$ are fixed. (We also assume that $\beta^{(4)} < 0$, which has been the case for most fibers used in OPA work in recent years; for $\beta^{(4)} > 0$ the analysis would need to be appropriately modified.) One can vary $\beta^{(2)}$ at will, in sign and magnitude, by adjusting λ_1 near λ_0. The graphs of $\Delta\beta$ versus u are parabolas passing through the origin of the axes, where their slopes are equal to $\beta^{(2)}$. They may or may not intersect horizontal lines drawn at $\Delta\beta_0$, depending on the parameters. Let us consider some important cases, illustrated in Fig. 5.10.

$\Delta\beta_0 = 0$

In this case the horizontal line is the horizontal axis. Since any parabola intersects that axis at the origin, Eq. (5.27) has a root at $u_1 = 0$, and complete pump depletion will always be achievable there.

If $\beta^{(2)} = 0$ the parabola remains close to the horizontal axis for a long distance before diverging from it. Thus we expect that this case will lead to a fairly flat gain spectrum, decreasing gradually from its maximum.

To obtain another interesting intersection, at $u_2 > 0$, we need $\beta^{(2)} > 0$. The larger $\beta^{(2)}$, the larger u_2 will be and the faster the parabola will depart from the horizontal axis near the origin. As a result the wider the spectrum, the more ripple it will have.

In the case $\beta^{(2)} < 0$ there is only one intersection at the origin. The more negative $\beta^{(2)}$ is, the faster the parabola diverges from the horizontal axis, which leads to a narrower gain spectrum. Therefore this regime is not useful for maximizing spectrum bandwidth.

$\Delta\beta_0 > 0$

The horizontal line is now above the horizontal axis. The origin is no longer on that line, for any parabola, and is thus no longer suitable for complete pump depletion.

There is always a particular positive value of $\beta^{(2)}$ for which the parabola is tangent to the line. In this case, we will have a u_1 value away from $u = 0$. As $\Delta\beta_0$ increases u_1 increases and the bandwidth around it decreases.

For larger β_2, there will be two intersection points, at u_1 and u_2. As $\beta^{(2)}$ increases, u_1 and u_2 will move apart but so will the maximum value of $\Delta\beta - \Delta\beta_0$. Hence again there will be a trade-off between gain bandwidth and gain ripple.

If $\beta^{(2)}$ is too small then there is no intersection and no opportunity to obtain an interesting spectrum.

The regime $\Delta\beta_0 > 0$ can therefore lead to interesting spectra, provided that $\Delta\beta_0$ is not too large.

$\Delta\beta_0 < 0$

In this case there is always a single intersection with a parabola, at $u_2 > 0$. For small $-\Delta\beta_0$ the conversion gain spectrum may be wide and flat, but for large $-\Delta\beta_0$ it exhibits a single narrow peak away from $u = 0$.

In summary, the investigation of the shape of $\Delta\beta$ in relation to $\Delta\beta_0$ shows that the best possibility for obtaining a wide gain spectrum with low ripple is to operate in a regime where $\Delta\beta_0 = 0$, i.e. $P_{30} = P_{10}/2$.

(b) Optimizing fiber length and signal power

While obtaining complete pump depletion is a desirable goal for a WC, there are several reasons why one should not necessarily strive to come as close to it as possible. They are as follows.

Fiber length

Complete pump depletion requires that the fiber length be infinite, which is not possible. One can approximate this condition by using a long fiber. This, however, would introduce several problems, as follows: (i) high cost; (ii) fiber loss; (iii) low SBS threshold. This is a very important point because generally, when L is much greater than the nonlinear length $L_{\text{NL}} = 1/(\gamma P_{10})$, SBS suppression becomes very difficult and may introduce additional problems such as broadening of the idler linewidth. (iv) Large dispersion fluctuations. In long fibers, there can be significant dispersion variations along the fiber. The presence of these fluctuations makes the WC performance different from that in a uniform fiber and difficult to predict. On the contrary, if a very short fiber can be used then dispersion fluctuations can become negligible and OPA performance can be reliably predicted on the basis of models assuming constant dispersion.

Conversion efficiency

Conversion efficiency is the ratio of desired output power and input power. For a WC the idler power P_4 is the desired output power. We define the input power as the total optical power entering the OPA; hence, we have the following definition of conversion efficiency:

$$\eta_{\text{eff}} = \frac{P_4}{P_{10} + P_{30}}. \tag{5.28}$$

This is a realistic definition for the optical conversion efficiency of the OPA itself, as it includes all the optical input power and not just the pump power. It is a quantity that one would want to maximize when designing an OPA-based WC.

In view of these considerations about fiber length and conversion efficiency, it is clear that designing a WC to approach complete pump depletion is not a very desirable strategy. It is better to design it so that it exhibits a large η_{eff} over a large wavelength-tuning range $\Delta\lambda$, using a small L, by suitable choice of the signal power P_{30}. Such design involves a trade-off between these goals, and the choice of an optimum design is possible only by introducing some rules for choosing a design that represents the best compromise.

In the following we examine in detail the trade-offs between η_{eff}, L, P_{30}, and $\Delta\lambda$.

η_{eff} versus L

We have seen that flat conversion gain spectra can be obtained when Eq. (5.27) has two positive roots. Then η_{eff} across the tuning range will be about the same as it is at the location of the roots. If there is only one root then η_{eff} will peak at the location of that root. At these points we have $\{\eta_1, \eta_3, \eta_4\} = \{-16P_{30}/7, P_{10}/2, P_{10}/2\}$ and the sn function becomes a tanh. We can then write η_{eff} as follows:

$$\eta_{\text{eff}} = -\frac{16P_{30}}{7(P_{10} + P_{30})}$$
$$\times \left\{ 1 - \left[1 - \frac{7P_{10}}{7P_{10} + 32P_{30}} \tanh^2\left(\frac{\gamma z}{4}\sqrt{(7P_{10} + 32P_{30})P_{10}}\right) \right]^{-1} \right\}. \tag{5.29}$$

η_{eff} versus P_{30}

The highest η_{eff} values are obtained for the lowest values of P_{30}. However, one needs a larger L to come close to these η_{eff} values. Actually, the plots for two different values of P_{30} cross at some value L_c of L. For $L < L_c$, η_{eff} for the larger P_{30} is larger, and this is reversed for $L > L_c$. This shows that, for a fixed L, increasing P_{30} may actually lead to a decrease in η_{eff}. It can be shown that this occurs even though here P_4 always increases with P_{30}. The decrease in η_{eff} for increasing P_{30} is related to the definition of η_{eff} by Eq. (5.28): if P_4 increases but $P_{10} + P_{30}$ increases faster then η_{eff} decreases.

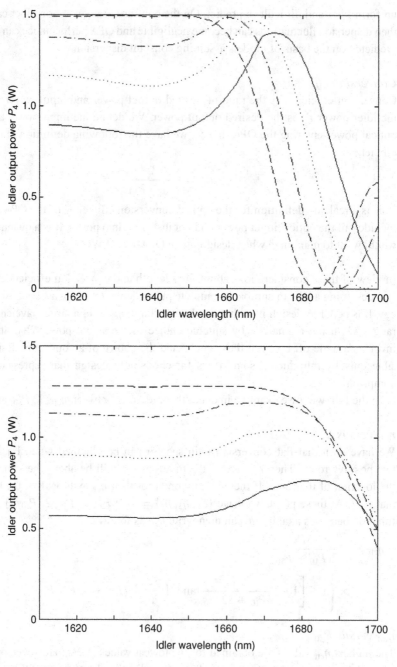

Fig. 5.11 Upper figure, plots of P_4 versus λ_4 for various values of P_{30}: solid line, 0.4 W; dotted line, 0.8 W; broken-and-dotted line, 1.2 W; broken line, 1.6 W. The other parameter values are $P_{10} = 3$ W, $\lambda = 1612.0$ nm, $\lambda_1 = 1611.8$ nm; $\beta^{(4)} = -1 \times 10^{-55}$ s^4 m, $D_\lambda = 0.024$ ps nm^{-2} km^{-1}, $\gamma = 12$ W^{-1} km^{-1}, $L = 80$ m. Lower figure, plots for the same values as in the upper figure except that $L = 40$ m.

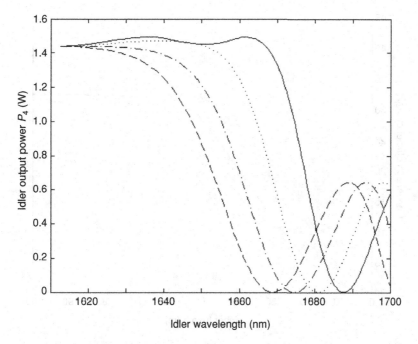

Fig. 5.12 Plots of P_4 versus λ_4, for various values of λ_1: solid line, 1611.6 nm; dotted line, 1611.8; broken-and-dotted line, 1612.0 nm; broken line, 1612.2 nm. The other parameter values are $P_{10} = 3$ W, $P_{30} = 1.8$ W, $L = 80$ m, $\lambda_0 = 1612.0$ nm, $\beta^{(4)} = -1 \times 10^{-55}$ s^4 m^{-1}, $D_\lambda = 0.024$ ps nm^{-2} km^{-1}, and $\gamma = 12$ W^{-1} km^{-1}.

$\Delta\lambda$ *versus* L

Let us now consider how $\Delta\lambda$ varies as L is reduced. From the preceding discussion, we expect that the maximum η_{eff} in the tuning range will be reduced. In addition, since a smaller L also implies smaller nonlinear phase shifts, we anticipate that η_{eff} across the tuning range will be more uniform and that $\Delta\lambda$ itself will be larger.

5.6.3 Numerical examples

We now present results of Matlab simulations carried out by using the exact Jacobian elliptic solution, for designs guided by the preceding qualitative observations. The function sn was calculated from its power series expansion, truncated after 10 terms; the elliptic integrals K and K' were obtained by numerical integration. The three roots of $g(x)$ were obtained by means of a Cardano solution.

We first consider an example with a pump power of $P_{10} = 3$ W. We assume that an HNLF is used, with parameters $\gamma = 12$ W^{-1} km^{-1}, $\lambda_0 = 1612.0$ nm, $\beta^{(4)} = -1 \times 10^{-55}$ s^4 m^{-1}, $D_\lambda = 0.024$ ps nm^{-2} km^{-1}; L, P_{30}, and λ_p are variable parameters.

(a) Conversion gain spectra for different values of P_{30} and L

The left-hand panel in Fig. 5.11 shows gain plots for $\lambda_1 = 1611.8$ nm, $L = 80$ m; P_{30} takes various values from 0.4 W to 1.6 W. We see that P_4 never exceeds 1.5 W, as expected.

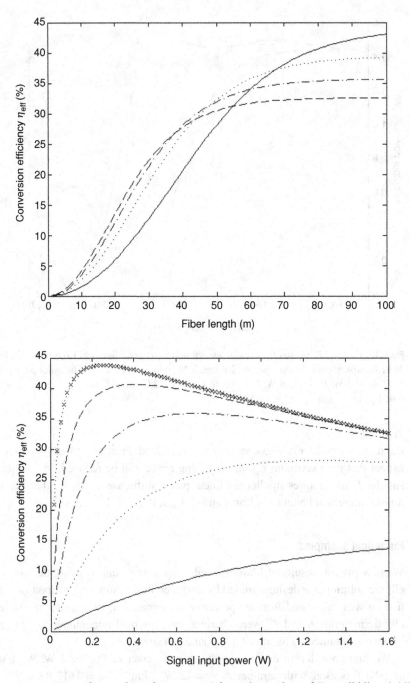

Fig. 5.13 Upper figure, plots of η_{eff} versus L for various values of P_{30}: solid line, 0.4 W; dotted line, 0.8 W; broken-and-dotted line, 1.2 W; broken line, 1.6 W. Lower figure, plots of η_{eff} versus P_{30}, for various values of L: solid line, 20 m; dotted line, 40 m; broken-and-dotted line, 60 m); broken line, 80 m; dotted line with crosses, 100 m. The fixed parameters were $P_{10} = 3$ W and $\gamma = 12$ W^{-1} km^{-1}.

In agreement with the discussion in the first part of subsection 5.6.2, for $P_{30} \ll 1.5$ W the spectrum has a single narrow peak far from the origin. For $P_{30} = 1.6$ W we have a flat spectrum that is quite wide. The spectra for $P_{30} = 1.2$ W and 1.6 W each provide good combinations of large tuning range and small deviation from the maximum.

The parameter values in the right-hand panel in Fig. 5.11 are the same as those in the left-hand panel except that $L = 40$ m. The shapes of the curves are similar but the idler power levels are lower and the tuning ranges are larger. This agrees with the discussion in subsection 5.6.2(a).

(b) Conversion spectra for different values of λ_1

Figure 5.12 shows gain plots for $P_{30} = 1.8$ W, $\lambda_0 = 1612.0$ nm; λ_1 takes various values from 1611.6 to 1612.2 nm. For $\lambda_1 = 1612.0$ nm the idler spectrum has a maximum at $\lambda_4 = 1612.0$ nm and decreases monotonically; it is not as wide as the widest monotonically decreasing spectrum in the left-hand panel of Fig. 5.11. For $\lambda_1 = 1612.2$ nm the spectrum has a similar shape but drops even faster. For $\lambda_1 = 1611.8$ nm the gain spectrum has a maximum near $\lambda_4 = 1640.0$ nm and is quite flat. The widest gain spectrum is obtained for $\lambda_1 = 1611.6$ nm; it has two maxima.

(c) Conversion efficiency versus L

When L is the variable, it is possible to choose the parameters so that they satisfy the condition for total pump depletion and then to make use of Eq. (5.29); this has been done in Fig. 5.13. The left-hand panel shows plots of the conversion efficiency η_{eff} versus L, for v values of P_{30}. In all cases longer fibers lead to increased η_{eff}. However, after a certain point one obtains diminishing returns by further increasing L.

(d) Conversion efficiency versus P_{30}

When P_{30} is the variable, it is still possible to meet the complete pump depletion condition and to use Eq. (5.29) for every P_{30}, but then a different parameter must be used for each P_{30}. One can in particular keep P_{10} fixed and vary $\Delta\beta$. This was done to produce the right-hand panel in Fig. 5.13, which again shows plots of η_{eff} versus P_{30} for various values of L. In agreement with the first part of subsection 5.6.2, for each value of L the plot passes through a maximum at a particular value of P_{30}. If it is important to maximize η_{eff} then this value of P_{30} should be considered to be the optimum.

(e) Wavelength conversion with $P_{10} = 300$ W

We now consider a wavelength conversion (WC) with 300 W of pump power. Since single-mode fibers lasers can deliver CW powers in excess of 1 kW [15], it is probable that in the future there will be some interest in exploring the use of such sources for wavelength conversion.

The pump power is 100 times larger than in the preceding example, and we can use the considerations of Sections 3.4 and 5.2 to investigate how the WC output would scale compared with the 3-W-pump case. If we also increase the signal input power by 100, to about 150 W, we will obtain a very flat conversion spectrum, with an idler power of the order of 150 W. This is true regardless of the type of fiber used.

The bandwidth, however, does depend on the fiber. If we use the same HNLF as before, γ and $\beta^{(4)}$ are the same and therefore the bandwidth scales as $(P_{10})^{1/4}$. Hence it is larger by $(100)^{1/4} = 3.16$, i.e. it is about 200 nm. This could be attractive for some applications.

The fiber would be about 100 times shorter than before, i.e. just 80 cm long. This fiber would have to dissipate 100 times the power of the preceding one in $1/100$ of the length. Therefore the power to be dissipated per unit length would be 10^4 times larger than in the first example. This could be excessive for a fiber with such a small core; water cooling might be required, which could be undesirable.

To reduce the heat dissipation problem, a dispersion-shifted fiber (DSF) could considered, with a γ value that is about $1/10$ that of HNLF. This would increase L to 8 m and therefore reduce the heat load per unit length by the same factor. Also the core size would be larger, which would help the fiber to withstand the power.

However, DSF has a $\beta^{(4)}$ value about 10 times larger than that of HNLF. Since the gain bandwidth scales as $(\gamma P_{10}/\beta^{(4)})^{1/4}$, we see that a 300 W pump WC made from DSF would have exactly the same bandwidth as a 3 W pump WC made from HNLF.

Finally, we could consider using a large-core fiber, commonly used in fiber lasers with kW output power [15]. Such fibers have a γ value much smaller than DSF, and so the required L would be much longer than 8 m, possibly up to 80 m, which would be similar to the fiber lengths used for fiber lasers, which can handle these high powers. Such fibers have the advantage of not requiring water cooling. However, a smaller γ will reduce the bandwidth. For such fibers $\beta^{(4)}$ is not well known and so it is not possible to predict the bandwidth.

(f) Discussion

One-pump OPAs cannot provide very flat gain spectra in the case where the pump is undepleted (particularly when $\beta^{(4)} < 0$) [2], while two-pump OPAs can [10]. Here we have shown, however, that one-pump OPAs can provide very flat gain spectra when the pump is strongly depleted. Thus two-pump OPAs do not necessarily have an advantage over one-pump OPAs for providing flat gain in the case of strong pump depletion.

We have defined the conversion efficiency as $\eta_{\text{eff}} = P_4/(P_{10} + P_{30})$. In the preceding discussion we have seen that P_{30} should be close to $P_{10}/2$ in order to obtain a flat gain spectrum; in that case η_{eff} will be close to 33%. Hence the design of OPA WCs with flat gain spectra involves a trade-off between gain bandwidth and conversion efficiency.

5.7 Fibers with longitudinal dispersion variations

In the preceding sections we studied the gain spectra of fibers with properties that are uniform along their entire lengths. This allowed us to find out the types of gain spectra can be obtained under such ideal conditions. We have seen in particular that, in the small-signal gain regime, two-pump OPAs have the potential for providing spectra that are very wide and flat while one-pump OPAs have spectra that have more ripple in the passband.

In this section we turn our attention to OPAs that have dispersion variations along their length. These variations may be either intentional, or they may occur naturally as a result of the manufacturing process.

Intentional dispersion variations may be used for modifying the gain spectrum in desirable ways. Naturally occurring random variations are actually present in all fibers, however, and we need to understand how they modify the ideal spectra associated with uniform fibers.

In this section we investigate these two types of variation. We approximate an arbitrary dispersion variation by a series of fiber sections with uniform dispersion. For each such section we know from Chapter 3 that we can express the relation that exists between the input and output fields by means of a matrix (in the absence of pump depletion). Therefore, we can obtain the transmission matrix for the whole fiber as the product of the matrices of the individual segments. We use this matrix approach throughout the present section as it has the advantages of being relatively simple to implement and of being computationally efficient. (The method breaks down when the pump depletion is not negligible; one must then resort to numerical integration techniques.)

5.7.1 Matrix approach

This is considered in [16–19]. We showed in Chapters 3 and 4 that in the small-signal regime the input and output fields (signal and idler) of a uniform fiber section are related by a 2×2 matrix, which depends on the fiber properties, the pump fields, and $\Delta\beta$. In a series of segments, one must calculate the pump fields by taking into account the cumulative effect of the preceding sections on their amplitudes and phases. Once that is done one can calculate the matrix for each section and then for the entire fiber by matrix multiplication, as mentioned above.

We will limit our calculations to lossless fibers, for which the matrices can be simply expressed in terms of exponentials, or equivalently in terms of sinh and cosh functions. It is worth noting, however, that the method can be extended to lossy fibers. In that case the pump powers decrease exponentially and the matrices are expressed in terms of Bessel or hypergeometric functions, for one- or two-pump OPAs respectively (see Chapter 3).

It is important to note that, since fiber OPAs are fundamentally phase sensitive, the phases of all the fields must be taken into account properly at all stages. In particular, one should remember that, in the derivations of the gain matrices, changes of field variables were made, involving phase transformations. Thus it is important to use the appropriate field variables and the correct expressions for the pump fields in the matrices.

It turns out that for the one- and two-pump OPAs studied in Chapter 3 the phase transformations used for the signal and idler fields are actually the same. Therefore the matrices arrived at in that chapter can be used without any change to relate the signal and idler input and output fields.

The pump fields entering the first matrix are $A_1(0)$ and $A_2(0)$, which are the original complex field envelopes at the fiber input. It can be shown that the same fields should be used to write down the matrix for any segment.

Therefore the transfer matrix for a fiber made up of N_s segments is given by

$$M = \prod_{k=1}^{N_s} M_k = \prod_{k=1}^{N_s} \begin{bmatrix} m_{33}^k & m_{34}^k \\ \left(m_{34}^k\right)^* & \left(m_{33}^k\right)^* \end{bmatrix}, \qquad (5.30)$$

where

$$m_{33}^k = \cosh(g_k L_k) + \frac{i\kappa_k}{2g_k}\sinh(g_k L_k) \qquad \text{for either type of OPA,}$$

$$m_{34}^k = \frac{2i\gamma A_{10} A_{20}}{g_k}\sinh(g_k L_k) \qquad \text{for a two-pump OPA,}$$

$$m_{34}^k = \frac{i\gamma (A_{10})^2}{g_k}\sinh(g_k L_k) \qquad \text{for a one-pump OPA.}$$

The length of the kth segment is L_k; the appropriate expressions for g_k and κ_k for the two types of OPA are given in Chapter 3.

5.7.2 One-pump OPA gain flattening

We have seen that it is difficult to obtain flat gain spectra with one-pump OPAs. This is particularly true when $\beta^{(4)} < 0$, which is valid for most available fibers, in which case we have a spectrum that has two symmetric peaks or a single peak, depending on the location of the pump with respect to the ZDW; this is shown in Figs. 5.3 and 5.4. It is natural, therefore, to consider placing two fibers in series, one with a two-peak gain spectrum and the other one with a one-peak spectrum; one would expect that the combined spectrum would be approximately the sum of the two and would therefore be fairly flat. This reasoning is only approximate because OPAs are phase sensitive, and therefore the gain in the second stage depends on the relative phases of the waves emerging from the first stage. Hence in general the gain of the combined sections is not simply the sum (in decibels) of the individual gains (both calculated in the usual manner, by assuming that there is no idler at the input).

Nevertheless, guided by this intuition one can implement this procedure, assuming that the same pump passes through two consecutive segments that have different ZDWs. In principle this could be achieved by splicing together different fibers or by using a single fiber and modifying the dispersion over part of it by using a different temperature [20] or stress [21].

It is found that this method indeed works reasonably well and that two such segments can lead to a relatively flat spectrum, by proper choice of lengths and ZDWs. However, there are ripples due to the phase-sensitive nature of the OPA and to reduce these it is necessary to introduce additional fiber segments. Thus one is led to extend the method beyond the two segments originally envisioned. One finds in practice that using four segments provides a sufficient reduction in ripple [18].

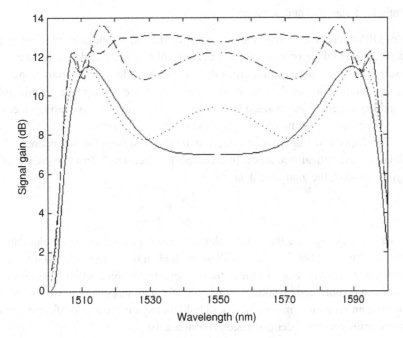

Fig. 5.14 Gain spectrum of a four-segment one-pump fiber OPA. Solid line, segment 1; dotted line, segments 1 and 2; broken-and-dotted line, segments 1, 2, and 3; broken line, segments 1, 2, 3, and 4.

Figure 5.14 shows a gain spectrum that can be obtained in this manner, with a four-segment arrangement. The parameters were the same as in [18], i.e. $L_1 = 200$ m, $L_2 = 77$ m, $L_3 = 118$ m, $L_4 = 27$ m, $\lambda_1 = 1550.00$ nm, $\lambda_{01} = 1550.00$ nm, $\lambda_{02} = 1551.49$ nm, $\lambda_{03} = 1548.47$ nm, $\lambda_{04} = 1546.00$ nm, $\beta_1^{(2)} = 0$, $\beta_2^{(2)} = 5.89 \times 10^{-2}$ ps^{2}km^{-1}, $\beta_3^{(2)} = -6.07 \times 10^{-2}$ ps^2 km^{-1}, $\beta_4^{(2)} = -16.5 \times 10^{-2}$ ps^2 km^{-1}, $P_1 = 0.5$ W, $\gamma = 20$ W^{-1} km^{-1}, and $\beta^{(4)} = -2.85 \times 10^{-4}$ ps^4 km^{-1}. The ZDW of the kth segment is λ_{0k}.

The spectra obtained by increasing the number of segments are plotted. This shows that all four segments are needed to obtain the desired flatness. The final spectrum is a little higher and not quite as flat as in [18], because the splice loss of 0.6 dB between adjacent sections assumed in that work has not been included here.

The results can be verified by using a numerical Runge–Kutta method to calculate the output fields without using the matrices.

The parameters in [18] were arrived at by trial and error. The search for optimum conditions could possibly be rendered more systematic by using some form of simulated annealing technique [22, 23].

While such a design for a flat-gain one-pump OPA is appealing, to our knowledge it has not been experimentally demonstrated. One reason is probably the difficulty of finding four fiber pieces with the desired ZDWs. This could in principle be achieved by using a controlled distribution of temperature [20] or stress [21] along a single fiber segments.

5.7.3 Periodic dispersion compensation

The OPA matrix theory was first used for investigating the possibility of using a periodic distribution of dispersion along the length of a fiber [16]. The idea is that if two fibers are available with dispersion properties that are not ideal for OPA operation, perhaps some linear combination of their properties would be more appropriate. In principle such a combination could be obtained by making a composite fiber, by cutting pieces from the two fibers and reassembling them with alternating fiber types.

The dispersion properties of a fiber can be obtained from the dependence of the group delay on the optical frequency. In a composite fiber, made from segments of fibers of types A and B, the total time delay is

$$\tau = \frac{L_A}{v_{gA}} + \frac{L_B}{v_{gB}}, \tag{5.31}$$

where v_{gA} and v_{gB} are the group velocities and L_A and L_B are the total lengths of the two fiber types. (This is true regardless of the lengths of the individual fiber segments and where they are located.) Since the average dispersion coefficients of the composite fiber can be obtained by taking derivatives of τ with respect to ω, we see that the same linear combination as in Eq. (5.31) will hold for any dispersion coefficient. For example, the average second-order dispersion coefficient is

$$\beta^{(2)}_{ave} = \frac{L_A \beta^{(2)}_A + L_B \beta^{(2)}_B}{L_A + L_B}. \tag{5.32}$$

Concerning nonlinear effects, simple averaging is generally not possible because of the lack of linearity of the equations. One exception is for self-phase modulation (SPM); it can be shown that the SPM phase shift of a pump can be calculated by using an average value of the nonlinearity coefficient calculated from Eq. (5.32), where the $\beta^{(2)}$'s are replaced by γ's.

Concerning the OPA equations, if the fiber segments are very short then it is clear that the composite fiber will behave as a homogeneous fiber, with averaged properties. Then the usual OPA results for a uniform fiber can be used with these coefficients. However, if the fiber segments are not very short, it will be necessary to represent each by a matrix and multiply the matrices to represent the entire fiber. Intuitively, we expect that for a moderate number of fibers the gain spectrum will approximate that of a uniform fiber with average values of the parameters.

In practice, if two different types of fiber are used, say DSF and HNLF, the inevitable splice loss between consecutive segments will limit the number of segments. In [16] only three segments of each type of fiber were used. In spite of this limited number, the characteristics of the resulting gain spectrum already approached those of the ideal fiber with averaged coefficients, which represented a significant improvement over the spectra that could be obtained with either type of fiber used by itself.

Of course, when losses are introduced by splices the matrix for each segment should be calculated by using the local pump power in each segment.

One way to avoid splice losses is to create variations in the dispersion properties within a single fiber, by using a temperature distribution [20]. This restricts how much the fiber properties can be varied, compared to using pieces of different fibers. However, the fact that no splices are needed implies that a very large number of short segments can be used. By so doing, it is possible to investigate the regime where the composite fiber effectively behaves as a uniform fiber with the averaged parameters.

5.7.4 Fibers with random longitudinal dispersion variations

The preceding examples in this chapter show that the shape of the gain spectra can be strongly affected if the pump wavelength varies, in some conditions by as little as 0.1 nm. Equivalently, if the pump wavelength is fixed but the fiber ZDW drifts by such an amount then the spectrum will also be affected. All practical fibers have random ZDW variations along their length, and this will modify their gain spectra from the ideal spectra calculated for perfectly uniform fibers.

A common model to investigate the effect of such variations on the spectrum is to approximate the continuous ZDW variation actually found by one where the ZDW has different constant values in a large number of small segments. The number and lengths of the segments, as well as the random values of the dispersion, are then chosen in such a way as to approximate the ZDW profile in the fiber.

We can use the matrix method for such a model [19]. We write down the corresponding 2×2 matrix for each segment with constant ZDW and then multiply the matrices to obtain the overall transfer matrix.

Figure 5.15 shows examples of one-pump OPA gain spectra for an HNLF, for various longitudinal variations of λ_0. Typical parameters are $P_1 = 0.5$ W, $L = 0.5$ km, $\gamma = 10$ W^{-1} km^{-1}, $\lambda_1 = 1550$ nm, $\beta^{(4)} = -5 \times 10^{-55}$ s^4 m^{-1}. The fibers were divided into 50 m segments, and in each segment the ZDW was chosen at random by adding a random amount $\Delta\lambda_0$ to the average value, 1550 nm; $\Delta\lambda_0$ is a uniformly distributed random variable with zero mean and maximum value $\Delta\lambda_{0,\max}$. The three curves correspond to the spectra for three fibers, with $\Delta\lambda_{0,\max} = 0, 2, 3$ nm. As expected, we see that as the dispersion variations increase the spectrum becomes more distorted. Nevertheless, we note that the distortions are not very large even though the λ_0 variations are substantial (possibly up to 6 nm for $\Delta\lambda_{0,\max} = 3$ nm). We therefore conclude that for this particular choice of parameters the gain spectrum is actually fairly robust with respect to λ_0 variations.

However, it should be noted that it is possible to find examples that are significantly more sensitive. For example, the very wide and flat two-pump gain spectrum of Fig. 5.7 is very sensitive to variations in the pump wavelengths, a change of 0.1 nm making a significant difference; this is an indication that similar variations in λ_0 may also lead to significant distortions.

Theoretical investigations of the effect of ZDW variations have been performed by a number of authors, for both one-pump [19, 24] and two-pump [25] OPAs, with conclusions similar to those arrived at above.

In [26] the changes in spectral shape due to ZDW variations as well as due to changes in pump wavelength were exploited to extract a fiber dispersion map. This was done by using a polynomial of order P to fit the dispersion curve and then using spectrum

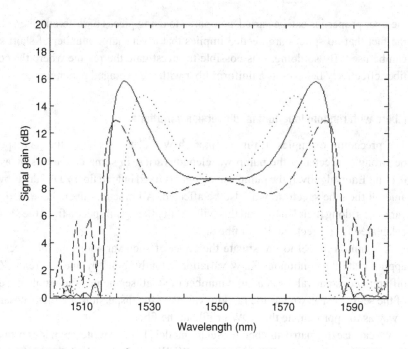

Fig. 5.15 Gain spectra of two-pump fiber OPAs with different amounts of ZDW variation $\Delta\lambda_{0,\max}$: solid line, 0 nm; broken line, 2 nm; dotted line, 3 nm.

measurements for different pump powers to generate a set of equations for the unknown coefficients of the polynomial. By this method it was possible to get dispersion maps for fibers as short as 300 m. The result of the work reported in [26] revealed that HNLFs could have ZDW variations as large as 1 nm over 100 m, which makes it difficult to obtain gain spectra approaching those of uniform fibers.

An interesting approach for dealing with random dispersion variations was introduced by Inoue [27].

5.8 Fibers with constant linear birefringence

In the preceding discussion we considered the gain spectra of OPAs made from isotropic fibers. For a one-pump OPA we saw that it is possible to create a narrowband gain region far from the pump, whose location can be adjusted by tuning the pump near the ZDW. Birefringence provides another means for creating narrow gain regions away from the pump. This approach was in fact used in some of the earliest fiber OPA work, to generate light at new wavelengths.

Here we are going to derive expressions for the location and width of the gain peak and discuss the potential for this approach.

In subsection 4.3.1 we showed that

$$\Delta\beta = \Delta\beta^x + 2\Delta n\,\omega_1/c. \tag{5.33}$$

5.8 Fibers with constant linear birefringence

Expanding $\Delta\beta^x$ to the fourth order, in the usual fashion, we obtain

$$\Delta\beta = \beta^{x(2)}(\Delta\omega_s)^2 + \beta^{x(4)}(\Delta\omega_s)^4/12 + 2\Delta n\omega_1/c \tag{5.34}$$

This shows that the birefringence adds a constant to the expression for $\Delta\beta^x$ in an isotropic fiber. The sign of this constant can be changed by exchanging the SOPs of pump and signal; its magnitude is given by the fiber birefringence, which can be as large as 10^{-2} in some fibers. Weak birefringence can also be introduced by applying stress to the fiber, for example by winding it on a cylinder. Thus the constant in Eq. (5.34) can be viewed as a parameter that can be chosen, within a wide range, to achieve particular effects.

It is interesting to note that the form of Eq. (5.34) is actually the same as that for $\Delta\beta$ for a two-pump OPA in an isotropic fiber, in which the constant (i.e. the part that does not depend on $\Delta\omega_s$) is $-\beta^{x(2)}(\Delta\omega_p)^2 - \beta^{x(4)}(\Delta\omega_p)^4/12$ and can be adjusted by changing the pump spacing. It thus appears that birefringence could in principle be used to obtain with a one-pump OPA the same types of flat gain spectra that can be obtained with a two-pump OPA, as described in Section 5.3. However, to obtain Eq. (5.34) we assumed that the two polarization axes have the same dispersion except for the presence of Δn, and this is only an approximation. Thus for shaping wide gain spectra with birefringence one needs to re-examine the assumptions made and to use accurate dispersion properties for the two axes.

If, however, one is interested in obtaining narrow gain regions away from the pump, the spectral shape is always about the same and the main quantity of interest is the distance of the gain region from the pump, $\Delta\omega_g$, which does not depend critically on the dispersion differences between the two axes. To calculate this distance, we let $\Delta\beta = 0$ in Eq. (5.34); this leads to

$$(\Delta\omega_g)^2 = \frac{6}{\beta^{(4)}}\left\{-\beta^{(2)} \pm \sqrt{(\beta^{(2)})^2 - \frac{4\pi}{3}\frac{\beta^{(4)}\Delta n}{c\lambda_1}}\right\}, \tag{5.35}$$

where we have omitted the superscript x for simplicity. We now examine the dependence of $\Delta\omega_g$ on $\beta^{(2)}$ (which can be adjusted by tuning the pump) and on Δn (which can be set during manufacture, or perhaps by inducing stress).

If $\beta^{(2)} = 0$, we simply have

$$\Delta\omega_g = 2\left(-\frac{3\pi\Delta n}{\lambda_1\beta^{(4)}}\right)^{1/4}. \tag{5.36}$$

In this case Δn and $\beta^{(4)}$ need to have opposite signs for a meaningful solution to be obtained.

As an example, for $\Delta n = 10^{-4}$, $\lambda_1 = 1500$ nm (i.e. 200 THz) and $\beta^{(4)} = -3.3 \times 10^{-55}$ s^4 m^{-1}, we have $\Delta v_g = \Delta\omega_g/2\pi = 66.6$ THz, which is about one-third of the pump frequency itself. This would correspond to gain peaks at 1125 nm and 2250 nm. Therefore with such an arrangement we could obtain wavelength conversion with very large shifts. It should be noted that this is true even though the values of Δn and $\beta^{(4)}$ in this example are quite moderate and easily realizable.

However, since the slope of $\Delta\beta$ versus $\Delta\omega_s$ is quite large as $\Delta\beta$ passes through zero, the width of the gain region will be quite small. This is quite similar to what occurs when

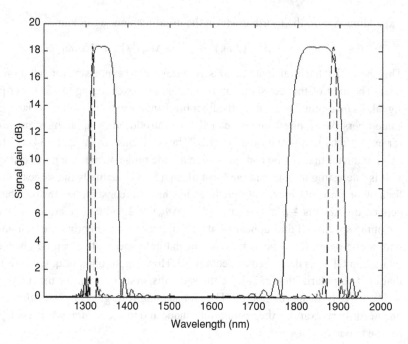

Fig. 5.16 Gain spectra corresponding to two one-pump birefringent fiber OPAs with opposite signs for the birefringence: solid line, $\Delta n = -2 \times 10^{-5}$; broken line, $\Delta n = 2 \times 10^{-5}$.

tuning the pump to obtain a narrow gain region with a one-pump OPA in an isotropic fiber (Section 5.3).

To avoid this problem, one can arrange for the slope of $\Delta\beta$ versus $\Delta\omega_s$ to vanish as $\Delta\beta$ passes through zero. This occurs for a finite $\beta^{(2)}$ when the two roots of Eq. (5.35) are the same. Now Δn and $\beta^{(4)}$ need to have the same sign to have a meaningful solution. It can be shown that the location of $\Delta\beta = 0$ is again given by Eq. (5.36), except that the negative sign is absent. Hence with the same conditions as in the preceding example we would obtain the same large shifts but now with a much larger gain bandwidth. The exact value of that bandwidth depends on how the shape of the gain spectrum is optimized in that region.

Figure 5.16 shows signal gain spectra obtained by following the procedures just described. The parameters were $P_1 = 700$ W, $L = 0.6$ m, $\gamma = 20$ W^{-1} km^{-1}, $\lambda_0 = 1550$ nm, and $\beta^{(4)} = -5 \times 10^{-55}$ s^4 m^{-1}. The narrow gain peaks (broken lines) correspond to $\Delta n = 2 \times 10^{-5}$ and $\lambda_1 = \lambda_0 = 1550$ nm ($\beta^{(2)} = 0$), while the wider gain peaks (solid lines) correspond to $\Delta n = -2 \times 10^{-5}$ and $\lambda_1 = 1507$ nm. It can be seen that the optimized gain spectra display flat-topped gain regions that are at least one order of magnitude wider than the others. In fact the gain region near 1310 nm is tens of nanometers wide, which could be of interest for the amplification of communication signals in that important wavelength range. (However, the high power required to obtain this bandwidth for this example renders it impractical.)

In early OPA work Stolen investigated the use of birefringent fibers in the case where the pump is not very close to λ_0, which makes $\beta^{(2)}$ relatively large; he did not include the effect of $\beta^{(4)}$ [28]. With these assumptions, one finds that

$$\Delta\omega_g = \left(-\frac{4\pi \Delta n}{\lambda_1 \beta^{(2)}}\right)^{1/2}, \tag{5.37}$$

which is valid as long as $\Delta\omega_g$ is not too large, so that the effects of $\beta^{(4)}$ can be neglected.

In other work, it has been proposed that the longitudinal variation in the magnitude of the birefringence should be controlled, in order to tailor the gain spectrum as well as to achieve complete pump depletion. The idea is to bend the fiber around a purpose-built shape in order to create a longitudinal variation of the fiber's local radius of curvature, which in turn creates a longitudinal variation in the birefringence. To be successful, this approach requires a fiber that has very little intrinsic birefringence in the absence of intentional bends [29].

5.9 Few-mode fibers

Most work with fiber OPAs uses waves that are in the fundamental mode of propagation, sometimes with different SOPs. However, it is sometimes desirable to consider the possibility that some or all the waves do not propagate in the same transverse mode of the fiber. Since different transverse modes have different transverse field profiles, it is then necessary to re-examine the calculation of γ_{jklm} in Chapter 3. The form of Eq. (5.32) remains correct, as it allows for the possibility of having different functions ψ_k that represent the mode profiles. When different profiles enter into Eq. (5.32), the resulting coefficients for the various interactions (self-phase modulation, cross-phase modulation, four-wave mixing) may all be different, and it is therefore necessary to calculate them all.

Another difference from the case of a single mode is that the expression for $\Delta\beta$ must also be re-examined. Specifically, we now have

$$\Delta\beta = \beta_3^c + \beta_4^d - \beta_1^a - \beta_2^b, \tag{5.38}$$

where the labels a, b, c, d refer to the modes at the frequencies $\omega_1, \omega_2, \omega_3, \omega_4$, respectively; these modes may also have different SOPs in the most general case. In this most general case, where all four modes are different, all their dispersion properties will need to be known in order to calculate $\Delta\beta$.

Since the use of different modes in fiber OPAs significantly complicates the analysis of their gain spectra, and few advantages have so far been identified, the use of higher-order modes has been very limited in the OPA field.

One possible area of interest is that different modes can have large differences between their dispersion curves, and this can be used to obtain phase matching for widely separated frequencies. In other words, it can be used for wavelength conversion with large frequency shifts. (Our treatment of gain spectra using different SOPs can be viewed as a particular case of this.)

In fact one reason for considering different modes when large frequency shifts are required is that a fiber designed to be single-mode near the pump wavelength (say

1550 nm) may very well support additional modes at shorter wavelengths, say near 1000 nm. In this case one would need to consider the possibility that these modes may be efficiently excited for some specific pump wavelengths. This type of situation has been encountered with microstructured fibers designed with high pump powers in the fundamental mode: parametric amplified spontaneous emission (ASE) was observed at much shorter wavelengths, with radiation patterns that corresponded to higher-order modes [30, 31]

References

1. "Unified analysis of four-photon mixing, modulational instability, and simulated Raman scattering under various polarization conditions in fibers," Golovchenko, E. A., Pilipetskii, A. N. *J. Opt. Soc. Amer. B*; 1994; vol. 11, pp. 92–101.
2. "Broadband fiber optical parametric amplifiers," Marhic, M. E., Kagi, N., Chiang, T. K., Kazovsky, L. G. *Opt. Lett.*; 1996; vol. 21, pp. 573–5.
3. "Optimization of spectrally flat and broadband single-pump fiber optic parametric amplifiers," Floridia, C., Sundheimer, M. L., Menezes, L. ed. S., Gomes, A. S. L. *Optics Comm.*; 2003; vol. 223, pp. 381–8.
4. "Broadband wavelength conversion over 193-nm by HNL-DSF improving higher-order dispersion performance," Hirano, M., Nakanishi, T., Okuno, T., Onishi, M. In *Proc. 31st European Conf. on Optical Communication*, September 2005, Glasgow, UK; vol. 6, pp. 43–4.
5. "Broadband single-pumped fiber-optic parametric amplifiers," Torounidis, T., Andrekson, P. *IEEE Photon. Technol. Lett.*; 2007; vol. 19, in press.
6. G. Kalogerakis, private communication.
7. "Wideband tuning of the gain spectra of one-pump fiber optical parametric amplifiers," Marhic, M. E., Wong, K. K. Y., Kazovsky, L. G. *IEEE J. Select. Topics Quantum Electron.*; 2004; vol. 10, pp. 1133–41.
8. "Widely tunable optical parametric generation in a photonic crystal fiber," Chen, A. Y. H., Wong, G. K. L., Murdoch, S. G.; Leonhardt, R., Harvey, J. D., Knight, J. C., Wadsworth, W. J., Russel, P. St. J. *Opt. Lett.*; 2005; vol. 30, pp. 762–4.
9. "Widely tunable sub-30-fs pulses from a compact erbium-doped fiber source," Tauser, F., Adler, F., Leitenstorfer, A. *Opt. Lett.*; 2004; vol. 29, pp. 516–8.
10. "Broadband fiber-optical parametric amplifiers and wavelength converters with low-ripple Chebyshev gain spectra," Marhic, M. E., Park, Y., Yang, F. S., Kazovsky, L. G. *Opt. Lett.*; 1996; vol. 21, pp. 1354–6.
11. "Demonstration of two-pump fiber optical parametric amplification," Yang, F. S., Ho, M. C., Marhic, M. E., Kazovsky, L. G. *Electron. Lett.*; 1997; vol. 33, pp. 1812–3.
12. "Fiber optical parametric amplifiers with linearly- or circularly-polarized waves," Marhic, M. E., Wong, K. K. Y., Kazovsky, L. G. *J. Opt. Soc. Amer. B.*; 2003; vol. 20, pp. 2425–33.
13. "Tunable fiber parametric wavelength converter with 900 mW of CW output power at 1665 nm," Marhic, M. E., Williams, G. M., Goldberg, L., Delavaux, J. M. P. In *Proc. Conf. Photonics West*, San Jose CA, January, 2006; *Proc. SPIE*, vol. 6103, pp. 165–76.
14. "92% pump depletion in a CW one-pump fiber OPA," Marhic, M. E., Wong, K. Y., Ho, M. C., Kazovsky, L. G. *Opt. Lett.*; 2001; vol. 26, pp. 620–2.
15. "1.53 kW from a single Yb-doped photonic crystal fiber laser," Bonati, G., Voelckel, H., Gabler, T., Krause, U., Tünnermann, A., Limpert, J., Liem, A., Schreiber, T., Nolte, S., Zellmer, H. In *Proc. Conf. Photonics West*, San Jose CA, January, 2005; *Proc. of the SPIE*, vol. 5709-2a.

16. "High-nonlinearity fiber optical parametric amplifier with periodic dispersion compensation," Marhic, M. E., Yang, F. S., Ho, M. C., Kazovsky, L. G. *J. Lightwave Technol.*; 1999; vol. 17, pp. 210–5.
17. "Gain enhancement in cascaded fiber parametric amplifier with quasi-phase matching: theory and experiment," Kim, J., Boyraz, O., Lim, J. H., Islam, M. N. *J. Lightwave Technol.*; 2001; vol. 19, pp. 247–51.
18. "Broadband and flat parametric amplifiers with a multisection dispersion-tailored nonlinear fiber arrangement," Provino, L., Mussot, A., Lantz, E., Sylvestre, T., Maillotte, H. *J. Opt. Soc. Amer. B*; 2003; vol. 20, pp. 1532–7.
19. "Parametric amplification in presence of fluctuations," Farahmand, M., de Sterke, M. *Optics Express*; 2004; vol. 12, pp. 136–42.
20. "Temperature control of the gain spectrum of fiber optical parametric amplifiers," Wong, K. K. Y., Marhic, M. E., Kazovsky, L. G. *Optics Express*; 2005; vol. 13, pp. 4666–73.
21. "Experiment of zero dispersion tuning by stretching down-sized HNLF," Takahashi, M., Tadakuma, M., Hiroishi, J., Sugizaki, R., Yagi, T. In *Proc. European Conf. on Optical Communication*, Cannes, France, September 2006; vol. 4, paper Th1.5.1, pp. 41–2.
22. "Optimized design of two-pump fiber optical parametric amplifier with two-section nonlinear fibers using genetic algorithm," Gao, M., Jiang, C., Hu, W., Wang, J. *Optics Express*; 2004; vol. 12, pp. 5603–13.
23. "Two-pump fiber optical parametric amplifiers with three-section fibers allocation," Gao, M., Jiang, C., Hu, W., Wang, J. *Optics and Laser Technol.*; 2006; vol. 38, pp. 186–91.
24. "Modulational instability in optical fibers with variable dispersion," Abdullaev, F. K., Darmanyan, S. A., Kobyakov, A., Lederer, F. *Phys. Lett. A*; 1996; vol. 220, pp. 213–8.
25. "Impact of dispersion fluctuations on dual-pump fiber-optic parametric amplifiers," Yaman, F., Lin, Q., Radic, S., Agrawal, G. P. *IEEE Photon. Technol. Lett.*; 2004; vol. 16, pp. 1292–4.
26. "Zero-dispersion wavelength mapping in short single-mode optical fibers using parametric amplification," Mussot, A., Lantz, E., Durecu-Legrand, A., Simonneau, C., Bayart, D., Sylvestre, T., Maillotte, H. *IEEE Photon. Technol. Lett.*; 2006; vol. 18, pp. 22–4.
27. "Arrangement of fiber pieces for a wide wavelength conversion range by fiber four-wave mixing," Inoue, K. *Opt. Lett.*; 1994; vol. 19, pp. 1189–91.
28. "Phase matching in birefringent fibers," Stolen, R. H., Bosch, M. A., Lin, C. *Opt. Lett.*; 1981; vol. 6, pp. 213–5.
29. "Quasi-phase matching in an optical fiber with periodic birefringence," Murdoch, S. G., Leonhardt, R., Harvey, J. D., Kennedy, T. A. B. *J. Opt. Soc. Amer. B*; July 1997; vol. 14, pp. 1816–22.
30. "Nonlinear generation of very high-order UV modes in microstructured fibers," Efimov, A., Taylor, A., Omenetto, F., Knight, J., Wadsworth, W., Russell, P. *Opt. Express*; 2003; vol. 11, pp. 910–8.
31. "Phase-matched third harmonic generation in microstructured fibers," Efimov, A., Taylor, A., Omenetto, F., Knight, J., Wadsworth, W., Russell, P. *Opt. Express*; 2003; vol. 11, pp. 2567–76.

6 The nonlinear Schrödinger equation

6.1 Introduction

In the preceding chapters we have considered the nonlinear interactions between a small number N_f ($N_f \leq 6$) of frequencies in a nonlinear fiber. We wrote down sets of N_f coupled equations for their complex amplitudes that depended only on z. These sets of equations, and the resulting solutions, are very useful and with them we were able to investigate several important aspects of fiber OPAs, particularly their gain spectra.

However, there are situations where this approach is not sufficient to describe the behavior of fiber OPAs. Examples of such situations are as follows.

1. **The number of frequencies N_f becomes large** An important example of this is the case of multiple signal carriers passing through the OPA at the same time, such as in a wavelength-division multiplexing (WDM) system. Then, if we tried to follow the previous method, the number of equations would grow as N_f and the number of four-wave mixing (FWM) terms in each equation would grow faster than N_f, resulting in a very complex set of equations.
2. **The pumps are modulated at a very high rate or are very short** For moderate modulation frequencies and pulse lengths it may be possible to make use of the quasi-CW approximation as a first-order approximation (see Chapter 7). But when these become extreme the quasi-CW approximation breaks down, and one needs to use a model that accurately takes into account the broadband nature of the waves.

In situations such as these it becomes necessary to use a technique that can deal with essentially arbitrary optical spectra passing through the nonlinear fiber. To do this we need to go back to the basic wave equation and investigate what it yields for an input that consists of a continuous spectrum of frequencies about the center frequency ω_c. The approach is to consider first what happens in the absence of nonlinearity and to then add the nonlinear contribution as a small perturbation; this leads to the nonlinear Schrödinger equation (NLSE). As in Chapter 4, we first consider an isotropic fiber and then add birefringence.

We will start by looking at some well-known analytic solutions of the NLSE in uniform fibers. We then turn our attention to the split-step Fourier method (SSFM), which can be used for broad optical spectra as well as for fibers exhibiting arbitrary longitudinal variations of dispersion and birefringence.

6.2 Derivation of the NLSE for an isotropic fiber

In the absence of nonlinearity, the guided waves in an isotropic fiber are only subject to dispersion. We consider a relatively narrow range of frequencies ω about a center frequency ω_c; hence the difference frequencies $\Omega = \omega - \omega_c$ are relatively small. The propagation of the frequency component at ω is determined by the wavevector $\beta(\omega)$, which can be expanded in a power series of Ω as follows:

$$\beta(\omega) = \beta(\omega_c + \Omega) = \beta_c + \sum_{n=1}^{\infty} \frac{\beta^{(n)}}{n!}\Omega^n, \tag{6.1}$$

where

$$\beta_c = \beta(\omega_c) \quad \text{and} \quad \beta^{(n)} = \frac{d^n\beta}{d\omega^n} \quad \text{at } \omega_c. \tag{6.2}$$

In most situations the series in Eq. (6.1) can be truncated after a few terms. As we saw in Chapter 5, for fiber OPAs it is generally necessary to include all terms up to $\beta^{(4)}$ in order to obtain accurate results for the shape of the gain spectrum.

We assume that the transverse field dependence is handled as in Chapters 3 and 4, and so here we consider only functions of z and t. In parallel with the approach of Chapter 4, we let the real electric field vector $\vec{E}(z, t)$ be of the form

$$\vec{E}(z, t) = \frac{1}{2}\{\vec{A}(z, t)\exp[i(\beta_c z - \omega_c t)] + \text{c.c.}\}, \tag{6.3}$$

where $\vec{A}(z, t)$ is the complex vector phaser (slowly varying envelops, SVE), of $\vec{E}(z, t)$.

We now decompose $\vec{A}(0, t)$ into its frequency components by means of the Fourier transform

$$\vec{A}(0, t) = \int_{\Omega=-\infty}^{\infty} \vec{B}(\Omega)\exp(-i\Omega t)d\Omega, \tag{6.4}$$

where $\vec{B}(\Omega)$ is the complex amplitude of the Fourier component at Ω. Because we assume linear propagation, we have

$$\vec{A}(z, t) = \int_{\Omega=-\infty}^{\infty} \vec{B}(\Omega)\exp[i(\beta - \beta_c)z - i\Omega t]d\Omega. \tag{6.5}$$

Taking the derivative of Eq. (6.5) with respect to z yields

$$\frac{\partial \vec{A}(z, t)}{\partial z} = \int \vec{B}(\Omega) i(\beta - \beta_c)\exp[i(\beta - \beta_c)z - i\Omega t]d\Omega$$

$$= i \int \vec{B}(\Omega) \left(\sum_{n=1}^{\infty} \frac{\beta^{(n)}}{n!}\Omega^n\right) \exp[i(\beta - \beta_c)z - i\Omega t]d\Omega. \tag{6.6}$$

Taking the mth derivative of Eq. (6.5) with respect to t yields

$$\frac{\partial^m \vec{A}(z,t)}{\partial t^m} = \int \vec{B}(\Omega)(-i\Omega)^m \exp[i(\beta - \beta_c)z - i\Omega t] d\Omega. \tag{6.7}$$

Comparing Eqs. (6.6) and (6.7), we see that

$$\frac{\partial \vec{A}(z,t)}{\partial z} - \sum_{n=1}^{\infty} \frac{i^{n+1} \beta^{(n)}}{n!} \frac{\partial^n \vec{A}(z,t)}{\partial t^n} = 0. \tag{6.8}$$

This equation, which is strictly a consequence of the linear propagation equations, is exact.

If the fiber is lossy, Eq. (6.8) needs to be slightly modified. If the power loss coefficient is α, a term $\alpha \vec{A}(z,t)/2$ should be added to the left-hand side.

We now introduce the Kerr nonlinearity as a perturbation of Eq. (6.8). We follow the same approach as in Chapter 4. Substituting Eq. (6.3) into Eq. (4.6) and keeping on both sides only those frequencies that are close to ω_c, we find that the terms of interest on the right-hand side must be proportional to $[\vec{A}, \vec{A}^*, \vec{A}]$. Following the same arguments as in Chapter 4 concerning the coefficient that multiplies this term, we finally arrive at the equation

$$\frac{\partial \vec{A}}{\partial z} - \sum_{n=1}^{\infty} \frac{i^{n+1} \beta^{(n)}}{n!} \frac{\partial^n \vec{A}}{\partial t^n} = i\gamma [\vec{A}, \vec{A}^*, \vec{A}], \tag{6.9}$$

which is a form of the NLSE valid for isotropic fibers.

When $\vec{A}(z,t)$ is linearly polarized and the series is limited to $n = 2$, we obtain the familiar scalar form of the NLSE,

$$\frac{\partial A}{\partial z} + \beta^{(1)} \frac{\partial A}{\partial t} + i \frac{\beta^{(2)}}{2} \frac{\partial^2 A}{\partial t^2} = i\gamma |A|^2 A, \tag{6.10}$$

where A is the complex amplitude of $\vec{A}$.

6.3 Derivation of the NLSE for a birefringent fiber

We now consider propagation in a nonlinear fiber with constant linear birefringence, such as polarization-maintaining fiber (PMF). As before, we consider the dispersive properties (which now include birefringence) and the nonlinear effect, separately and then combine the contributions to form the complete equation.

As in Chapter 4 we let x and y denote the directions of the fast and slow polarization axes, and we decompose any incident field SOP into its linearly polarized x and y components. Linear propagation along the fiber can then be studied very simply, the $x(y)$ component having the corresponding wavevector $\beta^x(\beta^y)$.

We will present two slightly different derivations. The difference arises from the way in which the SVEs are defined: we may use a separate definition for each linear SOP, or a common definition. The resulting NLSEs are in principle equivalent, but they have features making them better suited for handling different situations.

6.3.1 First derivation

Here we use a separate definition for the SVE of each linear SOP, i.e. we let

$$E^x(z,t) = \frac{1}{2}\{A^x(z,t)\exp[i(\beta_c^x z - \omega_c t)] + \text{c.c.}\}, \tag{6.11}$$

$$E^y(z,t) = \frac{1}{2}\{A^y(z,t)\exp[i(\beta_c^y z - \omega_c t)] + \text{c.c.}\}, \tag{6.12}$$

where

$$\beta_c^x = \beta^x(\omega_c), \qquad \beta_c^y = \beta^y(\omega_c) \tag{6.13}$$

and A^x and A^y are the slowly varying envelopes of the two polarization components. Because of the way in which they are defined, A^x and A^y each have slow variations in time as well as in space [1].

These definitions lead to two scalar equations similar to Eq. (6.8), namely

$$\frac{\partial A^x}{\partial z} - \sum_{n=1}^{\infty} \frac{i^{n+1}\beta^{x(n)}}{n!}\frac{\partial^n A^x}{\partial t^n} = 0, \tag{6.14}$$

$$\frac{\partial A^y}{\partial z} - \sum_{n=1}^{\infty} \frac{i^{n+1}\beta^{y(n)}}{n!}\frac{\partial^n A^y}{\partial t^n} = 0. \tag{6.15}$$

We now consider the nonlinear effects alone. We have previously investigated the vector properties of the Kerr effect, using the notation introduced in Chapter 4. The treatment is very general, and it is suitable for arbitrary polarization states. Therefore, the x- and y-polarized components of the nonlinear polarization are proportional to $[\vec{A}, \vec{A}^*, \vec{A}, \hat{x}]$ and $[\vec{A}, \vec{A}^*, \vec{A}, \hat{y}]$, where $\vec{A} = A^x e^{i\beta_c^x z}\hat{x} + A^y e^{i\beta_c^y z}\hat{y}$ is the total electric field. We obtain the two equations

$$\frac{\partial A^x}{\partial z} - \sum_{n=1}^{\infty}\frac{i^{n+1}\beta^{x(n)}}{n!}\frac{d^n A^x}{dt^n}$$
$$= i\gamma[\vec{A}, \vec{A}^*, \vec{A}, \hat{x}]$$
$$= i\gamma\left[\left(P^x + \frac{2}{3}P^y\right)A^x + \frac{1}{3}(A^y)^2 A^{x*} e^{-2i\Delta\beta z}\right], \tag{6.16}$$

$$\frac{\partial A^y}{\partial z} - \sum_{n=1}^{\infty}\frac{i^{n+1}\beta^{y(n)}}{n!}\frac{\partial^n A^y}{\partial t^n}$$
$$= i\gamma[\vec{A}, \vec{A}^*, \vec{A}, \hat{y}]$$
$$= i\gamma\left[\left(P^y + \frac{2}{3}P^x\right)A^y + \frac{1}{3}(A^x)^2 A^{y*} e^{2i\Delta\beta^{xy} z}\right], \tag{6.17}$$

where

$$\Delta\beta^{xy} = \beta_c^x - \beta_c^y = \beta^x(\omega_c) - \beta^y(\omega_c). \tag{6.18}$$

Because A^x and A^y appear in products in both equations, they are nonlinearly coupled.

Under some circumstances it can be argued that the last terms in the x and y equations can be neglected, because of the phase factor $e^{\pm 2i\Delta\beta^{xy}z}$. This term oscillates rapidly due to the difference in phase velocity between the two axes and will essentially average out to zero over several beat lengths. One situation where this is applicable is the case where the birefringence is strong and the fiber is many beat lengths long. Another example is when the birefringence is weak but varies randomly along the length, as in most communication fibers. In such situations we then keep only those terms on the right-hand sides that are synchronous. We then obtain the forms [2, 3]

$$\frac{\partial A^x}{\partial z} - \sum_{n=1}^{\infty} \frac{i^{n+1}\beta^{x(n)}}{n!}\frac{\partial^n A^x}{\partial t^n} = i\gamma\left(P^x + \frac{2}{3}P^y\right)A^x, \qquad (6.19)$$

$$\frac{\partial A^y}{\partial z} - \sum_{n=1}^{\infty} \frac{i^{n+1}\beta^{y(n)}}{n!}\frac{\partial^n A^y}{\partial t^n} = i\gamma\left(P^y + \frac{2}{3}P^x\right)A^y. \qquad (6.20)$$

The main disadvantage of these equations is that they have been obtained by neglecting the coherent-coupling, i.e. ellipse-rotation, terms (see the next subsection for definitions of these terms), which may not always be a good approximation.

6.3.2 Alternative derivation

Here we use a common definition for the SVE of each linear SOP, i.e. we now let

$$E^x(z,t) = \frac{1}{2}\{C^x(z,t)\exp[i(\beta_c z - \omega_c t)] + \text{c.c.}\}, \qquad (6.21)$$

$$E^y(z,t) = \frac{1}{2}\{C^y(z,t)\exp[i(\beta_c z - \omega_c t)] + \text{c.c.}\}, \qquad (6.22)$$

where

$$\beta_c = \frac{1}{2}\left(\beta^x(\omega_c) + \beta^y(\omega_c)\right). \qquad (6.23)$$

Comparing with Eqs. (6.11), (6.12), we see that the new SVEs C^x and C^y are simply related to the previous ones A^x and A^y by phase factors:

$$C^x = A^x e^{-i\Delta\beta^{xy}z/2}, \qquad C^y = A^y e^{i\Delta\beta^{xy}z/2}. \qquad (6.24)$$

When $\Delta\beta^{xy}$ is small, as in communication fibers with random birefringence, the C's can be said to be slowly varying envelopes of the electric fields, like the A's, but one should bear in mind that "slowly varying" has a slightly different meaning for the two types of variable.

These definitions lead to two equations similar to Eqs. (6.14), (6.15), namely

$$\frac{\partial C^x}{\partial z} - i\frac{\Delta\beta^{xy}}{2}C^x - \sum_{n=1}^{\infty}\frac{i^{n+1}\beta^{x(n)}}{n!}\frac{\partial^n C^x}{\partial t^n} = 0, \qquad (6.25)$$

$$\frac{\partial C^y}{\partial z} + i\frac{\Delta\beta^{xy}}{2}C^y - \sum_{n=1}^{\infty}\frac{i^{n+1}\beta^{y(n)}}{n!}\frac{\partial^n C^y}{\partial t^n} = 0. \qquad (6.26)$$

We now consider the nonlinear effects alone. The x- and y-polarized components of the nonlinear polarization are proportional to $[\vec{C}, \vec{C}^*, \vec{C}, \hat{x}]$ and $[\vec{C}, \vec{C}^*, \vec{C}, \hat{y}]$, where $\vec{C} = (C^x \hat{x} + C^y \hat{y}) e^{i\beta_c z}$ is the total electric field envelope. Therefore, we obtain the two equations

$$\frac{\partial C^x}{\partial z} - i \frac{\Delta \beta^{xy}}{2} C^x - \sum_{n=1}^{\infty} \frac{i^{n+1} \beta^{x(n)}}{n!} \frac{d^n C^x}{dt^n}$$

$$= i\gamma \left[\vec{C}, \vec{C}^*, \vec{C}, \hat{x} \right]$$

$$= i\gamma \left[\left(P^x + \frac{2}{3} P^y \right) C^x + \frac{1}{3} (C^y)^2 C^{x*} \right], \quad (6.27)$$

$$\frac{\partial C^y}{\partial z} + i \frac{\Delta \beta^{xy}}{2} C^y - \sum_{n=1}^{\infty} \frac{i^{n+1} \beta^{y(n)}}{n!} \frac{\partial^n C^y}{\partial t^n}$$

$$= i\gamma \left[\vec{C}, \vec{C}^*, \vec{C}, \hat{y} \right]$$

$$= i\gamma \left[\left(P^y + \frac{2}{3} P^x \right) C^y + \frac{1}{3} (C^x)^2 C^{y*} \right]. \quad (6.28)$$

Now we see the advantage of using this alternate formulation: the right-hand sides of Eqs. (6.27)–(6.28) no longer explicitly contain exponentials of the form $e^{\pm 2i \Delta \beta^{xy} z}$. As a result, they contain only nonlinear terms (with no explicit dependence on dispersion or birefringence), while the left-hand sides contain only linear terms due to dispersion and birefringence (with no explicit dependence on fiber nonlinearity). This clear separation of the basic physical phenomena is very helpful for formulating the solution of the NLSE by means of the split-step Fourier method (SSFM), as will be shown in Section 6.6. The fact that Eqs. (6.27), (6.28) were obtained without making any approximations implies that the results of the associated SSFM should be accurate under all circumstances.

6.4 Analytic solutions of the scalar NLSE

6.4.1 No dispersion, no loss

It is sometimes useful to assume that the frequency components of interest experience essentially relative delays as they travel through a fiber; this can occur in practice because the frequencies are close and/or because the fiber is short. This assumption is equivalent to neglecting fiber dispersion entirely, i.e. letting $\beta^{(n)} = 0$ for all $n > 1$ in Eq. (6.10). Then Eq. (6.10) reduces to

$$\frac{\partial A}{\partial z} + \beta^{(1)} \frac{\partial A}{\partial t} = i\gamma |A|^2 A. \quad (6.29)$$

It can be shown that the general solution of Eq. (6.29) is

$$A(z, t) = A\left(0, t - \frac{z}{v}\right) \exp \left\{ i\gamma z \left| A\left(0, t - \frac{z}{v}\right) \right|^2 \right\} \quad (6.30)$$

where $v = 1/\beta^{(1)}$ is the group velocity. This solution is particularly useful when implementing numerical techniques for integrating the NLSE.

(a) Expression for K input waves

(Note that in this section the subscripts denoting pump signal and idler are not the same as in the rest of the book, in order to accommodate the standard notation for series.)

The preceding particular solution can be used to study a number of nonlinear phenomena when a finite number K of waves are present at the fiber input. We then assume that

$$A(0, t) = \sum_{k=1}^{K} A_k e^{i\omega_k t}, \quad (6.31)$$

where A_k is the complex amplitude of the Fourier component of frequency ω_k. Then

$$A(z, t) = A(0, t - \beta^{(1)}z) = \sum_{k=1}^{K} A_k e^{i\omega_k t'}, \quad (6.32)$$

where $t' = t - \beta^{(1)}z$. Consequently,

$$|A(0, t')|^2 = \sum_{k=1}^{K} |A_k|^2 + \sum_{k \neq l=1}^{K} \sum_{l=1}^{K} A_k^* A_l e^{i\omega_{l,k} t'}$$

$$= P_0 + 2 \sum_{k>l=1}^{K} \sum_{l=1}^{K} |A_k A_l| \cos(\omega_{l,k} t' + \phi_{l,k}), \quad (6.33)$$

where $P_0 = \sum_{k=1}^{K} |A_k|^2$ is the total input power, $\omega_{l,k} = \omega_l - \omega_k$, $\phi_{l,k} = \phi_l - \phi_k$, and ϕ_k is the phase angle of A_k. Then

$$\exp(i\gamma z |A(0, t')|^2) = e^{i\gamma z P_0} \prod_{k>l=1}^{K} \prod_{l=1}^{K} \exp\{2i\gamma z |A_k A_l| \cos(\omega_{l,k} t' + \phi_{l,k})\}. \quad (6.34)$$

Using the Jacobi–Anger identity

$$\exp(i\rho \cos \theta) = \sum_{p=-\infty}^{\infty} i^p J_p(\rho) \exp(i p \rho), \quad (6.35)$$

where J_p is the Bessel function of the first kind of order ρ, we obtain

$$\exp(i\gamma z |A(0, t')|^2) = e^{i\gamma z P_0} \prod_{k>l=1}^{K} \prod_{l=1}^{K} \left\{ \sum_{p_{l,k}=-\infty}^{\infty} J_{p_{l,k}}(2\gamma z |A_k A_l|) \right.$$

$$\left. \times \exp\left[i p_{l,k} \left(\omega_{l,k} t' + \phi_{l,k} + \frac{\pi}{2}\right)\right] \right\}, \quad (6.36)$$

where $p_{l,k}$ is a running integer; there is a different sum over $p_{l,k}$ for every distinct ordered pair $\{l, k\}$. Finally,

$$A(z, t) = e^{i\gamma z P_0} \sum_{m=1}^{K} A_m e^{i\omega_m t'}$$

$$\times \prod_{k>l=1}^{K} \prod_{l=1}^{K} \left\{ \sum_{p_{l,k}=-\infty}^{\infty} J_{p_{l,k}}(2\gamma z |A_k A_l|) \right.$$

$$\left. \times \exp\left[i p_{l,k} \left(\omega_{l,k} t' + \phi_{l,k} + \frac{\pi}{2}\right)\right] \right\}. \quad (6.37)$$

(b) Case of two input waves

We now consider the simplest non-trivial case, that of only two waves at the input, with angular frequencies ω_1 and ω_2. This case is interesting for several reasons.

If one input wave (the pump) is strong compared with the other one (the signal), this situation will correspond well to the input conditions of a one-pump fiber OPA when the signal is close to the pump. Then we mainly expect one new wave (the idler) to emerge at the output. By calculating the signal and idler amplitudes at the output, we will obtain new expressions for the signal and idler gains for a one-pump fiber OPA.

However, if the two input waves have about the same amplitude then they can give rise to a large series of evenly spaced FWM terms, i.e. they can give rise to frequency-comb generation.

We now develop the formalism common to these two cases and then study them individually. To simplify the notation, we now write down the field without the phase term common to all the components, namely we set

$$C(z, t) = A(z, t) e^{-i\gamma P_0 z}. \tag{6.38}$$

Then

$$\begin{aligned}
C(z, t) &= \left[A_1 e^{i\omega_1 t'} + A_2 e^{i\omega_2 t'}\right] \\
&\quad \times \sum_{n=-\infty}^{\infty} J_n(2\gamma z |A_1 A_2|) \exp\left[in\left(\omega_{1,2} t' + \phi_{1,2} + \frac{\pi}{2}\right)\right] \\
&= \sum_{n=-\infty}^{\infty} i^n J_n(2\gamma z |A_1 A_2|) \left[A_1 e^{i\omega_1 t'} + A_2 e^{i\omega_2 t'}\right] \exp[in(\omega_{1,2} t' + \phi_{1,2})] \\
&= \sum_{n=-\infty}^{\infty} i^n J_n(2\gamma z |A_1 A_2|) \left(A_1 \exp\{i\left[(n+1)\omega_1 - n\omega_2\right] t' + n\phi_{1,2}\} \right. \\
&\quad \left. + A_2 \exp\{i[n\omega_1 - (n-1)\omega_2] t' + n\phi_{1,2}\}\right). \tag{6.39}
\end{aligned}$$

We can use Eq. (6.39) to find the complex amplitudes of the various frequency components. The general term has frequency $\omega_1 + m(\omega_1 - \omega_2)$, where m is an integer. Its amplitude is proportional to

$$\begin{aligned}
i^m J_m A_1 e^{im\phi_{1,2}} &+ i^{m+1} J_{m+1} A_2 e^{i(m+1)\phi_{1,2}} \\
&= i^m e^{im\phi_{1,2}} (J_m A_1 + i J_{m+1} A_2), \tag{6.40}
\end{aligned}$$

where the argument of all the Bessel functions is $2\gamma z |A_1 A_2|$. For the original frequencies ω_1 and ω_2 the amplitudes are respectively proportional to $J_0 A_1 + i J_1 A_2$ and $J_0 A_2 + i J_1 A_1$.

The total power at $\omega_1 + m(\omega_1 - \omega_2)$ is given by

$$P_m = P_{10} J_m^2 + P_{20} J_{m+1}^2, \tag{6.41}$$

where $P_{10} = |A_1|^2$ and $P_{20} = |A_2|^2$ are the input powers at ω_1 and ω_2.

(We can extend this result to the case of lossy fibers by means of the double exponential transformation of Eq. (6.48) below. The result is similar to that obtained by Waarts et al. [4].)

One-pump OPA

We now assume that one wave is initially much stronger than the other (as e.g. if there were a pump at ω_1 and a signal at ω_2). Then the main new frequency component in the initial phase of the process (i.e. for small L values L) corresponds to $m = 1$ and is at $2\omega_1 - \omega_2$, the idler frequency. From Eq. (6.41) we obtain the idler power as

$$P_i = P_{10}J_1^2 + P_{20}J_2^2. \tag{6.42}$$

We can then define the signal-to-idler power conversion gain (or idler gain) as

$$G_i = \frac{P_i}{P_{20}} = \frac{P_{10}}{P_{20}}J_1^2 + J_2^2. \tag{6.43}$$

Since, for small x, $J_1(x) \approx x/2$ and $J_2(x) \approx x^2/8$, at the beginning of the process G_i is well approximated by the first term only, i.e.

$$G_i \approx \frac{P_{10}}{P_{20}}J_1^2(2\gamma z|A_1 A_2|) \approx \frac{P_{10}}{P_{20}}(\gamma z|A_1 A_2|)^2 = (\gamma P_{10}z)^2. \tag{6.44}$$

This result is identical to the gain of a one-pump fiber OPA when the signal is close to the pump and there is no pump depletion [5].

We can also obtain the conversion gain for the case of a lossy fiber OPA by using the double exponential transformations of Section 3.7 (see also subsection 6.4.2). The result is

$$G_i \approx e^{-\alpha L}\left(\gamma L_{\text{eff}} P_{10}\right)^2, \tag{6.45}$$

as in [6]. An interesting feature of this expression is that it predicts that, if L keeps increasing, G_i will eventually reach a maximum value and then decrease monotonically. This is of course a consequence of the fiber loss, which eventually attenuates all waves in the fiber. Therefore there is an optimum length L_{opt} for which maximum conversion efficiency is reached, for given α, γ, and P_{10}. It is obtained by setting the derivative of G_i with respect to L equal to zero, i.e.

$$\frac{dG_i}{dL} = \frac{d}{dL}\left[e^{-\alpha L}(\gamma L_{\text{eff}} P_{10})^2\right]$$
$$= -\alpha e^{-\alpha L}(\gamma L_{\text{eff}} P_{10})^2 + e^{-\alpha L} 2\gamma P_{10} e^{-\alpha L} \gamma L_{\text{eff}} P_{10} = 0. \tag{6.46}$$

Equation (6.46) leads to the value $L_{\text{opt}} = (\ln 3)/\alpha \approx 1.1/\alpha$, for which we obtain the maximum conversion gain

$$G_{i,\text{max}} = \frac{2}{9}\frac{\gamma P_{10}}{\alpha} = 0.222\frac{\gamma P_{10}}{\alpha}. \tag{6.47}$$

The form of these equations indicates that, for a given fiber, increasing the pump power will result in a proportional increase in the maximum conversion efficiency, which will always occur at the same location.

As an example, consider DSF with $\alpha = 0.06$ nepers km^{-1}, $\gamma = 2$ W^{-1} km^{-1}, and $P_{10} = 1$ W. We find that $L_{\text{opt}} = 18.3$ km and $G_{i,\max} \approx 8.9$. Hence we can get nearly 10 dB of conversion gain, but a rather long fiber is required.

An advantage of the present approach is that it automatically handles pump depletion by the nonlinear process, when the Bessel functions are used instead of their approximations; by contrast, the methods of [5] and [6] do not take depletion into account.

We may also define the signal gain as $G_s = P_2/P_{20}$. It can then be shown that

$$G_s = G_i + e^{-\alpha L} \left[J_0^2(\nu) - J_2^2(\nu) \right], \tag{6.48}$$

where $\nu = 2\gamma L_{\text{eff}} |A_1 A_2|$. Equation (6.48) constitutes an extension of the usual relationship between G_s and G_i in lossy fiber OPAs when only the signal, pump and idler are considered.

Two strong input waves

The solution derived for two waves at the input can also be used in the situation where these two waves are of about equal strength and the fiber is long. In this case, since all FWM interactions are assumed to be phase matched, the efficient generation of high-order FWM terms can occur. In fact, this situation in principle offers the possibility of generating many equally spaced frequencies, i.e. a frequency-comb. Frequency-comb generation has received a great deal of attention in recent years since such sources are required for WDM communication systems, which use a number of evenly spaced carriers located on the ITU grid.

Other considerations indicate that the number of frequencies generated in this manner is of the same order as the nonlinear phase shift $\Phi = \gamma P_{10} L$. While it appears that one could obtain an arbitrary number of harmonics by increasing Φ, stimulated Brillouin scattering (SBS) will set a practical limit. Also, the assumption that dispersion is negligible will eventually break down for high-order harmonics.

Another consequence of FWM due to two strong input waves is that it is not fruitful to try to model a standard two-pump fiber OPA with no dispersion by the preceding approach. The reason is that the two pumps would then immediately interact and generate strong spurious FWM terms, which would quickly draw power away from the pumps. This of course would be very undesirable in practice and in fact is not observed if the pump spacing is large enough to introduce significant dispersion for the pumps.

Thus while a one-pump OPA can be modeled reasonably well by neglecting dispersion when the signal is close to the pump, the same is not true for a two-pump OPA with a realistic pump spacing, regardless of where the signal is located.

There is one situation where this two-strong-waves model is useful for OPAs. This is the case of a one-pump OPA where the pump power is rapidly modulated in sinusoidal fashion. (This arrangement will be studied in subsections 7.5.3 and 12.2.1.) In practice, this can be obtained by placing two CW pumps close to each other. In such situations the output spectrum of the pump consists of a series of evenly spaced tones, which can be viewed as either resulting from FWM between the two original waves or from SPM acquired by a carrier with sinusoidal intensity modulation. Either way, the amplitudes of the various tones are then obtained from Eq. (6.42).

6.4.2 No dispersion, loss

In this case Eq. (6.29) becomes

$$\frac{\partial A}{\partial z} + \beta^{(1)} \frac{\partial A}{\partial t} + \frac{\alpha}{2} A = i\gamma |A|^2 A. \tag{6.49}$$

One can still find the general solution. It is obtained by making a double exponential transformation of the solution in the lossless case. More precisely,

$$A(z,t) = A\left(0, t - \frac{z}{v}\right) e^{-\alpha z/2} \exp\left\{i\gamma L_{\text{eff}} \left|A\left(0, t - \frac{z}{v}\right)\right|^2\right\}. \tag{6.50}$$

We note that this transformation is very similar to that obtained in Section 3.7. The present transformation is somewhat more general, as time is explicitly included; also it is applicable to a waveform with a continuous spectrum rather than just discrete frequency components.

6.4.3 Soliton solutions of the NLSE

Probably the best-known solutions of the NLSE correspond to solitons. Solitons are isolated short pulses that can travel in lossless fibers while retaining their original shape at all times, as a result of a precise balance between dispersion and fiber nonlinearity. This remarkable property is of obvious interest in the field of optical communication. For this reason, solitons have been the object of intense research over the past few decades. At the present time it appears, however, that the limitations encountered in typical communication fibers, namely loss, dispersion, and amplifier noise, considerably reduce the attractiveness of solitons in long-haul communication. Thus systems based on the original pure soliton concept may not be competitive with systems based on different schemes. Nevertheless, solitons remain a very important example of NLSE solutions.

Solitons require fairly high peak powers and travel over long distances. Therefore they can induce strong nonlinear effects in other co-propagating waves. In fact it has been recognized that under certain circumstances they can provide parametric gain for noise components [7].

Generally speaking, however, the soliton regime is not very closely related to the conditions generally sought for making fiber OPAs, and for that reason we will not dwell on this subject here. The interested reader can consult the extensive literature on the subject; see for example [8].

6.5 Including the Raman gain in the NLSE

The preceding formulation of the NLSE does not contain any term corresponding to the Raman gain. As a result, it can only provide first-order approximations for wave

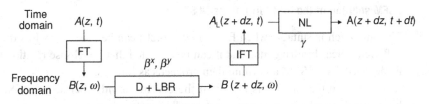

Fig. 6.1 Schematic representation of the structure of the SSFM algorithm in an elementary fiber segment. D, dispersion; LBR, linear birefringence; NL, nonlinearity; FT, Fourier transform; IFT, inverse FT. Here $dt = \beta^{(1)} dz$.

propagation in optical fibers. These will be quite accurate when the width of the optical spectrum is small compared with the Raman shift, in which case the Raman gain is small. However, if the optical spectrum has a width comparable with the Raman shift then the Raman gain may play an important role, particularly if the parametric gain is not well phase matched. In such situations, important effects such as the Raman self-frequency shift may occur. In order to model these effects properly, it is necessary to modify the NLSE by adding terms corresponding to the Raman gain.

Since the Raman gain does not play a large role in well-phase-matched fiber OPAs, we will not present the modified form of the NLSE here. The interested reader can consult [8] to see how to introduce Raman gain into the NLSE.

6.6 Numerical solutions of the NLSE by the split-step Fourier method

When analytic solutions are not available, the NLSE is generally numerically integrated by the split-step Fourier method (SSFM). The structure of the numerical split-step method is related to the method that we used to obtain the NLSE itself in Sections 6.2 and 6.3, in the sense that dispersion and nonlinearity are treated separately. The numerical calculation makes use of exponential solutions respectively associated with dispersion and nonlinearity.

Turning now to the structure of the numerical integration itself, its most basic implementation is represented schematically in Fig. 6.1. The fiber is divided into a large number N_z of equal elementary segments with length $dz = L/N_z$. In each segment, the elementary contributions due to dispersion (including birefringence if present) and nonlinearity are calculated sequentially. For this reason, this is known as a split-step procedure. If birefringence and dispersion have random variations along the fiber, these can be introduced by varying the parameters in consecutive segments (as well as introducing a random rotation of the birefringence axes, as explained in Chapter 4).

As seen in Section 6.3, A and C formulations are possible for the NLSE, with significant differences in the resulting nonlinear terms. These will in turn have important implications regarding the solutions of the respective NLSEs by the SSFM. For these reasons, we will deal separately with the corresponding SSFM formulations in the following subsections.

6.6.1 SSFM solution of the first form of the NLSE

This approach is suitable when Eqs. (6.19), (6.20) can be used, i.e. when the fiber has sufficient linear birefringence that it can be assumed that the ellipse rotation terms can be dropped. The SSFM algorithm then proceeds as follows.

We assume that $\vec{A}(z, t)$ is known at the end of the nth fiber segment. Since linear birefringence and dispersion are best handled in the frequency domain, $\vec{B}(z, \Omega)$, the (temporal) Fourier transform of $\vec{A}(z, t)$, is calculated by means of a fast Fourier transform (FFT). We then have

$$B^x(z+dz, \Omega) = B^x(z, \Omega) \exp[i(\beta^x - \beta_c^x)dz], \qquad (6.51)$$

$$B^y(z+dz, \Omega) = B^y(z, \Omega) \exp[i(\beta^y - \beta_c^y)dz]. \qquad (6.52)$$

To calculate the effect of the nonlinearity, we wish to work in the time domain, in order to make use of the results of Section 6.4. We therefore take an inverse fast Fourier transform (IFFT) to go from $\vec{B}(z+dz, \Omega)$ to $\vec{A}_L(z+dz, t)$, where the subscript L means that this intermediate field includes only the linear effects. Since the nonlinear section is dispersion-free and birefringence-free, the expression for the final output field can be written down using Eq. (6.31):

$$A^x(z+dz, t+dt) = A_L^x(z+dz, t) \exp\left[i\gamma\left(P^x + \frac{2}{3}P^y\right)dz\right] \qquad (6.53)$$

$$A^y(z+dz, t+dt) = A_L^y(z+dz, t) \exp\left[i\gamma\left(P^y + \frac{2}{3}P^x\right)dz\right], \qquad (6.54)$$

where $dt = \beta^{(1)}dz$.

A random rotation of the SOP on the Poincaré sphere is also introduced, as in Chapter 4. The whole procedure is repeated for each successive segment.

The FFT used in the SSFM uses a finite number N of evenly spaced frequencies, occupying a finite bandwidth $\Delta\omega$. Thus the SSFM neglects everything outside this bandwidth, which is generally taken to be at least 2.5 times larger than the bandwidth occupied by all the input waves. In practice, one should choose a calculation bandwidth large enough to include all the major spurious FWM terms.

In the time domain, the total field envelope is also calculated at N instants, spaced dt apart. Hence the total time window has duration $\Delta\tau = Ndt$; $\Delta\omega$ and $\Delta\tau$ satisfy $\Delta\omega\Delta\tau = 2\pi$ because of the FT relationship.

The input field can be specified as a time-domain function that describes the variation in the envelope during the time window. This is convenient when one deals with pulsed signals with an envelope defined by some function of time, such as a Gaussian. This is also useful when some of the input waves are modulated by communication signals, with patterns specified in the time domain.

Alternatively, the input field can be specified in terms of its frequency components. For OPA work this is often convenient because all input waves may be monochromatic; an example is a two-pump OPA into which is fed a wavelength-domain multiplexing (WDM) spectrum consisting of a number of evenly spaced monochromatic carriers.

Specifying such a spectrum would for example be useful in investigating effects such as FWM between WDM signals in a fiber OPA (see Chapter 13).

For a one-pump fiber OPA, it is convenient to take ω_c as the pump frequency. For a two-pump OPA, ω_c can be taken as halfway between the two pump frequencies. The initial field $A(0, t)$ is then the sum of the waves present at the input; typically it consists of the pump(s) and one or more signals. As the field propagates, new frequency components are generated by FWM (see Chapter 13).

Another possibility is to present broadband noise at the OPA input. This is accomplished by having the same power level at all frequencies, possibly with random phases.

One reason for doing this is to calculate the OPA amplified spontaneous emission (ASE), which can be viewed as resulting from the amplification of vacuum fluctuations and can be modeled as white noise with a specific power density (−55 dB m in a 0.1 nm bandwidth).

Another reason is that the shape of the ASE output spectrum will also provide a measure of the gain spectrum, if it is symmetric. However, with this approach the output power at ω_a comes not only from input power at ω_a but also from input power at $2\omega_c - \omega_a$. Assuming that the gain spectrum is symmetric and that the output powers coming from different input frequencies add up incoherently (on a power basis), then the ratio of the output spectral power density at ω_a and the input spectral power density at ω_a is $2G_s(\omega_a) - 1$, i.e. about twice the signal gain when the gain is large. This method should be used with care if there is any indication that the gain spectrum is not symmetric.

A simple modification solves the latter problem. If one injects white noise over one-half the spectrum, say for $\omega < \omega_c$, then one can calculate the signal gain and idler conversion efficiency for that part of the spectrum, without any overlap. By repeating this for $\omega > \omega_c$, the other half of the gain spectrum is obtained. Combining the results for the two halves yields the correct spectra for the signal gain and the idler conversion gain.

Alternatively, one can inject noise only at every even frequency on the positive side and at every odd frequency on the negative side. The resulting idlers then fall into slots that were initially empty, one per slot. It is then a simple matter to extract the signal and idler gains.

The advantage of using this approach to calculate the gain, instead of scanning the frequency of a monochromatic input signal, is that the entire gain spectrum can be calculated faster, since the gain at all frequencies is calculated simultaneously.

6.6.2 SSFM solution of the alternative NLSE

Let $\vec{D}$ denote the FT of $\vec{C}$. Examination of the linear propagation equations show that the D's propagate as follows in the linear section:

$$D^x(z+dz, \Omega) = D^x(z, \Omega)\exp\left\{i\left(\beta - \beta_c + \frac{\Delta n\Omega}{2c}\right)dz + i\Phi_0\right\}, \quad (6.55)$$

$$D^y(z+dz, \Omega) = D^y(z, \Omega)\exp\left\{i\left(\beta - \beta_c - \frac{\Delta n\Omega}{2c}\right)dz - i\Phi_0\right\}. \quad (6.56)$$

These equations are obtained by assuming that

$$\beta = (\beta^x + \beta^y)/2, \tag{6.57}$$

$$\beta^x - \beta^y = \Delta n(\omega_c + \Omega)/c. \tag{6.58}$$

Here Φ_0 is given by $\Delta n\,\omega_c dz/(2c) = \pi \Delta n dz/\lambda_c$; $\beta - \beta_c$ can be expanded as usual in terms of the desired orders of dispersion, generally up to the fourth order for fiber OPAs. Note that $\beta^{(1)}$ is not be present in this expansion, just as it is not present in the NLSE itself.

With these substitutions, one can then use Eqs. (6.55) and (6.56) to propagate the D's through the linear fiber section.

Let us now turn to propagation through the nonlinear fiber section. We showed in Section 6.3 that the ellipse rotation terms could be kept, without exponential factors, by using the C's instead of the A's to represent the SVEs. However, since the structure of the nonlinear terms has been modified, a different integration procedure must be used for them. The ellipse rotation terms can easily be handled by using a basis of orthogonal circularly polarized SOPs. The corresponding unit Jones vectors are $\hat{r} = (\hat{x} - i\hat{y})/\sqrt{2}$ and $\hat{l} = (\hat{x} + i\hat{y})/\sqrt{2}$. At the input of the nonlinear section the fields are converted from the linear basis to the circular basis by

$$C^r = \frac{C^x + iC^y}{\sqrt{2}}, \qquad C^l = \frac{C^x - iC^y}{\sqrt{2}}. \tag{6.59}$$

In the absence of birefringence and dispersion, the equations describing the Kerr effect are

$$\frac{dC^r}{dz} = i\gamma \left[C^r\hat{r} + C^l\hat{l}, C^{r*}\hat{l} + C^{l*}\hat{r}, C^r\hat{r} + C^l\hat{l}\right] \cdot \hat{l} = \frac{2i\gamma}{3}(P^r + 2P^l)C^r, \tag{6.60}$$

$$\frac{dC^l}{dz} = i\gamma \left[C^r\hat{r} + C^l\hat{l}, C^{r*}\hat{l} + C^{l*}\hat{r}, C^r\hat{r} + C^l\hat{l}\right] \cdot \hat{r} = \frac{2i\gamma}{3}(P^l + 2P^r)C^l. \tag{6.61}$$

It can readily be shown that P^r and P^l remain constant. Therefore Eqs. (6.53) and (6.54) can be integrated, with the following results:

$$C^r(z + dz) = C^r(z) \exp\left\{\frac{2i\gamma}{3}(P^r + 2P^l)dz\right\} = C^r(z)e^{i\theta_r}, \tag{6.62}$$

$$C^l(z + dz) = C^l(z) \exp\left\{\frac{2i\gamma}{3}(P^l + 2P^r)dz\right\} = C^l(z)e^{i\theta_l}, \tag{6.63}$$

where $\theta_r = \frac{2\gamma}{3}(P^r + 2P^l)dz$ and $\theta_l = \frac{2\gamma}{3}(P^l + 2P^r)dz$. This shows that the circular field components simply experience phase shifts θ_r and θ_l when they pass through the nonlinear section. When the phase difference between them, $\Delta\theta = \theta_r - \theta_l = 2(P^l - P^r)\gamma dz/3$, is not equal to zero, this leads to the well-known effect of ellipse rotation, whereby the major axis of the ellipse of the input SOP is rotated by $\delta\theta = \Delta\theta/2$ [9].

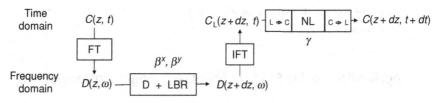

Fig. 6.2 Schematic representation of the SSFM algorithm, modified to keep the ellipse rotation terms. In comparison with Fig. 6.1, the main difference is the introduction of changes in the basis Jones vectors, from linear to circular and back; they are respectively represented by $L \Rightarrow C$ and $C \Rightarrow L$.

The field at the output of the nonlinear section can then be expressed in terms of its x and y components by

$$C^x = \frac{C^r + C^l}{\sqrt{2}}, \qquad C^y = \frac{C^r - C^l}{i\sqrt{2}}. \tag{6.64}$$

They are then ready to use as inputs for the next fiber segment.

Figure 6.2 shows in schematic form the SSFM algorithm, modified to keep the ellipse rotation terms.

As an alternative, it is also possible to do all the calculations of the Kerr effect in terms of C^x and C^y, without any approximation. Specifically, one can show that

$$\begin{bmatrix} C^x(z+dz) \\ C^y(z+dz) \end{bmatrix} = e^{i\gamma P_0 dz} \begin{bmatrix} \cos(\delta\theta) & \sin(\delta\theta) \\ -\sin(\delta\theta) & \cos(\delta\theta) \end{bmatrix} \begin{bmatrix} C^x(z) \\ C^y(z) \end{bmatrix}, \tag{6.65}$$

where $P_0 = P_z + P_y = P_r + P_l$ is the total power, and

$$\delta\theta = \frac{\gamma dz}{3}(P_l - P_r) = \frac{i\gamma dz}{3}(C_x C_y^* - C_x^* C_y). \tag{6.66}$$

Equation (6.62) says that the field after the section of length dz is obtained by rotating the input field by $\delta\theta$ and multiplying the result by a phase factor that is calculated as an SPM phase shift, which involves the total power regardless of how it is distributed among the SOPs. These are the effects of ellipse rotation.

In practice, then, there is a choice between two approaches for implementing ellipse rotation. (i) One can go from LP to CP states, and back; then all that is needed is to calculate powers and to introduce two phase shifts in the CP basis. (ii) Alternatively, one can stay in the LP basis; now it is also necessary to calculate powers, the SPM phase shift, and the ellipse rotation angle and then to rotate the input fields accordingly.

Either way, these operations are all relatively simple compared with those required for the FFT and so either approach can be implemented without introducing significant computational overheads.

This modified SSFM algorithm has the advantage of keeping all the terms in the nonlinear interaction. Therefore it can provide accurate results even when the ellipse rotation terms should be kept, as for isotropic fibers or low-birefringence fibers. In view of the potential benefits of the modified algorithm, its simple formulation, and the small computational overheads associated with it, it is worth implementing in many situations.

6.7 Applications of the SSFM to fibers with longitudinal variations

As described in subsection 5.7.4, the impact of random longitudinal ZDW variations can be studied by means of a matrix method. The same could be done for random birefringence. An alternate approach would be to use a Runge–Kutta-type method to solve the differential equations coupling the main waves involved in the interactions (there are four or six such waves in the preceding chapters). These methods can provide accurate results as long as the additional waves arising from FWM among the basic waves are not very large. However, solving the NLSE by the SSFM is not subject to these limitations and can therefore be used almost universally. In the following sections we present examples of the spectral distortions, created by random variations of either zero-dispersion wavelength (ZDW) or birefringence, obtained when the NLSE is solved by means of the SSFM.

6.7.1 Random ZDW variations

In subsection 5.7.4 we showed examples of the effect of ZDW variations obtained by a matrix multiplication method [10]. The same effect was also investigated using a different mathematical technique [11]. Here, we present results obtained by means of the NLSE. We use the scalar equation and the SSFM to solve it. Figure 6.3 shows examples of gain spectra for a two-pump OPA with parallel pumps, obtained for different values of the ZDW standard deviation. The parameter values were $\gamma = 10\,\text{W}^{-1}\,\text{km}^{-1}$, $\beta^{(4)} = -1 \times 10^{-55}\,\text{s}^4\,\text{m}^{-1}$, $D_\lambda = 0.0615\,\text{ps}\,\text{nm}^{-2}\,\text{km}^{-1}$, $L = 250\,\text{m}$, $\lambda_1 = 1502.4\,\text{nm}$, $\lambda_2 = 1599.4\,\text{nm}$, and $P_{10} = P_{20} = 1\,\text{W}$. The dispersion was modeled by assuming that the ZDW in consecutive segments of length $L_{cd} = 10\,\text{m}$ was a Gaussian random variable, with mean ZDW $\lambda_0 = 1550\,\text{nm}$, and standard deviation σ_d. Birefringence was not included.

We see that the spectrum for the uniform fiber ($\sigma_d = 0$) is quite flat, but the spectrum for the fiber with the largest random ZDW variations drops by more than 15 dB in the center. Note that the spectra for the non-uniform fibers each correspond to a single fiber (not to an average over an ensemble of fibers). If the simulation is run with different fibers having the same σ_d, different spectral shapes are obtained each time.

6.7.2 Random birefringence variations

In Section 4.4 we investigated in detail what occurs when there is sufficient randomization of the SOPs to be able to average the coefficients of the OPA equations. In particular, we made the assumption that all the waves had the same SOP at every z. While this assumption is expected to hold relatively well for small frequency spacings,

6.7 Applications of the SSFM to fibers with longitudinal variations

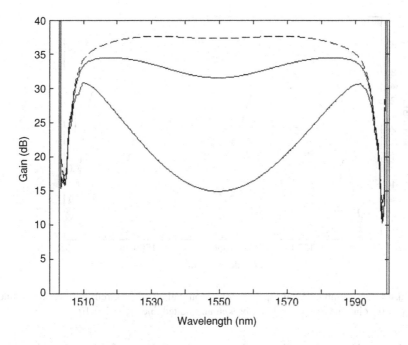

Fig. 6.3 Two-pump OPA gain spectra for a fiber with random ZDW variations, for the following values of σ_d: top line, 0 nm; middle line, 0.4 nm; bottom line, 0.8 nm. (The spikes at the pump wavelengths are artifacts and do not correspond to actual gain values.)

it becomes invalid as the frequency spacings become large. The reason for this can be understood from the waveplate model of polarization-mode dispersion (PMD). In a waveplate the phase retardation is given by $\Delta\Phi = \omega L_w \Delta n/c$, which is proportional to ω (L_w is the waveplate length). Hence waves at different frequencies experience different phase retardations. Therefore, if they enter the waveplate with the same SOP then they will exit it with slightly different SOPs. If we consider propagation over one beat length, corresponding to a nominal 2π phase shift, and we have waves centered at 1559 nm and separated by 150 nm, their retardations will differ by about 10%, or 36°. This is a considerable amount. Fortunately in non-PMF fibers the magnitude and sign of the birefringence varies randomly, so that the phases actually perform a random walk about 0, which prevents the difference from accumulating linearly. Instead, the growth in the phase difference can be modeled as a diffusion process: one can associate with it a characteristic diffusion length L_d, which is the distance after which the SOPs of the most distant wavelengths are uncorrelated. If $L > L_d$, the assumption that all SOPs remain the same at any z breaks down, and it is no longer possible to obtain an accurate description of the spectrum simply by using equations with modified coefficients.

An expression has been derived for L_d [12]. It is

$$L_d = \frac{3}{(D_p \Delta \omega)^2}, \qquad (6.67)$$

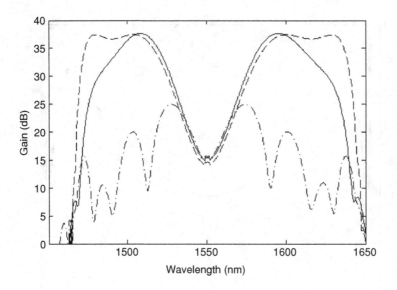

Fig. 6.4 One-pump OPA gain spectra, for various values of the birefringence: broken line, $\sigma_b = 0$; solid line, $\sigma_b = 1 \times 10^{-6}$; broken-and-dotted line, $\sigma_b = 2 \times 10^{-6}$.

where D_p is the polarization-mode dispersion (PMD) coefficient and $\Delta\omega$ is the frequency range of interest. As an example, if $D_p = 0.05$ ps km$^{-1/2}$ and the two wavelengths are 30 nm apart then $L_d = 200$ m. If $L < L_d$, there will be little difference between the SOPs at the various wavelengths and the averaged equations should provide a good approximation to the spectrum.

If $L > L_d$, however, the SOPs will quickly become uncorrelated, and the averaged equations will not be suitable. In that situation, each particular fiber will have a different evolution for the SOPs and thus a different longitudinal distribution for the gain, since the local gain depends on the relative orientation of the SOPs. As a result, the fibers will all have different gain spectra, in general, which can only be calculated by numerical means.

In such situations it is not possible to obtain closed-form solutions even by averaging, and one needs to resort to numerical integration. The effects of PMD on one- and two-pump fiber OPAs have been investigated by various techniques [13, 14]. Here we present similar results, obtained by means of an SSFM specifically tailored for modeling fiber OPAs [15]. The fiber is divided into sections of length L_{cb}, the birefringence coherence length. The magnitude of the birefringence Δn in each segment is represented by a random variable, of mean zero, which is often chosen to be a Gaussian. The orientation of the axes of polarization in each segment is also assumed to be random. It is represented by an angle, which is another random variable, with a uniform probability density function (p.d.f.).

Figure 6.4 shows one-pump OPA gain spectra obtained for parameter values: $\gamma = 10$ W^{-1} km^{-1}, $\beta^{(4)} = 1 \times 10^{-55}$ s^4 m^{-1}, $D_\lambda = 0.0615$ ps nm^{-2} km^{-1}, $L = 250$ m, $\lambda_1 = 1550.2$ nm, and $P_{10} = 2$ W. The pump and signal SOPs at the input were linear, and the same. There were no dispersion fluctuations. Random birefringence was present with

6.7 Applications of the SSFM to fibers with longitudinal variations

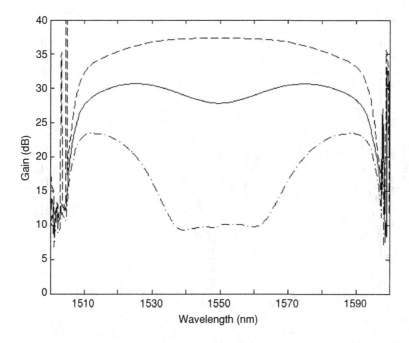

Fig. 6.5 Two-parallel-pump OPA gain spectra, for various values of the birefringence: broken line, $\sigma_b = 0$; solid line, $\sigma_b = 1 \times 10^{-6}$; broken-and-dotted line, $\sigma_b = 2 \times 10^{-6}$.

standard deviation σ_b for Δn (a Gaussian random variable) and birefringence coherence length $L_{cb} = 10$ m.

For relatively low birefringence, $\sigma_b = 1 \times 10^{-6}$, the shape of the gain spectrum is not drastically altered compared with that for an isotropic fiber, but there is an overall reduction in gain. For $\sigma_b = 2 \times 10^{-6}$ the distortion becomes very severe. In addition, for large σ_b, fibers that have nominally the same parameters can exhibit individual gain spectra that are quite different. Clearly this could pose significant difficulties for the design and fabrication of fiber OPAs with specified gain spectra. The deterioration of the gain spectra at high σ_b is in good qualitative agreement with what we expect when the diffusion length becomes shorter and shorter, since the gain is reduced when the SOPs of the three waves do not remain identical to each other everywhere.

Figure 6.5 shows gain spectra calculated by the SSFM for two-pump fiber OPAs with co-polarized pumps, for various values of σ_b; similar results have been obtained by different means [14]. The parameter values were $\gamma = 10\,\text{W}^{-1}\,\text{km}^{-1}$, $\beta^{(4)} = -1 \times 10^{-55}\,\text{s}^4\,\text{m}^{-1}$, $D_\lambda = 0.0615\,\text{ps}\,\text{nm}^{-2}\,\text{km}^{-1}$, $L = 250$ m, $\lambda_1 = 1502.4$ nm, $\lambda_2 = 1599.4$ nm, $P_{10} = P_{20} = 1$ W, $L_{cb} = 10$ m, and $\lambda_0 = 1550$ nm. As in the one-pump case, for $\sigma_b = 1 \times 10^{-6}$ the spectral shape is not greatly affected. For larger values of σ_b, however, the gain remains high only near the pumps; it drops significantly near the center frequency, which is far from both pumps.

Figure 6.6 shows gain spectra for two-pump fiber OPAs with orthogonally polarized pumps, for various values of the angle θ of the signal SOP and for that of the short-wavelength pump. The parameters were the same as for Fig. 6.5, except as indicated.

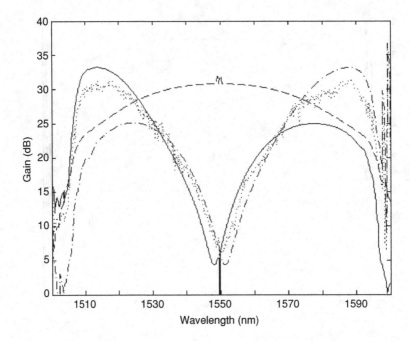

Fig. 6.6 Two-orthogonal-pump OPA gain spectra, for various values of the angle θ between the signal SOP and the x-axis. The broken line shows the spectrum for an isotropic fiber. The other three lines are for a birefringent fiber with $\sigma_b = 2 \times 10^{-6}$ and θ values as follows: solid line, $\theta = 0$; broken-and-dotted line, $\theta = 90°$; dots, $\theta = 45°$. The same fiber was used for all three spectra.

The main difference from the case of co-polarized pumps is that the spectra become strongly asymmetric for certain values of θ. For example, for $\theta = 0$ the gain is large near the short-wavelength pump and is significantly lower near the other pump. This effect is in agreement with the expected evolution of the various SOPs for large σ_b [14]. The fact that the gain spectrum varies with θ shows that such an OPA will not perform adequately in an optical communication system, where it is important to have a gain that does not vary with the SOP of the incident signal. This may reduce the appeal of the two-pump OPA with orthogonal pumps for making polarization-insensitive OPAs.

6.8   Sources of SSFM software

Several commercial software packages implementing the SSFM are available for investigating the propagation of communication signals in fibers. Many research groups have developed their own versions of the SSFM for similar purposes. A few groups have also developed SSFM software more specifically for OPAs.

The Matlab SSFM software developed by the author is available upon request. The ellipse rotation (coherent coupling) terms are kept in this code.

6.9 Conclusion

In this chapter we have introduced the NLSE and its solution by the SSFM method, and we have applied this approach to several situations for which the analytical solutions developed in Chapters 3 and 4 are not applicable. These are cases where the fiber exhibits significant longitudinal variations of the ZDW and/or linear birefringence that varies randomly along the length. In both cases we have seen that the performance of fiber OPAs is generally degraded, in the sense that the flat gain spectrum achievable in a perfectly uniform fiber can become greatly distorted. Since these longitudinal variations are always present to some extent in practical fibers, it is clear that they pose a challenge for the design and operation of fiber OPAs with flat gain spectra. In addition, the fact that the gain in two-pump OPAs with perpendicular pumps varies with the signal SOP indicates that this type of OPA may not be well suited for polarization-independent operation when very large bandwidths are desired.

An important point, however, is that the type of simulation presented here is pessimistic, in the sense that the calculated spectra are not optimized whereas spectra that can be obtained experimentally with a given fiber can be optimized by slightly tuning the pump wavelengths. While this can be done in near-real time in the laboratory, it would be quite time consuming to perform numerically. It is therefore encouraging to see that excellent experimental results have been obtained with fibers with significant ZDW variations and PMD, by careful optimization of parameters even though non-optimized simulations predict worse results [16, 17]. This indicates that the performance of fiber OPAs with imperfect fibers may be somewhat better than expected from the necessarily limited numerical simulations.

In the next chapter we will apply the SSFM to the case of OPAs with pulsed pumps, and in Chapter 13 we will use it to investigate crosstalk between WDM signals.

References

1. "Nonlinear pulse propagation in birefringent optical fibers," Menyuk, C. *IEEE J. Quantum Electron.*; 1987; vol. QE-23, pp. 174–6.
2. "Modulational instability for normal dispersion," Rothenberg, J. E. *Phys. Rev. A*; 1990; vol. 42, pp. 682–5.
3. "Cross-phase modulational instability in high-birefringence fibers," Drummond, P. D., Kennedy, T. A. B., Dudley, J. M., Leonhardt, R. Harvey, J. D. *Opt. Commun.*; 1990; vol. 78, pp. 137–42.
4. "Nonlinear effects in coherent multichannel transmission through optical fibers," Waarts, R. G., Friesem, A. A., Lichtman, E.; Yaffe, H. H., Braun, R. P. *Proc. IEEE*; 1990; vol. 78, pp. 1344–68.
5. "Parametric amplification and frequency conversion in optical fibers," Stolen, R. H., Bjorkholm, J. E. *IEEE J. Quantum Electron.*; 1982; vol.QE-18, pp. 1062–72.
6. "Parametric instability of optical amplifier noise in long-distance optical transmission systems," Lorattanasane, C., Kikuchi, K. *IEEE J. Quantum Electron.*; 1997; vol. 33, pp. 1068–74.

7. "Parametric noise amplification inherent in the coherence of fundamental optical soliton sequence propagating in fiber," Inoue, T., Namiki, S. *IEEE J. Selected Topics in Quantum Electron.*; 2004; vol. 10, pp. 900–5.
8. *Nonlinear Fiber Optics*, Agrawal, G. P. 1989; Academic, San Diego.
9. "Intensity-dependent changes in the refractive index of liquids," Maker, P. D., Terhune, R. W., Savage, C. M. *Phys. Rev. Letters*.; 1964; vol. 12, pp. 507–9.
10. "Parametric amplification in presence of fluctuations," Farahmand, M., de Sterke, M. *Optics Express*; 2004; vol. 12, pp. 136–42.
11. "Impact of dispersion fluctuations on dual-pump fiber-optic parametric amplifiers," Yaman, F., Lin, Q., Radic, S., Agrawal, G. P. *IEEE Photon. Technol. Lett.*; 2004; vol. 16, pp. 1292–4.
12. "Autocorrelation function of the polarization-mode dispersion vector," Karlsson, M., Brentel, J. *Opt. Lett.*; 1999; vol. 24, pp. 939–41.
13. "Effects of polarization-mode dispersion on fiber-based parametric amplification and wavelength conversion," Lin, Q., Agrawal, G. P. *Opt. Lett.*; 2004; vol. 29, pp. 1114–6.
14. "Effects of polarization-mode dispersion in dual-pump fiber-optic parametric amplifiers," Yaman, F., Lin, Q., Agrawal, G. P. *IEEE Photon. Technol. Lett.*; 2004; vol. 16, pp. 431–3.
15. "Accurate modeling of fiber OPAs with nonlinear ellipse rotation terms in the split-step Fourier method," Marhic, M. E., Kazovsky, L. G., Rieznik, A. A., Fragnito, H. L. In *Proc. Conf. on Optical Amplifiers and Their Applications*, Whistler BC, Canada, June 2006; paper JWB35.
16. "Nearly 100 nm bandwidth of flat gain with a double-pumped fiber optic parametric amplifier," Marconi, J. D., Chavez Boggio, J. M., Fragnito, H. L. In *Proc. Optical Fiber Communication Conf.*, Anaheim CA, March 2007; paper OWB1.
17. "Broadband single-pumped fiber-optic parametric amplifiers," Torounidis, T., Andrekson, P. *IEEE Photon. Technol. Lett.*; 2007; vol. 19, in press.

7 Pulsed-pump OPAs

7.1 Introduction

In the preceding chapters we investigated in detail the gain spectra of fiber OPAs with CW pumps. This is an important case for several reasons. From a practical standpoint, CW pumps are a requirement in applications such as amplifiers for optical communication. Also, experiments with CW pumps are simple to set up and are thus widely used. From a theoretical standpoint, the assumption of a CW pump leads to the relatively simple sets of equations and analytic solutions that we have studied in previous chapters.

It is also often desirable, however, to use pumps that do not have constant intensities, for a variety of reasons. One reason is that pumps that are pulsed with a low duty cycle can generate high peak powers while having a modest average power; this is a common way of generating peak powers of several watts or tens of watts using relatively inexpensive EDFAs with sub-watt average output power. Another reason is to generate OPA gain only during specified periods of time, to select particular parts of a signal waveform in order to implement some signal processing function. Examples of this are optical sampling for displaying optical pulse shapes, optical sampling for generating multiple replicas of optical signals, and optical switching for demultiplexing optical pulse trains.

Here we present alternative techniques for calculating the output fields of pulsed-pump OPAs. We begin by describing the quasi-continuous-wave (quasi-CW) regime approach, which has the merit of providing good approximations for long pulses in a relatively simple manner; for short pulses, it is necessary to resort to numerical methods such as the SSFM in order to obtain accurate results. We then present examples of calculations for selected cases.

7.2 The quasi-CW regime

When a pump is pulsed, it no longer consists of a single monochromatic wave and therefore the equations developed in the preceding chapters in principle can no longer be used, as they are based on monochromatic pumps. Intuitively, however, it is clear that if the pump-intensity variation is very slow then the CW equations should work quite well over time intervals during which the pump can be assumed to be constant. This use of CW equations to describe slowly varying phenomena is common in many areas of optics and is referred to as the quasi-CW (or quasi-steady-state) approximation. In this case one simply uses solutions of the CW equations for which some system parameters (here the pump intensities) are slowly varying functions of time.

The use of a quasi-CW solution is helpful because it allows us to use all the knowledge that we have derived from the CW case to obtain a first-order approximation to the system's behavior under pulsed operation. However, one should bear in mind that the accuracy of the quasi-CW solution will deteriorate when it is used for pump variations that are actually not very slow.

An important question is then to determine whether a pump variation is slow or fast in the context of a fiber OPA. To do so we must identify some characteristic time constant or some particular frequency of the OPA which separates the two regimes. The fact that the fiber nonlinearity responds on a femtosecond time scale does not mean that this is the appropriate time scale; additional considerations come into play.

A major limitation comes from phase modulation. The time-dependent pump intensity causes time-dependent phase modulation of the pump, signal, and idler, which in turn cause frequency modulation of the three waves. Since the OPA gain characteristics depend on the frequencies of the waves, they may be significantly modified if the frequency excursions are large. Since the latter are relatively larger for short pulses, we see that, as anticipated, the CW approximation will begin to break down for short pulses.

To quantify this effect, let us calculate the pump frequency shift due to self-phase modulation (SPM) (we consider a one-pump OPA for simplicity). We assume that the pump pulse has a Gaussian temporal shape which does not change along the fiber length. The peak power is P_0, and so the maximum self-phase shift is $\Phi = \gamma P_0 L$. Then the frequency excursion due to SPM is of the order of $\Delta v_{SPM} \approx \Phi \Delta v$, where Δv is the half-width of the pump input spectrum and $\Delta v \approx 1/\tau_p$, where τ_p is the pulse half-width. As an example, if τ_p is of the order of 1 ns and $\Phi = 3$, $\Delta v_{SPM} \approx 3$ GHz. For many practical OPAs this is not a very large frequency excursion, and generally it will not cause drastic changes in gain. Under these circumstances we can use the CW gain expressions with P_0 replaced by $P_0(t)$, but assuming that $\Delta \beta$ is constant and therefore can be calculated in terms of the input frequencies.

If we now consider τ_p values of the order of 10 ps, however, and $\Phi = 3$ then $\Delta v_{SPM} \approx 300$ GHz (about 2.5 nm in terms of wavelength). This is now a considerable shift, which in general will strongly affect the gain via the corresponding change in $\Delta \beta$. Under these circumstances, assuming that $\Delta \beta$ remains constant would lead to results that would be likely to be very inaccurate.

These examples indicate that for typical fiber OPAs the quasi-CW approximation should provide a reasonable first-order approximation when the pulse durations are of the order of nanoseconds or longer but that the accuracy will progressively get worse as the pump pulses become much shorter.

It is interesting to note that the SPM-related frequency shift is zero at the extrema of the pump pulse; for a Gaussian pulse there is one such point, the maximum at the pulse center. At such points $\Delta \beta$ is the same as if there were no SPM, and therefore the gain can still be calculated with good accuracy simply by replacing P_0 by $P_0(t)$ in the CW gain expression. In contrast, at times where the pump power varies rapidly, i.e. near inflection points of $P_0(t)$, the frequency deviation will be high and the same gain equation will be subject to a large error.

Another significant limitation of the quasi-CW approach is due to pulse walkoff. This refers to the fact that, since the pump and signal pulses have different nominal frequencies, they travel at different group velocities and therefore their relative overlap may vary as they pass through the OPA. The shorter the pulses, the more important this may be. When the walkoff is significant the pump and signal effectively interact only over a fraction of the length L, and therefore we expect a gain reduction in comparison with the no-walkoff case.

If $\lambda_1 \approx \lambda_0$ then the time delay between the pump and signal over the OPA length is $\Delta\tau = D\Delta\lambda_s L/2$. For the quasi-CW method to be easily applicable, we need to have $\Delta\tau \ll \tau_p$. As an example, for an a highly nonlinear fiber (HNLF) with $D_\lambda = 0.02$ ps nm^{-2} km^{-1} and $L = 100$ m, if we use $\Delta\lambda_s = 50$ nm then we have $\Delta\tau = 0.1$ ps. Hence in this case we should not expect the quasi-CW approach to work well for subpicosecond pulses. (But, as we have seen, pump SPM may set a limit for longer pulses anyway.)

Finally, if the signal pulses are so short that their shape is affected by dispersion as they propagate through the OPA, the quasi-CW approach breaks down simply because it is based on the assumption of a fixed pulse shape. This, however, will occur only after SPM and walkoff have become problems. In general, the two effects of SPM and walkoff will need to be taken into account when considering whether the quasi-CW method is applicable.

To use the quasi-CW expression for the signal and idler gains of a one-pump OPA, we take Eqs. (3.77) and (3.78), with the appropriate parameters for a one-pump OPA given in Section 3.6.2, and replace P_0 by $P_0(t)$ everywhere. Hence we have explicitly

$$h_3(t) = \left\{\cosh[g(t)L] + i\frac{\kappa(t)}{2g(t)}\sinh[g(t)L]\right\}\exp\left\{i\left[\gamma P_0(t) - \frac{\Delta\beta}{2}\right]L\right\}, \qquad (7.1)$$

$$h_4(t) = i\frac{\gamma P_0(t)}{g(t)}\sinh[g(t)L]\exp\left\{i\left[\gamma P_0(t) - \frac{\Delta\beta}{2}\right]L\right\}, \qquad (7.2)$$

where

$$\kappa(t) = \Delta\beta + 2\gamma P_0(t) \qquad (7.3)$$

and

$$g^2(t) = -\frac{\Delta\beta}{4}[\Delta\beta + 4\gamma P_0(t)]. \qquad (7.4)$$

Here $\Delta\beta$ is the wavevector mismatch, which depends only on $\Delta\omega_s = \omega_3 - \omega_1$ and is assumed to be constant.

Equation (7.2) for h_4 contains the time-varying phase term $\exp\{i[\gamma P_0(t)]\}$, which constitutes the SPM experienced by the pump. Also, it can be shown that $\sinh[g(t)L]/g(t)$ is always real, regardless of whether $g(t)$ itself is real or imaginary. Therefore the idler is subject to the same phase modulation (and hence frequency modulation) as the pump itself. This may seem counter-intuitive, because in normal cross-phase modulation (XPM) (where parametric interaction is not involved), it is well known that the XPM is twice as large as the SPM; the reason for the difference has to be attributed to the complexity of the parametric interaction.

Equation (7.1) for h_3 also contains the same phase term, $\exp\{i[\gamma P_0(t)L]\}$. However, the expression in the braces multiplying this phase term is generally a complex function of time. The total time-dependent phase angle of the signal is

$$\theta_3(t) = \gamma P_0(t)L + \tan^{-1}\left\{\frac{\kappa(t)}{2g(t)}\tanh[g(t)L]\right\}. \tag{7.5}$$

(This is valid for $g(t)$ real or imaginary.) The first term ranges between 0 and $\Phi_{\max}$, which is itself generally between 2 and 6; the second term can vary between $-\pi/2$ and $\pi/2$. Hence the second term can make a significant contribution under certain circumstances.

The preceding equations can be used for studying the small-signal gain in OPAs with pulsed pump when circumstances warrant it.

There is, however, nothing in our preceding reasoning that limits it to the small-signal case. In fact, it applies just as well to the case where pump depletion is significant. One could therefore also write down quasi-CW solutions to the gain equations, involving Jacobian elliptic functions (see subsections 3.5.1, 3.5.2, and 3.6.1), simply by introducing the time dependence of the pump(s) explicitly [1].

This is very useful, particularly when it comes to designing OPAs where it is desirable to obtain strong pump depletion. The conditions for complete pump depletion are still the same, but they can only be satisfied at particular times. Clearly the use of pulsed pumps with a Gaussian-like pulse shape will complicate the optimization process in such a situation, because maximizing the conversion efficiency at a particular time may not yield the highest conversion efficiency for the pulse as a whole. In addition, strong pulse distortions may occur if the peak power of the pulse is larger than that required for complete depletion. Also, the inevitable XPM phase shifts will introduce spectral broadening, which will make it challenging to obtain nearly-transform limited pulses in this manner.

Another difficulty with using the Jacobian elliptic function solution is that it yields only expressions for the output powers of the waves. To our knowledge closed-form solutions for the phases are not available. Therefore it is not possible to write down expressions for the fields, which limits the usefulness of this approach.

7.3   The split-step Fourier method

We have used this method previously for obtaining accurate values for the gain of fiber OPAs with continuous-wave (CW) pumps but random variations in zero-dispersion wavelength (ZDW) and/or birefringence. Even though the individual input waves were CW, the total input field, being the sum of discrete tones, had an envelope which exhibited temporal variations. These variations were handled in the manner described in Section 6.6, involving the FFT and its inverse to switch repeatedly between the time domain and the frequency domain.

The situation here is quite similar. The total input field again displays temporal variations, not only because of the different nominal frequencies of the waves but also because of the intensity modulation of the pump. The SSFM method can deal with this situation just as well as the previous one. The only difference is that the output waves, instead of being continuous, will exhibit temporal intensity (and phase) modulation and will therefore have individual spectra with non-zero width.

As stated earlier, the quasi-CW method becomes increasingly inaccurate as the pulse duration is reduced. It then becomes necessary to resort to the SSFM method. In principle it should provide accurate results for pump pulses as short as one picosecond, which are of interest for a variety of applications.

If, however, relatively long pulses are used (say longer than 1 ns), the SSFM method may become difficult to use. This is particularly true if it is necessary to cover a very wide wavelength range, such as when studying a fiber OPA with a 400 nm bandwidth. In that case, in order to sample the narrow pump spectrum accurately and to cover the entire frequency span, one may need to use about 10^6 frequency points. Such a large number of points would lead to long computation times, making this approach unattractive. Since the quasi-CW method should be quite accurate for time scales in excess of 1 ns, it will then be preferable to use it.

Overall, we see that the quasi-CW approach and the SSFM approach play complementary roles. The former is best suited to long pump pulses while the latter is best suited to short pulses. In the intermediate region, the quasi-CW method can provide a simple first-order approximation, with all the physical insight associated with the CW solutions; the SSFM method can then provide greater accuracy.

7.4 Important pulse shapes

Rectangular pulses

The case of a rectangular pulse is interesting because the pump SPM at the top of the pulse is constant and therefore does not shift the pump frequency. Thus, according to the preceding reasoning, one can use the CW expressions to calculate accurately the gain at the top of the pulse, even if the pulse is very short.

At the pulse edge the SPM phase is discontinuous, and this leads to an infinite shift in pump frequency with signal gain unity. However, the duration of this event is zero and therefore the contribution of the discontinuity can be ignored.

While the rectangular pulse provides an interesting theoretical model, it should be recognized that it is an idealized pulse shape and can only be approximated in practice, since all physical pulses will actually have finite rise times and fall times, limited by the physical characteristics of the modulators used to generate them. The best approximation to a rectangular pulse is obtained by taking a CW source and turning it on and off by means of a high-speed modulator. Such modulators can have sub-nanosecond rise times (with bandwidths of tens of gigahertz). In this manner one can generate pulses with a flat top lasting several nanoseconds using sub-nanosecond transitions; these pulses have a nearly rectangular appearance. This method is often used in practice to measure the gain that would be available from a CW OPA if a CW pump had the same power as the peak power of the rectangular pulse. For example, one could generate 5 ns pulses with 10 W peak power, with a duty cycle of 10^{-2} and therefore an average power of only 0.1 W; then, measuring the gain near the center of the pulse will yield an accurate measurement of the gain that could be obtained with a constant pump power of 10 W.

Cosine-squared pulses
These pulses can be obtained in practice simply by linearly combining two unmodulated monochromatic waves with radian frequencies ω_1 and ω_2. If they have the same power then the result can be viewed as a carrier with frequency $(\omega_1 + \omega_2)/2$ and intensity proportional to $\cos^2[(\omega - \omega_2)t/2]$. While such pulse shapes are easy to obtain, they are quite rounded. They do not have a flat top, and the intensity is a maximum only at discrete times. Maximum gain is achieved only at these times, which would make it difficult to measure the peak gain.

Raised-cosine pulses
To increase the duration of the regions of maximum (and minimum) gain, one could hypothetically modify cosine-squared pulses as follows: leave the S-shaped transitions between maxima and minima unchanged, but stretch the duration of the extrema from 0 to some arbitrary duration t_{rc}. As t_{rc} is increased, this type of raised-cosine waveform looks more and more like a rectangular wave. One could in principle choose t_{rc} to be large enough that an accurate measurement of the peak gain could be made during that time.

This type of waveform is frequently used for modeling communication signals and has received its name in that context. A raised-cosine model would be a good fit for the experimental flat-top pulses with finite transition times described above (suitably modified to have different on and off durations, if the duty cycle is not 50%).

Gaussian and super-Gaussian pulses
A model which is useful to describe the output of some pulsed lasers is that of Gaussian pulses. Such a pulse has a field envelope of the form

$$f(t) = e^{-(t/T_p)^2} \tag{7.6}$$

where T_p is the half-width, measured from the center to where the amplitude drops by $1/e$.

Gaussian functions are very important in optics, because they have the unique property that their uncertainty product, defined as in quantum mechanics, has the smallest possible value.

However, the spectral and temporal quality of laser pulses is generally defined in a different manner, by means of the time–bandwidth product (TBP) $\Delta\nu\Delta\tau$, where $\Delta\tau$ is the full width at half maximum (FWHM) of the intensity (field-squared) waveform and $\Delta\nu$ is the FWHM of the optical power spectrum [2]. With this definition, the TBP for a Gaussian pulse is equal to 0.441. Because Gaussians are minimum-uncertainty functions in quantum mechanics, one might expect that they also have a minimum TBP.

This, however, is not true because the definition adopted above for the TBP is different from that used for the uncertainty product (which is based on moments, rather than FWHMs). With this definition, pulses with a TBP as low as 0.11 have been observed [3], and it is in fact possible to construct functions with a vanishingly small TBP.

Gaussian pulses have a bell-shaped appearance, with a parabolic peak like the cosine-squared pulse. As such, they are a poor approximation to flat-top pulses. To obtain a better representation of such pulses, one can use so-called super-Gaussian pulses, which are simply obtained by replacing the power 2 in Eq. (7.6) by a power $n = 4, 6, 8$, etc. The larger n is, the closer the pulse approches a rectangular shape and the larger its TBP is.

Sech pulses

An important type of pulse in nonlinear optics is the sech (hyperbolic secant) pulse. It can be shown that some types of pulse can propagate without changing shape in optical fibers, because of a balance between fiber nonlinearity and dispersion. The temporal shape of the amplitude of a *soliton* pulse is given by a function of the form sech $= 1/\cosh$. The shape is close to that of a Gaussian but does not drop as fast away from the center.

Soliton pulses are very important because of their possible use in optical communication. In addition, fiber light sources generating periodic trains of short soliton pulses (a few picoseconds long) have been developed, and these could be used as sources for nonlinear fiber optic interactions.

In addition to the sech envelope, soliton pulses may also exhibit a linear frequency variation, or chirp. In the absence of chirp, soliton pulses have a TBP of 0.315, smaller than for Gaussian pulses [2].

Periodic pulses

So far we have discussed pulses as if they occurred in isolation. In practice, however, it is very common to encounter light sources which generate periodic pulse trains, which for all practical purposes can be considered to be of infinite duration. An example would be the output of a mode-locked laser: it consists of identical pulses, spaced by the round-trip propagation time in the cavity. Another would be a CW light wave modulated by an electro-optic modulator, driven by a periodic electrical waveform provided by an electrical waveform generator.

The time-domain analysis of the performance of an OPA driven by a periodically modulated pump is essentially the same as that for an isolated cycle of the pump. The reasons for this are that in an OPA all waves travel with nearly the same speed and the nonlinearity is memoryless, therefore what takes place at one instant of time is not influenced by what occurred before or what is coming next. For example, applying the quasi-CW approximation to any period in a train of pulses will yield the same result as if this period were isolated. Therefore the total response to a pulse train can be inferred simply by replicating the response to a particular period.

When viewed in the frequency domain, the response to a periodic pulsed pump train leads to some important conclusions. Consider for example the case where a fiber OPA operates in the linear regime. Let the time-dependent signal gain due to a single pulse

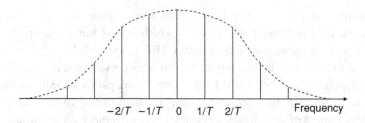

Fig. 7.1 Relationship between the spectrum of a single pulse (broken line) and a periodic train of pulses (solid lines). The pulse repetition frequency is $1/T$.

be $h_3(t)$. Then the gain due to a periodic train of identical pulses is

$$h_{3,\text{per}}(t) = \sum_{n=-\infty}^{\infty} h_3(t - nT) = h_3(t) \otimes \sum_{n=-\infty}^{\infty} \delta(t - nT)$$

$$= h_3(t) \otimes \frac{1}{T} \text{comb}\left(\frac{t}{T}\right), \quad (7.7)$$

where T is the period, $\delta(t)$ is Dirac's delta function, comb is the comb function, and $\otimes$ denotes convolution. The Fourier transform of $h_{3,\text{per}}(t)$ is therefore of the form

$$FT\{h_{3,\text{per}}(t)\} = FT\left\{\sum_{n=-\infty}^{\infty} h_3(t - nT)\right\} = FT\{h_3(t)\} \text{comb}(fT), \quad (7.8)$$

where FT denotes the Fourier transform operator and f is the variable in the frequency domain. Thus the spectrum of the pulse train consists of an infinite series of δ-functions spaced by the frequency interval $1/T$. The amplitudes of these frequency components are obtained from the complex amplitude of the Fourier transform of a single pulse, which is generally defined continuously at all frequencies. Thus we may say that the continuous spectrum of a single pulse acts as an envelope for the discrete spectrum of an infinite pulse train. This is illustrated in Fig. 7.1.

A slightly different perspective is obtained by introducing the notion of a Fourier *series*. This is done by saying that since $h_{3,\text{per}}(t)$ is periodic it can be written as an infinite sum of harmonics, with frequencies that are multiples of the modulation frequency $f_m = 1/T$, i.e.

$$h_{3,\text{per}}(t) = \sum_{n=-\infty}^{\infty} a_n \exp(i2\pi n f_m t), \quad (7.9)$$

where the a_n are the Fourier series coefficients. They are calculated by integrating the product of $h_{3,\text{per}}(t)$ and a harmonic over one period, i.e.

$$a_n = \frac{1}{T} \int_0^T h_{3,\text{per}}(t) \exp(-i2\pi n f_m t) \, dt. \quad (7.10)$$

This is equivalent to saying that the Fourier transform $H_3(f)$ of $h_{3,\text{per}}(t)$ consists of a series of delta functions, namely

$$H_3(f) = \sum_{n=-\infty}^{\infty} a_n \delta(f - n f_m). \quad (7.11)$$

7.4 Important pulse shapes

While this approach is strictly equivalent to the preceding one, it has the disadvantage of obscuring the role of the continuous FT of a single interval and the nature of the envelope of the Fourier coefficients. But from a calculation standpoint it is direct and efficient.

The interpretation of $|a_n|^2$ is that it is the power gain associated with the conversion from the original signal to a new signal, displaced in frequency from the original one by nf_m. A parallel reasoning can be used for the periodic idler gain, which can be written in terms of Fourier series coefficients b_n, i.e. $h_{4,\text{per}}(t) = \sum_{n=-\infty}^{\infty} b_n \exp(i2\pi n f_m t)$. The modulus squared $|b_n|^2$ is then the power gain associated with the conversion from the original signal to a new idler, displaced in frequency from the original one by nf_m.

An interesting consequence of this approach is that both total signal and idler spectra should be symmetric with respect to their respective central frequencies if $P_3(t)$ is an even function of time, i.e. we should have $a_{-n} = a_n$ and $b_{-n} = b_n$ for all n.

The discrete nature of the spectrum of a train of pulses may be important in practice and depends on the characteristics of the instruments used to display or measure the spectrum. For example, it is common to look at the output optical spectrum of a fiber OPA with an optical spectrum analyzer (OSA) having a resolution of 0.1 nm (about 12.5 GHz near 1550 nm wavelength). Thus if the pulse repetition frequency is smaller than 12.5 GHz then the displayed optical spectrum will appear to be continuous, like that of a single pulse. In other words, under such conditions an OSA cannot show that the optical spectrum is due to a periodic waveform or what its pulse repetition frequency (PRF) is. By contrast, if the PRF is larger than 12.5 GHz then the discrete nature of the spectrum will be revealed and it will be possible to infer the PRF from the spacing between the peaks.

The discrete nature of the gain spectrum of an OPA driven by a periodic pulse train can be exploited for practical applications. If the OPA input is an optical communication signal then the output signal spectrum consists of several carriers, as explained above, but now each carrier is also modulated by the input signal. The result is that at the output several replicas of the input signal are obtained. If the pulse PRF is high enough, the spectra of these individual replicas have negligible overlaps and can therefore be selected by means of optical filters. Therefore this approach can be used for generating multiple copies of an input signal, i.e. for multicasting [4].

Transform-limited pulses

Frequently encountered when dealing with short pulses is a transform-limited (TL) pulse or spectrum. Essentially this refers to a pulse which does not have any chirp, i.e. which corresponds to a monochromatic carrier modulated only in amplitude. The spectrum of such a pulse is generally narrower than that of the same pulse with chirp. As a result, TL pulses have a smaller TBP than pulses with the same temporal envelope but with chirp. This can easily been verified for sech pulses with varying degrees of chirp [2].

The importance of this concept is experimental. In practice it is relatively easy to measure the temporal pulse and frequency spectrum, but not the instantaneous variation in phase or frequency, i.e. the chirp. The existence of chirp can be ascertained as follows, however. One calculates the FT of the real pulse envelope (without chirp) and compares the result with the measured spectrum. If the two are very close then there is little chirp but if the calculated spectrum is much narrower than the measured one then the pulse has a large amount of chirp.

7.5 Examples of pulsed-pump OPAs

7.5.1 TDM demultiplexing with Gaussian pulses

One of simplest applications of pulsed OPAs is in demultiplexing signals which were formatted by time-division multiplexing (TDM). In TDM one divides time into equal-duration frames, each of which contains N equal consecutive time slots. There are N communication channels; the nth channel is assigned the nth time slot in each frame. Binary information is sent through each channel by sending either a pulse or no pulse in each of its time slots.

To retrieve the information in the nth channel at the receiver one must use a device which can discriminate between the incoming time slots and select only the information in the desired slots. A pulsed-pump OPA can be used for this purpose. Assuming that a suitable clock can be provided at the receiver by standard techniques, it can be used to generate pump pulses with about the duration of a time slot, well synchronized so that each pulse is centered in a time slot and coincides only with the nth slot in each frame. Under these circumstances, the pulsed OPA will generate gain for the signal in that time slot, and nowhere else; it will also generate a pulse at the idler wavelength. The pulse trains coming out of the OPA, at either wavelength, then contain primarily the pulses of interest and little or no power corresponding to the other time slots. Hence these pulse streams can now be detected with slow detectors (having a response time of the order of a frame duration), and the outputs of the latter will contain only the binary information associated with the nth channel. This arrangement will perform the desired time demultiplexing operation.

If the idler wavelength is used as the output then there will be no leakage of power to the undesired time slots, because the pump will be off during them. This is true regardless of how large the OPA gain is. Hence this mode of operation will provide the best discrimination against undesired pulses. However, it changes the wavelength and this may be considered as an undesirable feature in some applications.

If one desires to retain the original wavelength, one needs to use the signal output. The drawback is that undesired pulses pass through the OPA with unit transmittance when the pump is off, and therefore the discrimination between consecutive pulses is not as good as if the idler is used as output. To maximize the discrimination, it is necessary to have a high signal gain. For example, if $N = 100$ and if one wants the desired channel energy to exceed the total energy of the others in one frame by 30 dB then one will need a gain of about 50 dB. This corresponds to a pump self-phase shift $\Phi \approx 2\pi$, which is considerable and can lead to a substantial broadening of the spectrum of the received pulses.

We modeled this situation by means of the SSFM method, using identical Gaussian pulses for the pump and the signal. The powers were chosen to have values appropriate

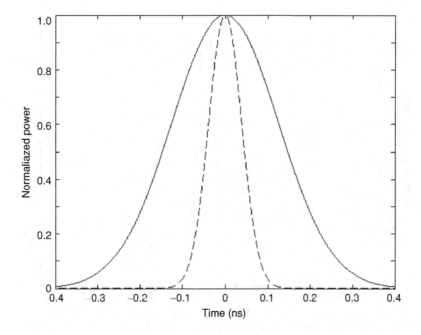

Fig. 7.2 Normalized signal pulse shapes before (solid line) and after (broken line) the OPA.

for a communication system. The fiber parameters were as follows: $L = 535$ m; $\lambda_0 = 1550$ nm; $\gamma = 10$ W^{-1} km^{-1}; $D_\lambda = 0.02$ ps nm^{-2} km^{-1}; $\beta^{(4)} = -5 \times 10^{-56}$ s^4 m^{-1}. For the pump we used $T_p = 0.25$ ns, $\lambda_1 = 1553$ nm, and a peak power $P_{1,\text{peak}} = 1$ W. The numbers were chosen to yield $\Phi_{\text{peak}} = 5.35$ rad and about 40 dB of maximum signal gain.

For the signal, we chose a peak power $P_{3,\text{peak}} = 0.01$ mW and the same pulse shape as the pump (the pulse shape would be appropriate for 10 Gb/s return-to-zero (RZ) signals). We used $\lambda_3 = 1531$ nm, which is fairly close to the pump value, in order to avoid the numerical problems associated with large frequency spacings (see Section 7.3). The computation parameters were: total wavelength range 1450–1650 nm; 1000 longitudinal steps; 2^{15}-point DFT.

Figure 7.2 shows the normalized signal pulse shapes before and after the OPA. It can be seen that the pulse width shrinks considerably: the FWHM is 290 ps at the input but only 90 ps at the output, i.e. it drops by more than a factor 3.

The reason for this is that, according to the quasi-CW picture, the OPA gain depends almost exponentially on the instantaneous pump power and so it takes little variation of pump power for the gain to vary strongly. This leads to output pulses that are generally much narrower than the pump pulses. To see this qualitatively, we assume that the gain depends exponentially on pump power, i.e. $G_3 \approx e^{2\gamma P_1 L}/4$. Maximum gain is obtained for maximum pump power, $P_{1,\text{peak}}$. The gain drops by one half for a power P_1' such that

$$G_3' \approx \frac{e^{2\gamma P_1' L}}{4} = \frac{G_3}{2} = \frac{e^{2\gamma P_{1,\text{peak}} L}}{8}. \tag{7.12}$$

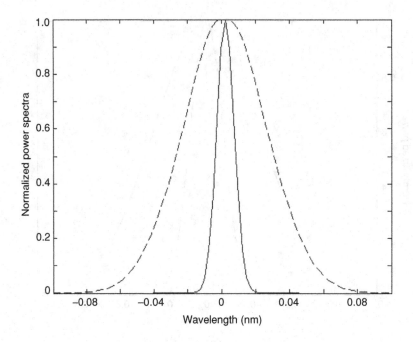

Fig. 7.3 Signal spectra before (solid line) and after (broken line) the OPA.

This leads to

$$\frac{\Delta P_1}{P_{1,\text{peak}}} = \frac{P_{1,\text{peak}} - P_1'}{P_{1,\text{peak}}} = \frac{\ln 2}{2\gamma P_{1,\text{peak}} L} = \frac{\ln 2}{2\Phi_{\text{peak}}}. \quad (7.13)$$

For a Gaussian, the instantaneous power varies in time as

$$P_1(t) = e^{-2(t/T_p)^2}. \quad (7.14)$$

Hence

$$\frac{\Delta P_1(t)}{P_{1,\text{peak}}} = 1 - e^{-2(t/T_p)^2}. \quad (7.15)$$

Equating the right-hand sides of Eqs. (7.13) and (7.15), we can solve for t, which is the time at which the gain drops by one-half. Consequently, if the input signal is CW, the output is a pulse with FWHM

$$2t = 2T_p \sqrt{-\frac{1}{2} \ln \left(1 - \frac{\ln 2}{2\Phi_{\text{peak}}}\right)}. \quad (7.16)$$

With the above parameters we find that the FWHM so calculated is 91 ps, in close agreement with the 90 ps obtained numerically by the FT method. Similar compression ratios have been observed in practice. Specifically, the signal pulse length was reduced from 93 ps to 20 ps in an OPA with gain 29 dB [5].

Figure 7.3 shows the signal spectra at the input and output. It can be seen that considerable spectral broadening takes place within the OPA: the FWHM increases by more than

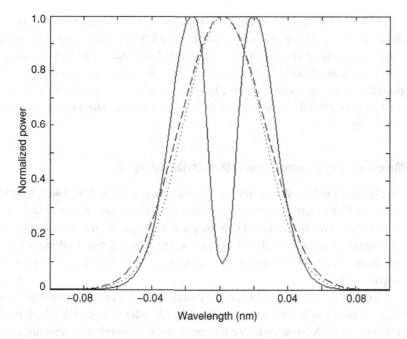

Fig. 7.4 Normalized spectra at the OPA output. Solid line, pump; broken line, signal; dotted line, idler.

a factor 5. The narrowing of the temporal pulse width alone would lead to a broadening of the spectrum by about a factor 3. The fact that we observe a larger broadening is most likely attributable to nonlinear frequency shifts, which are substantial for $\Phi_{peak} = 5.35$ rad.

While this is not directly relevant to the performance of the demultiplexer, it is interesting to calculate the TBP of the signal output pulses. From the measurements of the FWHMs in the time and wavelength domain, relative to the TL input pulse, we see that the TBP for the signal output pulse is larger by about $5/3 = 1.66$. Thus, even though the pulses exhibit a great deal of spectral broadening, their TBP deterioration is not very large.

The fact that the output pump and idler have the same phase variation does not imply that they have the same spectra. As can be seen from Fig. 7.4, the output pump and idler in fact have quite different spectra. The reason for this is that the idler pulse, like the signal output pulse, is about three times shorter than the pump pulse.

By contrast, the output signal and idler have very similar spectra, even though they have different phase variations. This is due to the fact that for high gain they have essentially the same pulse shape (power versus time), because they each gain one photon from the pump at the same time.

The double-humped shape of the pump output spectrum is typical of the spectrum of a Gaussian pulse which acquires a maximum self-phase modulation (SPM) phase shift of about 2π, which is close to the value 5.35 in this example [6].

The increase in signal bandwidth introduced by this demultiplexer, which uses the output signal, could lead to performance degradation in some detection systems. By contrast, if the idler is used as the output then lower gains can be used since there is no leakage from undesired channels and therefore pulse broadening can be reduced. Spectral broadening could also be reduced by using pump pulses with a flatter top than Gaussian pulses but this is difficult to do at the high speeds where optical demultiplexing is attractive.

7.5.2 Efficient wavelength conversion with Gaussian pulses

In subsection 5.6.2 we studied the conditions for obtaining efficient wavelength conversion with a CW pump and signal. We showed that for a pump power of several hundred watts, and short fibers, one could in principle obtain efficient wavelength conversion that is tunable over several hundred nanometers. We also discussed difficulties associated with the use of such high powers, namely stimulated Brillouin scattering (SBS) power dissipation, etc.

These difficulties can be alleviated by using a pulsed pump with a high peak power and a low duty cycle. This approach should yield idler pulses with a high peak power, tunable over a wide range, which could be useful for a number of applications.

In order for this approach to be practical it would be desirable to have the following features: (i) high conversion efficiency; (ii) nearly TL output pulses.

Here we model a wavelength converter designed for this purpose. We use the SSFM because of the difficulties associated with the quasi-CW method in this case. We assume again that the pump pulses are Gaussian, with $T_p = 0.25$ ns. This type of pulse can readily be generated by erbuim-doped fiber lasers (EDFLs).

The fiber parameters are as follows: $\lambda_0 = 1550$ nm; $\gamma = 10$ W^{-1} km^{-1}; $D_\lambda = 0.02$ ps nm^{-2} km^{-1}; $\beta^{(4)} = -5 \times 10^{-56}$ s^4 m^{-1}. The length is adjusted to optimize performance. For the pump we have $T_p = 0.25$ ns, $\lambda_1 = 1550$ nm, and peak power $P_{1,\text{peak}} = 100$ W. For the signal, we consider both the case where it is CW, with low power (100 mW), and the case where it is pulsed with a higher peak power (>10 W) and the same pulse shape as the pump. We use $\lambda_3 = 1548$ nm, which is close to the pump, in order to avoid the numerical problems associated with large frequency spacings (see Section 7.3).

The case of a the weak CW input signal is quickly found to be inadequate. The reason for this is that the signal needs to be amplified to a higher level in order to start depleting the pump significantly. This requires a long fiber, which leads to a large value for the cross-phase modulation (XPM) experienced by the signal and idler (this situation is very similar to that in the preceding subsection). As a result, the idler spectrum is found to be considerably broadened, with several peaks and consistent with a high level of phase modulation [6]. The TBP for the idler is several times larger than 0.441, and so the idler is far from being Gaussian.

With a pulsed input signal we can adjust its peak power to be large enough that pump depletion can start taking place from the beginning. As a result, a shorter fiber can be used to reach the same level of pump depletion as obtained with a weaker CW signal, and therefore XPM of the signal and idler by the pump is minimized.

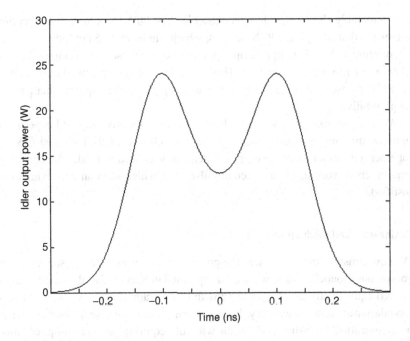

Fig. 7.5 Idler output pulse shape for $L = 3.5$ m.

Guided by the results of the CW investigation (Section 5.6), we will launch fairly high signal powers along with the pump. For $P_{3,\text{peak}} = 20$ W and $L = 2$ m we find that we can obtain an idler output pulse with nearly Gaussian temporal and spectral shapes, with a TBP only 1.57 times larger than the TL value. The peak idler power is $P_{4,\text{peak}} = 35$ W. The conversion efficiency and pulse quality obtained should be satisfactory for a number of applications.

Trying to increase L to improve the conversion efficiency while maintaining a near-TL idler pulse is counter-productive, as shown in Fig. 7.5. The conditions are the same as above, except that L has been increased from 2 m to 3.5 m. As a result, the peak of the signal pulse strongly depletes the pump before the fiber end, yielding maximum idler power then, but beyond that point the idler power at the center of the pulse decays by returning some power to the pump and also by giving up some power to the higher-order FWM terms. The conditions for strong depletion of the pump at L are now met at times on either side of the pulse center, where the initial launched power is smaller than at the center. Hence the idler pulse has a double-humped shape, which is not attractive for many applications.

All the preceding results can be scaled to pulses with the same duration and higher peak powers. For example, fiber lasers can now generate pulses with 1 kW peak power [7]. If we increase all the preceding powers by a factor 10 and reduce the length by the same factor then we expect to obtain a similar scaling of the idler peak power to about 350 W. This would be obtained with just 20 cm of highly nonlinear fiber (HNLF), which is considerably shorter than has been used in most OPA work to date. In fact this length is close to that of typical PPLN devices, which can be up to 5 cm long.

For short pulses, it is in principle possible to use pulse compression in an auxiliary fiber in order to improve the TBP. This, however, would be practical only for pulses that are a few picoseconds long. For nanosecond-long pulses the required fiber lengths would be prohibitive.

We note that extensive work has been done both theoretically and experimentally to optimize the conversion efficiency of pulsed fiber OPAs [1, 8, 9]. The quasi-CW approach, together with exact Jacobian elliptic functions, was used to model the conversion. The experimental work used nanosecond pulses, for which such an approximation is well justified.

7.5.3 Cosine-squared modulation

We now consider the case where the pump consists of two closely spaced equal-power continuous monochromatic waves. As explained in Section 7.4, the pair of waves can be viewed equivalently as a single wave with the average frequency and a cosine-squared modulation at the beat frequency. This case is interesting for several reasons: (i) it provides well-controlled intensity modulation without requiring an electro-optical modulator; (ii) the quasi-CW approach can be investigated by calculating Fourier series coefficients; (iii) the FT approach can be very efficient because of the simplicity of the input optical spectrum.

The fiber parameters are as follows: $\lambda_0 = 1561.14$ nm; $\gamma = 11$ W^{-1} km^{-1}; $D_\lambda = 0.03$ ps nm^{-2} km^{-1}; $\beta^{(4)} = -5.8 \times 10^{-56}$ s^4 m^{-1}; $L = 500$ m. The length L is adjusted to optimize performance. For the pumps we have $P_1 = P_2 = 0.2$ W, $\lambda_1 = 1563.13$ nm, and $\lambda_2 = 1563.93$ nm (i.e. the pumps are 0.8 nm or about 100 GHz apart). We consider a CW signal with $\lambda_3 = 1551.32$ nm and $P_3 = 0.85$ µW. At the peak power of the pump, the gain is about 13 dB. The low value of the signal power results in negligible pump depletion.

Figure 7.6. shows the OPA output spectrum obtained by means of the SSFM. The computation parameters were: total wavelength range 1450–1650 nm; 1000 longitudinal steps; 2^{14}-point DFT.

The spectrum consists of five lobes, each about 10 nm wide. Within each lobe there are several frequency components spaced by the modulation frequency (about 0.8 nm).

The central lobe corresponds to the pump. At the input it consists of just two frequencies, but this number is considerably larger at the output. The broadening is due to pump SPM.

The first lobe on the left (right) of the central lobe corresponds to the signals (idlers). The number of components in these lobes is due to the intensity modulation of the pump, which modulates the signal (idler) intensity and phase.

The smaller lobes near 1540 nm and 1590 nm are due to higher-order four-wave mixing, respectively between the signals and the pump and the idlers and the pump. Their presence is generally undesirable, as they draw power away from the signals and idlers. They are not taken into account in the quasi-CW calculation of the gain and therefore are a source of error for that method when they carry significant power.

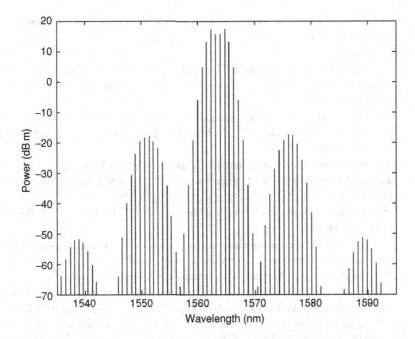

Fig. 7.6 Optical output spectrum of OPA with cosine-squared pump modulation.

The amplitudes of the various frequency components can be calculated alternatively by means of the FT of the amplified pulse shapes obtained using the quasi-CW approximation. The results are in excellent agreement with the results of the SSFM [4].

7.6 Conclusion

We have investigated aspects of pulsed fiber OPAs which are not apparent in the CW analysis presented in Chapters 4 and 5. In particular, we have shown that even if the quasi-CW analysis is applicable it reveals new effects such as pulse compression and a reduction in conversion efficiency. Together with Fourier analysis, it can also be used to provide a relatively simple picture of amplification and wavelength conversion by pumps with periodic modulation.

We have also shown how to use the SSFM in the case of short pulses, where the quasi-CW approach may not be applicable. This approach is generally accurate for pulses that are longer than one picosecond. For even shorter pulses [10], the NLSE itself may need to be modified to include Raman gain and other effects [11]

References

1. "Large-signal enhanced frequency conversion in birefringent optical fibers: theory and experiments," Seve, E., Millot, G., Trillo, S., Wabnitz, S. *J. Opt. Soc. Amer. B*; 1998; vol. 15, pp. 2537–51.

2. "Time-bandwidth product of chirped sech^2 pulses: application to phase–amplitude-coupling factor measurement," Lazaridis, P., Debarge, G., Gallion, P. *Opt. Lett.*; 1995; vol. 20, pp. 1160–2.
3. "Transform-limited, femtosecond WDM pulse generation by spectral filtering of gigahertz supercontinuum," Morioka, T., Kawanishi, S., Mori, K., Saruwatari, M. *Electron. Lett.*; 1994; vol. 30, pp. 1166–8.
4. "Multiple wavelength conversion with gain by a high-repetition-rate pulsed-pump fiber optical parametric amplifier," Kalogerakis, G., Marhic, M. E., Kazovsky, L. G., *J. Lightwave Technol.*; 2005; vol. 23, pp. 2954–60.
5. "Active optical pulse compression with a gain of 29.0 dB by using four-wave mixing in an optical fiber," Yamamoto, T., Nakazawa, M., *IEEE Photon. Technol. Lett.*; 1997; vol. 9, pp. 1595–7.
6. "Self-phase-modulation in silica optical fibers," Stolen, R. H., Lin, C. *Phys. Rev.*; 1978; vol. A 17, pp. 1448–53.
7. "1.53 kW from a single Yb-doped photonic crystal fiber laser," Bonati, G., Voelckel, H., Gabler, T., Krause, U., Tünnermann, A., Limpert, J., Liem, A., Schreiber, T., Nolte, S., Zellmer, H. In *Proc. Conf. Photonics West*, San Jose, January 2005; *Proc. SPIE*, vol. 5709–2a.
8. "Failure of phase-matching concept in large-signal parametric frequency conversion," Trillo, S., Millot, G., Seve, E., Wabnitz, S. *Appl. Phys. Lett.*; 1998; vol. 72, pp. 150–2.
9. "Strong four-photon conversion regime of cross-phase-modulation-induced modulational instability," Seve, E., Millot, G., Trillo, S., *Phys. Rev. E*; 2000; vol. 61, pp. 3139–50.
10. "Generation of femtosecond anti-Stokes pulses through phase-matched parametric four-wave mixing in a photonic crystal fiber," Konorov, S. O., Serebryannikov, E. E., Zheltikov, A. M., Zhou, P., Tarasevitch, A. P., von der Linde, D. *Opt. Lett.*; 2004; vol. 29, pp. 1545–7.
11. *Nonlinear Fiber Optics*, Agrawal, G. P. 1985; Academic, San Diego.

8 OPO theory

8.1 Introduction

If an optical gain medium is available, one can in principle convert it into an oscillator by providing optical feedback to form an optical resonator. If the gain medium is a fiber OPA, the resulting device is a fiber optical parametric oscillator (fiber OPO). There are many possibilities for designing the feedback system, and hence a variety of possible fiber OPOs. In the simplest versions the pump is supplied externally and is non-resonant, i.e. it passes through the OPO in one direction, interacting with the signal and idler within the nonlinear medium.

Let us assume that the optical cavity is resonant at the signal wavelength but not at the idler wavelength; we then have a singly resonant oscillator (SRO). If the cavity has a high quality factor Q, little or no signal escapes from it, and so the useful output of the SRO may have to be the idler. If the idler is not available, however, the output is a fraction of the signal, exiting through a partially transparent mirror or via a fiber coupler; in such a case the coupling introduces a signal loss, and so Q may not be very high. Most fiber OPOs that have been implemented to date are of the SRO type.

Figure 8.1 shows schematically various types of fiber OPO that have been implemented. Figure 8.1(a) corresponds to a Fabry–Pérot cavity, i.e. a linear cavity with mirrors at the ends; in this case the mirrors are implemented as fiber Bragg gratings (FBGs), and they reflect only the signal [1]. Figure 8.1(b) shows a unidirectional fiber ring, with couplers to inject the pump and to extract the output signal [2]. A unidirectional fiber ring is also used in Fig. 8.1(c), but this time a circulator is used to enforce unidirectional operation; a single FBG reflects the signal outside the ring, and a coupler is used for extracting part of the signal [3]. Figure 8.1(d) shows a rather different arrangement [4]. Here the HNLF and the 50/50 coupler form a loop mirror, within which light travels in both directions. An interesting feature is that the pump emerges from the same port by which it has entered, while the signal and idler use the other port; hence the pump is neatly separated from the signal and idler outside the loop. The signal and idler continue on to a diffraction grating, which is set to reflect only the signal; the idler is lost at the grating. Part of the signal is then extracted by the 10/90 coupler. Wavelength tuning is obtained by rotating the diffraction grating.

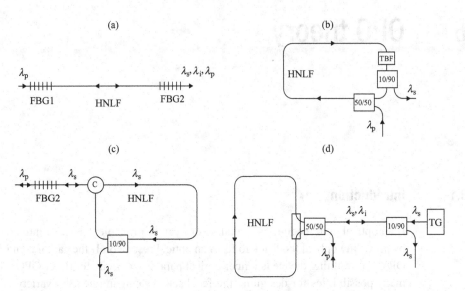

Fig. 8.1 Examples of fiber OPO implementations: (a) linear Fabry–Pérot using fiber Bragg gratings; (b) unidirectional ring using couplers and a tunable filter; (c) unidirectional ring using a circulator, a fiber Bragg grating and a tunable bypass filter; (d) bidirectional loop mirror using couplers and a tunable diffraction grating (TG).

8.2 Fiber OPO theory

Relatively little work has been done with fiber OPOs, as compared with fiber OPAs. The main reason is that $\chi^{(2)}$-based OPOs have been under development for many years, and are already at the commercial stages [1, 5, 6]. These devices have the following characteristics: (i) hundreds of milliwatts of average output power; (ii) high conversion efficiency; (iii) tunability over wide wavelength ranges, such as from 500 nm to 2 µm; (iv) single-frequency operation. At this time it is not clear that fiber OPOs can successfully compete in most of these areas or offer unique advantages.

Another reason for the relatively small amount of work on fiber OPOs is that, to a large extent, making an oscillator from a gain medium is a generic problem. If the physics of the particular medium is understood, calculating such things as the OPO threshold and output power can be done by using techniques that are largely independent of the gain medium. Hence these calculations for a fiber OPO are essentially the same as for a $\chi^{(2)}$-based OPO or a doped-fiber OPO. Only if a more detailed analysis is required, incorporating some fiber-specific features such as self-phase modulation (SPM), cross-phase modulation (XPM), etc., is the generic approach no longer sufficient. However, since as explained above there is little practical incentive for undertaking such detailed studies, little work has been done in this area to date.

Since the theory of fiber OPOs is not very advanced, we will discuss only the simplest types, bearing in mind that more complex versions may come to be of interest in the future. For example, we will consider only a single pump, but of course similar systems using two are feasible.

As seen in Fig. 8.1(a), one approach to making a fiber OPO is to use a straight fiber terminated by two FBGs, forming a Fabry–Pérot cavity. The signal is reflected upon itself by the FBGs and travels in both directions within the nonlinear medium. We will denote this SRO with bidirectional signal as SRO-B. However, since the pump travels in only one direction it generates OPA gain only in that direction. Hence the signal traveling backwards experiences no gain. If the signal power builds up to a high level in the cavity, as it should for an efficient OPO, the backward-traveling signal will affect the forward-traveling pump and idler by means of XPM and the forward-traveling signal by means of SPM. These phase terms are absent in the basic OPA theory of the preceding chapters, and their influence on OPA gain and OPO performance will have to be carefully taken into account.

Other arrangements avoid this problem by forcing the pump, signal, and idler only to co-propagate through the nonlinear medium; there is no counter-propagating signal. Two examples are given in Figs. 8.1(b) and 8.1(c). Then the externally incident pump travels in one direction through the loop, generating gain only in that direction. Spontaneous emission due to the gain is amplified only if it travels in the same direction, and this leads to a build-up of the signal in the resonant cavity formed by the loop. The idler is also generated in the same direction. We will denote this SRO with unidirectional signal as SRO-U.

8.2.1 Threshold

We will begin the analysis by studying the threshold condition for SROs. At threshold the signal power is negligible, and so there are no extra XPM and SPM terms due to the backward signal in SRO-B. Therefore the signal gains for SRO-U and SRO-B are calculated from the same expressions, and the threshold conditions will have similar forms. Also at threshold there is no pump depletion, and we can use the OPA gain derived in subsection 3.6.2.

In the absence of fiber loss, the threshold condition for either type of SRO is $G_3 T = 1$, where G_3 is the signal power gain in the fiber and T is the total transmittance for light making a round trip in the cavity with the pump off. Threshold is reached most easily if the signal is at the maximum gain wavelength, for which $G_3 = \cosh^2(\gamma P_0 L)$ (assuming a lossless fiber). The required pump power is then

$$P_0 = \frac{1}{\gamma L} \cosh^{-1}\left(\frac{1}{\sqrt{T}}\right). \tag{8.1}$$

As an example let us consider a 100 m HNLF with $\gamma = 20$ W^{-1} km^{-1} and $T = 0.98$ (0.1 dB round-trip loss). Then we find that $P_0 = 71$ mW. If we now consider a longer fiber with $L = 1$ km, $\gamma = 20$ W^{-1} km^{-1}, and $T = 0.05$ (13 dB round-trip loss) we find that $P_0 = 109$ mW, which is of the same order of magnitude. Therefore for both examples we have $P_0 \approx 100$ mW, a relatively low threshold power compared with other types of OPO. This indicates that it should be fairly easy to operate CW fiber OPOs.

In the presence of fiber loss, the threshold condition for SRO-U remains $G_3 T = 1$ while for SRO-B the threshold condition becomes $G_3 T e^{-\alpha L} = 1$. In neither case does

T include the fiber loss, but G_3 is now the gain in the lossy fiber; G_3 is expressed in terms of Bessel functions, and it is not possible to calculate the threshold power in closed form. However, if $\lambda_3 \approx \lambda_1$ then $\Delta\beta \approx 0$ and we have the simple form $G_3 \approx e^{-\alpha L}[1 + (\gamma P_0 L_{\text{eff}})^2]$.

For SRO-B we then have a threshold pump power

$$P_{0,\text{th}} = \frac{\alpha}{\gamma} \frac{(1/T - v^2)^{1/2}}{v(1-v)}, \qquad (8.2)$$

where $v = e^{-\alpha L}$. For SRO-U, we obtain

$$P_{0,\text{th}} = \frac{\alpha}{\gamma} \frac{(1/T - v)^{1/2}}{v^2(1-v)}. \qquad (8.3)$$

8.2.2 Conversion efficiency

It is generally desirable to operate OPOs with a high pump-to-signal conversion efficiency, which means that the signal power in the cavity is of the same order as the pump power and therefore pump depletion must take place. In that case, we must use the results of subsection 3.6.1, which do not take fiber loss into account.

(a) Unidirectional SRO

For SRO-U there is no need to consider XPM and SPM from the counter-propagating signal; hence the previous results are directly applicable.

Consider an OPO operating in the steady state. The saturated signal gain must be such that it satisfies the threshold condition. For a given P_{10}, there is a P_{30} such that the saturated gain has the correct value. Because of the complexity of the Jacobian-elliptic-function solution, it is not possible to calculate P_{30} in closed form.

Hence it is necessary to solve for the unknown power values by a numerical technique such as iteration or by a shooting method. Here we describe an iterative approach, as follows.

1. Choose a seed value for P_{30}; call it $P_{30}^{(1)}$.
2. Launch P_{10} and $P_{30}^{(1)}$ into the fiber, and calculate the signal power at the fiber output, $P_3^{(1)}(L)$, by using the solution from subsection 3.6.1.
3. The signal power reaching the fiber input after one round trip is then $P_{30}^{(2)} = T P_3^{(1)}(L)$.
4. If $|P_{30}^{(2)} - P_{30}^{(1)}| > \varepsilon$, where ε is a small number, the value is insufficiently accurate, and a new iteration is needed, so go back to step 2 and repeat steps 2–4, increasing all superscripts along the way.

With a suitable choice for $P_{30}^{(1)}$ the procedure will converge after a certain number of iterations. This will provide an accurate approximation for the steady-state signal power in the fiber. With this value, the idler output power and conversion efficiency can be calculated.

8.2 Fiber OPO theory

Maximum conversion efficiency
In subsection 3.6.1 it is shown that it is possible to obtain complete pump depletion in a one-pump OPA, with suitable values of $\Delta\beta$ and other parameters. The same is true in an OPO. If this happens, at the fiber end $P_{10}/2$ has been transferred from the pump to both signal and idler. The idler output power is then $P_{10}/2$, and the pump conversion efficiency (idler output power over pump input power) is 50%. This is the maximum conversion efficiency that can be achieved with this type of SRO. In real devices, the conversion efficiency will be lowered by losses in the components.

To find the $\Delta\beta$ value for which maximum conversion can be obtained (for a given pump power), we can use the condition $\Delta\beta = \gamma(P_{30} - P_{10}/2)$ (see subsection 3.6.1), which leads to complete pump depletion in a long fiber. Then, for long L, the conversion efficiency from pump to idler will approach 50%.

As an example, consider a Fabry–Pérot SRO-U with $\gamma = 18$ W^{-1} km^{-1} and $L = 200$ m. The pump, with power $P_{10} = 0.5$ W, first passes through an FBG with 100% reflectivity at the signal wavelength and then exits through an FBG with 70% reflectivity. A shooting method was used to solve for P_{30}: P_{30} was adjusted until $TP_3^{(1)}(L) = P_{30}$ was obtained after one round trip. The expression $\Delta\beta = \gamma(P_{30} - P_{10}/2)$ was used to calculate $\Delta\beta$, in order to ensure that near-complete pump depletion could be achieved. With $P_{30} = 0.58$ W the pump power was almost exhausted, and a conversion efficiency greater than 49% was obtained. Figure 8.2 shows the evolution of the three waves in the cavity. Outside the cavity, the signal and idler powers are both nearly 0.25 W, confirming that virtually 100% of the initial power of 0.5 W has been converted into signal and idler, in equal amounts. (The calculations were performed with the equation after Eq. (3.64)).

Note that the signal power is greater than the pump power everywhere, which indicates that stimulated Brillouin scattering (SBS) will be a greater concern for the signal than for the pump.

(b) Bidirectional SRO
With SRO-B, the effect of XPM and SPM due to the backward signal must now be considered. Since we are assuming no fiber loss the backward signal power is constant. The propagation equations for the forward waves in the preceding subsection must now be modified. The pump and idler equations will include an additional XPM term, while the signal equation will include an additional SPM term. The result is simply that $\Delta\beta/\gamma$ in Eq. (3.139) must be replaced by $\Delta\beta/\gamma - P_3^-$, where P_3^- is the backward signal power. In the iterative algorithm given above, for the kth iteration we let $P_3^- = R_2 P_3^{(k)}$ where R_2 is the reflectance of the output FBG. Otherwise, the algorithm remains unchanged. To ensure that near-complete pump depletion is obtained, we now set $\Delta\beta/\gamma = P_{30} - P_{10}/2 + R_2 P_3^{(k)}$; alternatively, for a shooting algorithm one uses $P_3^- = P_{30}/R_1$ and $\Delta\beta/\gamma = P_{30} - P_{10}/2 + P_{30}/R_1$. Under these circumstances, the results are the same as those obtained in the preceding subsection. The only difference is that the $\Delta\beta$s are different for SRO-B and SRO-U.

If the pump frequency is modulated to reduce SBS, while the signal frequency is not, then the idler frequency is also modulated. As a result, the backward and forward

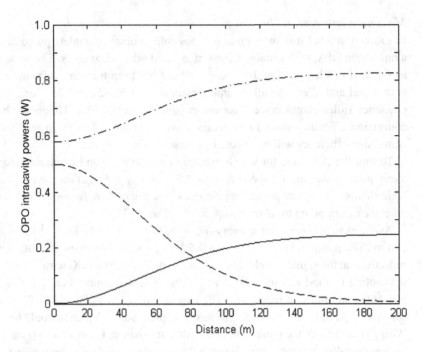

Fig. 8.2 Evolution of the intracavity powers of the various waves along a unidirectional SRO: solid line, idler; broken line, pump; broken-and-dotted line, signal.

signals have different frequencies at most points along the fiber. Thus the phase modulation induced by the backward wave is to be calculated as XPM, because two significantly different frequencies are involved. It can then be shown that this XPM, now common to all three forward waves, has no effect on $\Delta\beta/\gamma$ and so G_3 is completely unaffected by it. Therefore in this case the analysis for SRO-B is identical to that for SRO-U.

8.3 Conclusion

Little work has been done on the theory of fiber OPOs, in part because $\chi^{(2)}$-based OPOs already provide excellent devices for tunable wavelength generation. In addition, the basic fiber OPO theory (other than the gain theory) does not differ much from that of OPOs based on other media.

Some aspects of fiber OPOs may prove to be advantageous in the long run. In particular, fibers have the potential for withstanding high average powers and this could be exploited for making OPOs with high average power values for situations where high-power sources currently are not available. If activity in this area increases then it will become necessary to develop a more accurate theory of fiber OPOs, including the detailed impact of fiber-specific effects such as birefringence, ZDW fluctuations, SPM, and XPM as well as the small spacing between longitudinal modes.

References

1. "Continuous-wave fiber optical parametric oscillator," Marhic, M. E., Wong, K. K. Y., Kazovsky, L. G., Tsai, T. E. *Opt. Lett.*; 2002; vol. 27, pp. 1439–41.
2. "Broadband single-pumped fiber-optic parametric amplifiers," Torounidis, T., Andrekson, P. *IEEE Photon. Technol. Lett.*; 2007; vol. 19, in press.
3. "Continuous-wave, totally fiber integrated optical parametric oscillator using holey fiber," de Matos, C. J. S., Taylor, J. R., Hansen, K. P. *Opt. Lett.*; 2004; vol. 29, pp. 983–5.
4. "Tunable fiber-optic parametric oscillator," Serkland, D. K., Kumar, P. *Opt. Lett.*; 1999; vol. 24, pp. 92–4.
5. "Prospects for continuous-wave fiber optical parametric amplifiers and oscillators," Marhic, M. E., Wong, K. K. Y., Kazovsky, L. G. In *Proc. Conf. on Lasers and Electro-Optics*, Baltimore MD, May 2003; paper CTuA1, pp. 417–8.
6. "Optical parametric oscillators," Ebrahimzadeh, M.; Dunn, M. H. In *Optics IV*, Bass, M., Enoch, J. M., Van Stryland, E. W., Wolfe, W. L., eds. McGraw-Hill, New York; 2000; pp. 2201–72.

9 Quantum noise figure of fiber OPAs

9.1 Introduction

In previous chapters we studied wave propagation in fiber OPAs from a classical standpoint, i.e. starting from Maxwell's equations. That approach provides only a partial view of the rich nature of the electromagnetic field, namely its wave nature. However, a complete description of the electromagnetic field requires quantum mechanics. This quantum description of the electromagnetic field (called quantum electrodynamics, or QED) predicts fluctuations in measurable quantities due to the Heisenberg uncertainty principle and sets ultimate limits on the performance of devices such as optical amplifiers. A remarkable result of this theory is that the statistics and thus the noise performance of a system are affected by the quantum statistics of the source *and* by the physical measurement that is made. Thus a different analysis is needed depending on whether homodyne, heterodyne, or direct detection (where the field simply impinges on a photodiode) is used. Direct detection reveals the particle nature of the electromagnetic field. More specifically, energy arrives in discrete units, called photons, each of which may create an electron–hole pair in a semiconductor photodetector. In particular, the number of photons in an optical pulse can be thought of as a random variable having a mean and a variance. By deriving the mean and variance of the photocurrent that would be present if one performed direct detection of ideal laser light at the input of an optical amplifier and then a similar measurement at its exit, one can quantify the degradation of the signal-to-noise ratio (SNR) caused by the amplifier. This is defined as the noise figure (NF), a very important figure of merit in optical communication. Because direct detection is overwhelmingly preferred to homodyne or heterodyne detection in currently installed optical communication systems, we will consider only the NF for direct detection.

In this chapter we will start from the quantum-mechanical propagation equations in a nonlinear medium and calculate the NF for a non-degenerate phase-insensitive fiber OPA, neglecting Raman gain. Two expressions will hold, depending on whether the OPA is used as an amplifier or as a wavelength converter. We will then consider the NF for a frequency-degenerate phase-sensitive OPA. Finally we will include the effect of Raman gain and show that it leads to a deterioration of the NF.

9.2 Quantum-mechanical derivation of the OPA equations

In Chapter 3 we derived in a purely classical way the OPA propagation equations, in the general case where there is pump depletion. We found that the differential equations satisfy basic photon-number conservation and total-power conservation.

In subsection 3.5.3 we found that in the undepleted-pumps approximation the OPA propagation equations become linear and the solution is obtained in terms of a transfer matrix, which is quasi-unitary (Appendix 1). In the case of wavelength exchange the matrix is unitary. The forms of these classical matrices can be interpreted in quantum mechanics as expressing the fact that the field operators are Bose operators.

Thus we find that although these derivations are purely classical (except for the assumption that the photon energy is $\hbar\omega$), they are compatible with the requirements of quantum optics. This is simply the result of the fact that photon-number conservation is already built into the basic differential equations. To see this clearly, we now present the derivation of the propagation equations using the quantum-mechanical field operators and the Heisenberg description of the system's evolution. We confine ourselves to the scalar case.

9.2.1 The Hamiltonian for the nonlinear interactions

The starting point is the expression for the polarization density of the medium:

$$P = \varepsilon E = \varepsilon_0 \left(\chi^{(1)} E + \chi^{(3)} E^3 \right). \tag{9.1}$$

Where $\chi^{(1)}$ and $\chi^{(3)}$ are respectively the first- and third-order susceptibilities (we assume that the fiber has a negligible second-order susceptibility, which is well justified unless the fiber has been specially prepared); ε_0 is the permittivity of the vacuum; E is the electric field, which we assume to be linearly polarized parallel to the x-axis.

We also assume that $\chi^{(3)}$ is due only to non-resonant electrons and is therefore a constant (i.e. independent of frequency), which we can take to be equal to the *xxxx* component of the corresponding tensor). Light guiding in optical fibers can be accomplished by a variety of means, but all have in common the fact that the refractive index must vary in the direction transverse to the z-axis. In turn this refractive index is generally accompanied by a transverse variation in $\chi^{(3)}(x, y)$. As an example, some highly nonlinear fibers (HNLFs) have a small core heavily doped with germania, surrounded by a cladding of nearly pure silica; this results in a core with a nonlinear index about twice as large as the cladding. Other examples are hollow-core and holey fibers, which are made from a single material with hollow cylindrical regions parallel to the axis; in this case the nonlinear index of the holes is zero, while that of the surrounding material is finite.

The energy density associated with the field is then $U = U_L + U_{NL}$ where U_L is the quadratic part of the energy and is present even in the absence of the nonlinear term in $\chi^{(3)}$ and U_{NL} is due to the nonlinear interaction. We have [1]

$$U_L = -\varepsilon_0 \frac{1 + \chi^{(1)}}{2} E^2 \qquad U_{NL} = -\varepsilon_0 \frac{\chi^{(3)}}{4} E^4; \tag{9.2}$$

U is such that $P = -dU/dE$. (The negative signs in U_L and U_{NL} will now be dropped for simplicity and without significant consequences for the derivation and the results.)

The Hamiltonian is then obtained by integrating U over a suitable volume. When studying the resonance modes of a cavity, the volume is that of the cavity. Here we study traveling guided waves and perform the integration over a volume V bounded by two planes perpendicular to the z-axis and separated by a distance L_V that is small enough for the field envelopes to be considered constant over that distance.

We consider the interaction between four monochromatic waves with radian frequencies ω_k, $k = 1$–4. We first consider frequency assignments suitable for an OPA, i.e. satisfying $\omega_1 + \omega_2 = \omega_3 + \omega_4 = 2\omega_c$, where ω_c is the center frequency; later on we will see how the procedure needs to be modified for wavelength exchange. We assume for simplicity that all the waves are in the fundamental mode and with the same linear polarization. (Cases involving different modes and/or states of polarization can be studied by making simple modifications of the theory presented here.)

We make the transition to quantum mechanics by introducing annihilation and creation operators a_k and a_k' for the guided mode at ω_k; these operators satisfy the boson commutation relation $[a_k, a_k'] = a_k a_k' - a_k' a_k = 1$. We write the electric field for that mode as follows:

$$E_k(x, y, z, t) = \sqrt{\frac{\hbar \omega_k}{2\varepsilon_0 L_V R_k}} \left[a_k(z) \exp(i\beta_k z - i\omega_k t) \right.$$
$$\left. + a_k'(z) \exp(-i\beta_k z + i\omega_k t) \right] \psi_k(x, y), \quad (9.3)$$

where $\psi_k(x, y)$ is the transverse profile of the main electric field component of the mode, which we assume for simplicity to be real. Also,

$$R_k = \int_{x=-\infty}^{\infty} \int_{y=-\infty}^{\infty} [n_k(x, y) \psi_k(x, y)]^2 \, dx \, dy, \quad (9.4)$$

where $n_k = \sqrt{1 + \chi^{(1)}}$ is the refractive index at ω_k.

The constants in the square root in Eq. (9.3) have been chosen so that the linear Hamiltonian $H_k^{(1)}$ of the mode at ω_k takes on the canonical form

$$H_k^{(1)} = L_V \int \int \varepsilon_0 [n_k(x, y) E_k(x, y, z, t)]^2 \, dx \, dy = \hbar \omega_k a_k' a_k + \frac{1}{2}, \quad (9.5)$$

where only the low-frequency terms in $(E_k)^2$ have been kept.

The nonlinear energy density term due to the interaction of the four waves is then of the form

$$U_{NL}(x, y, z, t) = \frac{\varepsilon_0 \chi^{(3)}(x, y)}{4} E^4(x, y, z, t)$$
$$= \frac{\chi^{(3)}(x, y) \hbar^2}{64 \varepsilon_0 L_V^2} \left\{ \sum_{k=1}^{4} \sqrt{\frac{\omega_k}{R_k}} [a_k(z) \exp(i\beta_k z - i\omega_k t) \right.$$
$$\left. + a_k'(z) \exp(-i\beta_k z + i\omega_k t)] \psi_k(x, y) \right\}^4. \quad (9.6)$$

The nonlinear part of the Hamiltonian is then obtained by integrating U_{NL} over the volume V:

$$H^{(3)} = \int_V U_{NL} \, dV. \tag{9.7}$$

We look only for the terms with zero frequency after expansion of the product (which means that we will be looking only at the evolution of the slowly varying envelopes of the operators). Among them is a term that corresponds to non-degenerate four-wave mixing and is of the form

$$H_{FWM} = H_{1234} + H'_{1234}, \tag{9.8}$$

where

$$H_{1234} = 24 M_{1234} a_1 a_2 a'_3 a'_4 \exp[i(\omega_3 + \omega_4 - \omega_1 - \omega_2)t - i(\beta_3 + \beta_4 - \beta_1 - \beta_2)z]$$
$$= 24 M_{1234} a_1 a_2 a'_3 a'_4 \exp(-i \Delta \beta z) \tag{9.9}$$

and

$$M_{klmn} = \frac{\hbar^2}{16 L_V \varepsilon_0} \sqrt{\frac{\omega_k \omega_l \omega_m \omega_n}{R_k R_l R_m R_n}} \iint \chi^{(3)} \psi_k \psi_l \psi_m \psi_n \, dx \, dy. \tag{9.10}$$

The factor 24 in Eq. (9.9) is due to the fact that there are 4! ways of obtaining the correct frequency combination, by selecting one frequency in each factor within Eq. (9.7).

The point of making this detailed quantum mechanical derivation can now be seen clearly. The form of the Hamiltonian that corresponds to non-degenerate four-wave mixing (FWM) contains combinations of creation and annihilation operators that reveal at the most fundamental level how photons are exchanged between the various waves. In particular, the presence of $a_1 a_2 a'_3 a'_4$ in H_{1234} indicates that it corresponds to the simultaneous annihilation of one photon at ω_1 and one at ω_2, and the creation of one photon at ω_3 and one at ω_4. In H'_{1234} the roles of creation and annihilation are exchanged. Thus these two processes operate simultaneously in the non-degenerate FWM term H_{FWM}.

In either case two photons are created and two are annihilated, and so there is no net change in the number of photons. Thus for this type of FWM the total photon number is conserved. (Note that this is not true of all types of FWM; for example, if $\omega_1 + \omega_2 + \omega_3 = \omega_4$ then three photons are annihilated when one is created, or vice versa, which clearly does not conserve total photon number.)

The way in which the various photons are created and annihilated also indicates that the following relations must hold:

$$\frac{dN_1}{dz} = \frac{dN_2}{dz} = -\frac{dN_3}{dz} = -\frac{dN_4}{dz}, \tag{9.11}$$

where N_k is the photon number density at ω_k. This type of equation, expressing a relation between photon-number variations, is known as a Manley–Rowe relation. We came across Eqs. (9.11) in our classical approach in Chapter 3. At the time, we had no fundamental classical explanation for why these relations should hold; we now see that their origin is actually quantum mechanical.

Finally, since the energy of one photon at ω_k is $\hbar\omega_k$, with H_{1234} an amount of energy $\hbar(\omega_1 + \omega_2) = 2\hbar\omega_c$ is annihilated while an energy $\hbar(\omega_3 + \omega_4) = 2\hbar\omega_c$ is created. Since these energies are the same, there is conservation of total energy, as expected for this closed system of four frequencies.

The preceding considerations show that the basic properties concerning photon-energy conservation and photon-number relations hold at the most fundamental level of the formulation. They must a fortiori also be present in any global result, such as in a transfer matrix relating input and output quantities. The main properties of such matrices, such as unitarity or quasi-unitarity, are simply the result of these fundamental local conservation laws.

In parametric devices it can often be assumed that two of the waves are much larger than the other two and are not appreciably depleted by the interaction. Then their operators can be replaced by constants (c-numbers) for most of the calculations. It is clear from the preceding discussion, however, that no operators can be considered to be constant when considering photon or energy conservation, because doing so would lead to erroneous conclusions. For example, in an OPA with gain the energy received by the growing signal and idler must come from the pumps, even though in relative terms the pump amplitudes might experience little depletion.

9.2.2 The Heisenberg evolution equations

Using the Hamiltonian $H^{(3)}$ of Eq. (9.7) we can now set up equations for the evolution of the operators of interest, namely the various creation and annihilation operators. The basis for this is Heisenberg's evolution equation for any operator O, namely

$$i\hbar \frac{dO}{dt} = [O, H] \equiv OH - HO. \quad (9.12)$$

In our context, Eq. (9.12) needs to be slightly modified. Since we are looking at the evolution of fields as a function of z, dt needs to be replaced by

$$dt = \frac{dz}{v} = \frac{\bar{n}dz}{c},$$

where $\bar{n} = c\beta/\omega$ is the effective refractive index.

Let us now write down this equation for a_1. To do this we select the part $H_1^{(3)}$ of $H^{(3)}$ whose terms have zero frequency and contain a_1 or a_1' (the terms that contain neither a_1 nor a_1' will commute with a_1 and will therefore disappear from $[a_1, H_1^{(3)}]$). The terms corresponding to non-degenerate FWM are H_{1234} and H_{1234}'. There are additional terms corresponding to self-phase modulation (SPM) and cross-phase modulation (XPM), for which zero frequency is obtained from the frequency combinations $\omega_1 + \omega_1 - \omega_1 - \omega_1$ and $\omega_1 - \omega_1 + \omega_k - \omega_k$, $k > 1$. Altogether, we find that

$$H_1^{(3)} = H_{1,\text{SPM}}^{(3)} + H_{1,\text{XPM}}^{(3)} + H_{\text{FWM}}^{(3)}, \quad (9.13)$$

where

$$H^{(3)}_{1,\text{SPM}} = M_{1111}(a_1 a'_1 a_1 a'_1 + a_1 a'_1 a'_1 a_1 + a'_1 a_1 a'_1 a_1 \\ + a'_1 a_1 a_1 a'_1 + a_1 a_1 a'_1 a'_1 + a'_1 a'_1 a_1 a_1), \quad (9.14)$$

$$H^{(3)}_{1,\text{XPM}} = 6(a_1 a'_1 + a'_1 a_1)[M_{1122}(a_2 a'_2 + a'_2 a_2) + M_{1133}(a_3 a'_3 + a'_3 a_3) \\ + M_{1144}(a_4 a'_4 + a'_4 a_4)], \quad (9.15)$$

and $H^{(3)}_{\text{FWM}}$ is given by Eq. (9.8). After calculating the various Poisson brackets, we obtain

$$i\frac{\hbar c}{12 n_1}\frac{da_1}{dz} = \frac{1}{12}[a_1, H^{(3)}_1]$$
$$= M_{1111}(N_1+1)a_1 + [M_{1122}(2N_2+1) + M_{1133}(2N_3+1) \\ + M_{1144}(2N_4+1)]a_1 + 2M_{1234} a'_2 a_3 a_4 e^{i\Delta\beta z}, \quad (9.16)$$

where $N_k = a'_k a_k$ is the photon-number operator of the mode at ω_k.

For a_3 we obtain

$$i\frac{\hbar c}{12 n_3}\frac{da_3}{dz} = \frac{1}{12}[a_3, H^{(3)}_3]$$
$$= M_{3333}(N_3+1)a_3 + [M_{3311}(2N_1+1) + M_{3322}(2N_2+1) \\ + M_{3344}(2N_4+1)]a_3 + 2M_{1234} a_1 a_2 a'_4 e^{-i\Delta\beta z}. \quad (9.17)$$

We see that the form of these equations is similar to that of Eq. (3.34), which we obtained from Maxwell's equations by a classical argument. In particular, we recognize the SPM, XPM, and FWM terms, which appear in that order on the right-hand side. The SPM and XPM terms contain some extra terms, namely the 1's appearing within the parentheses, but these correspond to very small photon numbers and can be safely neglected when strong (classical) pumps and signals are present.

As a check, we can make the transition to the classical equations by assuming that all four waves are strong and dropping the "1" terms in Eq. (9.17). To obtain a more familiar form, we replace a_k by a classical complex amplitude A_k such that $A_k A^*_k = P_k$, the total power in the mode at z. This requires that

$$a_k = A_k \sqrt{\frac{\bar{n}_k L_V}{c\hbar\omega_k}}, \quad (9.18)$$

where $\bar{n}_k = \beta_k c/\omega_k$. Making this substitution in Eq. (9.16) we obtain

$$-i\frac{dA_1}{dz} = \gamma_{1111} P_1 A_1 + 2(\gamma_{1212} P_2 + \gamma_{1313} P_3 + \gamma_{1414} P_4) A_1 \\ + 2\gamma_{1234} A_3 A_4 A^*_2 e^{i\Delta\beta z}, \quad (9.19)$$

where

$$\gamma_{klmn} = \frac{3\mu_0 \omega_k}{4}\sqrt{\frac{\bar{n}_k \bar{n}_l \bar{n}_m \bar{n}_n}{R_k R_l R_m R_n}} \iint \chi^{(3)} \psi_k \psi_l \psi_m \psi_n\, dx\, dy. \quad (9.20)$$

The convention used for the subscripts in Eq. (9.19) is the same as in Eq. (3.31). We see that Eq. (9.19) has the same form as Eq. (3.31), which was obtained by purely classical means, starting from Maxwell's equations and the expression for the nonlinear polarization (and a particular choice of multiplying function for eliminating the x and y variations). The agreement between the two approaches shows that the classical method is essentially equivalent to our method based on a quantum-mechanical Hamiltonian and Heisenberg's evolution equation. In both cases the quantities of interest are the slowly varying amplitudes of the fields. The only significant difference between the resulting equations is the "1" terms in Eq. (9.16), which are purely quantum mechanical in origin and can be safely neglected for classical fields. At this level of approximation, the two approaches provide identical equations, but the interest in giving a quantum-mechanical derivation of the classical equations is that now we can explain why they satisfy the Manley–Rowe relations; no obvious explanation of this appears when they are derived from Maxwell's equations.

It is also interesting to note that the classical form of the Hamiltonian $H_1^{(3)}$ (in terms of the A_k and A_k^*) is such that if we use Hamilton's equation

$$\frac{dA_1}{dz} = -\frac{\partial H_1^{(3)}}{\partial A_1^*},$$

we obtain the correct classical equation for the evolution of A_1. This confirms that $H_1^{(3)}$ has the correct form, and it also shows that A_1 and A_1^* are conjugate variables with respect to this Hamiltonian.

9.3 Noise figure of non-degenerate fiber OPAs

Here we calculate at the quantum-mechanical level the NF of a fiber OPA. We will assume no pump depletion, and we will neglect the Raman gain. Since the form of the propagation equations for the a_k is the same as the equations for the classical A_k studied in Chapter 3, the solution has the same form, and therefore we can again make use of the OPA transfer matrix derived in Appendix 1. We leave out the exponential phase terms relating A_k and C_k as they have no impact on the rest of the discussion and can be restored if needed. With these assumptions, the operators $a_3(z)$ and $a_4'(z)$ at a distance z from the input are obtained form

$$\begin{aligned} a_3(z) &= m_{33}a_3 + m_{34}a_4', \\ a_4'(z) &= m_{34}^*a_3 + m_{33}^*a_4', \end{aligned} \tag{9.21}$$

where the subscripts 3 and 4 respectively refer to the signal and idler; a_3 and a_4 denote the annihilation operators at the fiber input.

This approach indicates that during the interaction the field operators are evolving while the system wavefunction remains fixed, i.e. we are using the Heisenberg representation for the interaction. Since at the input the states of the signal and the idler are independent, we assume that the total wavefunction of the system is of the form $\Psi_k \Psi_l$.

The wavefunction for the input signal is Ψ_k; we assume it to be a coherent state (CS). The wavefunction for the input idler is Ψ_l. We assume that there is no finite idler at the input and therefore that it is in a vacuum state (VS), which is the same as a CS with zero mean field. Thus we let $\Psi_l = u_0$, where u_0 is the wavefunction for the harmonic oscillator (HO) ground state with zero quanta in it. The wavefunction u_0 satisfies the relations $a_k u_0 = 0$ and $a'_k u_0 = u_1$, where u_1 is the one-photon HO wavefunction; u_1 satisfies the relations $a_k u_1 = u_0$ and $a'_k u_1 = \sqrt{2} u_2$. Since the HO wavefunctions form an orthonormal set we have $\langle u_0 | u_0 \rangle = \langle u_1 | u_1 \rangle = 1$ and $\langle u_0 | u_1 \rangle = 0$, where $\langle \Phi | \Psi \rangle$ denotes the Hermitian product of some wavefunctions Φ and Ψ.

We can now make use of the expressions for the output operators to calculate the mean and the variance of the signal photon numbers at the output, and hence the NF for signal amplification, by means of the signal photon number operator at z, $N_3(z) = a'_3(z) a_3(z)$. The mean signal photon number at z is

$$\begin{aligned}
\langle N_3(z) \rangle &= \langle a'_3(z) a_3(z) \rangle = \langle \Psi_3 \Psi_4 | a'_3(z) a_3(z) | \Psi_3 \Psi_4 \rangle \\
&= \langle \Psi_3 \Psi_4 | (m^*_{33} a'_3 + m^*_{34} a_4)(m_{33} a_3 + m_{34} a'_4) | \Psi_3 \Psi_4 \rangle \\
&= \langle \Psi_3 \Psi_4 | |m_{33}|^2 a'_3 a_3 + |m_{34}|^2 a_4 a'_4 + m^*_{33} m_{34} a'_3 a'_4 + m^*_{34} m_{33} a_4 a_3 | \Psi_3 \Psi_4 \rangle \\
&= G_s N_{30} + G_i (N_{40} + 1) + m^*_{33} m_{34} \langle \Psi_3 | a'_3 | \Psi_3 \rangle \langle \Psi_4 | a'_4 | \Psi_4 \rangle \\
&\quad + m^*_{34} m_{33} \langle \Psi_3 | a_3 | \Psi_3 \rangle \langle \Psi_4 | a_4 | \Psi_4 \rangle,
\end{aligned} \qquad (9.22)$$

where N_{30} (N_{40}) is the mean number of input signal (idler) photons. Since we assume that there is no idler power at the input, $N_{40} = 0$. Also

$$\langle \Psi_4 | a'_4 | \Psi_4 \rangle = \langle u_0 | a'_4 | u_0 \rangle = \langle u_0 | u_1 \rangle = 0$$

and

$$\langle \Psi_4 | a_4 | \Psi_4 \rangle = \langle u_0 | a_4 | u_0 \rangle = \langle u_0 | 0 \rangle = 0.$$

Hence, we simply have

$$\langle N_3(z) \rangle = G_s N_{30} + G_i. \qquad (9.23)$$

It is interesting to note that, even in the absence of signal power at the input, there is finite average power at the output ($\langle N_3(z) \rangle = G_s - 1$), which corresponds to the amplified spontaneous emission (ASE) of the OPA.

We note that when G_s is large the ASE power is essentially proportional to G_s. Hence, one can obtain information about the gain of a fiber OPA without injecting a signal into it, simply by observing the output ASE. One can then infer the magnitude of the gain at some wavelength by properly calibrating the detection apparatus; one can also easily obtain the relative shape of the gain spectrum, which does not require any calibration. This technique is commonly used in high-gain fiber OPAs.

The variance of $N_3(z)$ is $[\delta N_3(z)]^2 = \langle [N_3(z)]^2 \rangle - \langle N_3(z) \rangle^2$. We have

$\langle [N_3(z)]^2 \rangle = \langle \Psi_3 \Psi_4 | (|m_{33}|^2 a_3' a_3 + |m_{34}|^2 a_4 a_4' + m_{33}^* m_{34} a_3' a_4'$
$\qquad + m_{34}^* m_{33} a_4 a_3)^2 | \Psi_3 \Psi_4 \rangle$
$= \langle \Psi_3 \Psi_4 | \{ |m_{33}|^4 a_3' a_3 a_3' a_3 + |m_{34}|^4 a_4 a_4' a_4 a_4'$
$\qquad + |m_{33} m_{34}|^2 \times (a_3' a_3 a_4 a_4' + a_4 a_4' a_3' a_3)$
$\qquad + (|m_{33}|^2 a_3' a_3 + |m_{34}|^2 a_4 a_4')(m_{33}^* m_{34} a_3' a_4' + m_{34}^* m_{33} a_4 a_3)$
$\qquad + (m_{33}^* m_{34} a_3' a_4' + m_{34}^* m_{33} a_4 a_3)(|m_{33}|^2 a_3' a_3 + |m_{34}|^2 a_4 a_4')$
$\qquad + (m_{33}^* m_{34})^2 a_3' a_4' a_3' a_4' + (m_{34}^* m_{33})^2 a_4 a_3 a_4 a_3$
$\qquad + |m_{33} m_{34}|^2 (a_3' a_4' a_4 a_3 + a_4 a_3 a_3' a_4') \}^2 | \Psi_3 \Psi_4 \rangle$
$= G_s^2 \langle N_3(0)^2 \rangle + G_i + 2 G_s G_i N_{30} + G_s G_i (N_{30} + 1).$ (9.24)

Since we assumed that the input signal is in a CS, its photon-number density has a Poisson density function, for which is it known that the variance is equal to the mean; thus we have $[\delta N_3(0)]^2 = \langle N_3(0)^2 \rangle - N_{30}^2 = N_{30}$, which implies $\langle N_3(0)^2 \rangle = N_{30}^2 + N_{30}$. Hence

$[\delta N_3(z)]^2 = \langle [N_3(z)]^2 \rangle - \langle N_3(z) \rangle^2$
$\qquad = G_s^2 (N_{30}^2 + N_{30}) + G_i^2 + 2 G_s G_i N_{30} + G_s G_i (N_{30} + 1)$
$\qquad - (G_s N_{30} + G_i)^2$
$\qquad = G_s^2 (N_{30}^2 + N_{30}) + G_i^2 + 2 G_s G_i N_{30} + G_s G_i (N_{30} + 1)$
$\qquad - G_s^2 N_{30}^2 - G_i^2 - 2 G_s G_i N_{30} G_i$
$\qquad = G_s^2 N_{30} + G_s G_i N_{30} + G_s G_i.$ (9.25)

To define the signal noise figure F_s, we introduce the input and output signal-to-noise ratios. To translate into the language of signal detection, we divide the photon energies by time, so that we are dealing with optical power. We assume that the optical signal power is detected by a square-law detector, so that the electrical current is proportional to the incident optical power and hence the electrical signal power is proportional to the square of the optical signal power. The electrical noise power is then proportional to the variance of the optical power. Hence we have

$$F_s = \frac{SNR_{in}}{SNR_{out}} = \left[\frac{N_{30}}{\langle N_3(z) \rangle} \right]^2 \frac{[\delta N_3(z)]^2}{[\delta N_3(0)]^2}$$
$$= \left(\frac{N_{30}}{G_s N_{30} + G_i} \right)^2 \frac{G_s^2 N_{30} + G_s G_i N_{30} + G_s G_i}{N_{30}}. \quad (9.26)$$

We have $G_i = G_s - 1$; thus in the limit where the input CS is strong enough that $N_{30} \gg 1$, we find that

$$F_s \approx 1 + \frac{G_i}{G_s} = 2 - \frac{1}{G_s}, \quad (9.27)$$

which is independent of N_{30}. We note that $F_s = 1$ when $G_s = 1$, as should be the case, because $G_s = 1$ means that there is no gain and so the OPA reduces to a piece of passive lossless fiber that does not alter the quality of the input signal. For large $G_s = 1$, the noise figure F_s approaches 2 (i.e. 3 dB) from below. This is expected from fundamental principles applying to all phase-independent amplifiers (PIAs) [2, 3].

Idler noise figure

The calculation is similar to that for the signal noise figure F_s. The necessary steps are

$$\langle N_4(z) \rangle = \langle a_4'(z) a_4(z) \rangle = \langle \Psi_3 \Psi_4 | a_4'(z) a_4(z) | \Psi_3 \Psi_4 \rangle$$
$$= \langle \Psi_3 \Psi_4 | (m_{34}^* a_3 + m_{33}^* a_4')(m_{34} a_3' + m_{33} a_4) | \Psi_3 \Psi_4 \rangle$$
$$= \langle \Psi_3 \Psi_4 | |m_{34}|^2 a_3 a_3' + |m_{33}|^2 a_4' a_4 + m_{34}^* m_{33} a_3 a_4 + m_{33}^* m_{34} a_4' a_3' | \Psi_3 \Psi_4 \rangle$$
$$= G_i(N_{30} + 1) + G_s N_{40} = G_i(N_{30} + 1), \qquad (9.28)$$

$$\langle [N_3(z)]^2 \rangle = \langle \Psi_3 \Psi_4 | (|m_{34}|^2 a_3 a_3' + |m_{33}|^2 a_4' a_4 + m_{34}^* m_{33} a_3 a_4 + m_{33}^* m_{34} a_4' a_3')^2 | \Psi_3 \Psi_4 \rangle$$
$$= \langle \Psi_3 \Psi_4 | \{ |m_{34}|^4 a_3 a_3' a_3 a_3' + |m_{33}|^4 a_4' a_4 a_4' a_4 + 2|m_{33} m_{34}|^2 a_3 a_3' a_4' a_4$$
$$+ (|m_{34}|^2 a_3 a_3' + |m_{33}|^2 a_4' a_4)(m_{34}^* m_{33} a_3 a_4 + m_{33}^* m_{34} a_4' a_3')$$
$$+ (m_{34}^* m_{33} a_3 a_4 + m_{33}^* m_{34} a_4' a_3')(|m_{34}|^2 a_3 a_3' + |m_{33}|^2 a_4' a_4)$$
$$+ (m_{34}^* m_{33} a_3 a_4)^2 + (m_{33}^* m_{34} a_4' a_3')^2$$
$$+ |m_{33} m_{34}|^2 (a_3 a_4 a_4' a_3' + a_4' a_3' a_3 a_4) \}^2 | \Psi_3 \Psi_4 \rangle$$
$$= G_i^2 \langle [N_3(0) + 1]^2 \rangle + G_s G_i (N_{30} + 1)$$
$$= G_i^2 (N_{30}^2 + 3N_{30} + 1) + G_s G_i (N_{30} + 1), \qquad (9.29)$$

$$F_i = \frac{SNR_{in}}{SNR_{out}} = \left[\frac{N_{30}}{\langle N_4(z) \rangle} \right]^2 \frac{[\delta N_4(z)]^2}{[\delta N_3(0)]^2}$$
$$= \left(\frac{N_{30}}{G_i N_{30} + G_i} \right)^2 \frac{G_i^2 N_{30} + G_s G_i (N_{30} + 1)}{N_{30}}. \qquad (9.30)$$

In the limit where the input CS is strong enough that $N_{30} \gg 1$, we find that [4, 5]

$$F_i \approx 1 + \frac{G_s}{G_i} = 2 + \frac{1}{G_i} \qquad (9.31)$$

Hence an OPA which has 100% conversion efficiency from signal to idler ($G_i = 1$) has an NF of 3 (i.e. 5 dB). As G_i increases, F_i approaches 2, just as does F_s (but from above).

9.4 Wavelength exchange

For wavelength exchange we consider that ω_1 and ω_2 correspond to two strong pumps on the same side of ω_c; we are interested in how the signal and the idler, respectively at ω_3 and ω_4, exchange energy. The frequencies satisfy $\omega_1 - \omega_2 + \omega_3 - \omega_4 = 0$, which is not the same as for an OPA. The Hamiltonian is still given by Eq. (9.13) but the four-wave mixing (FWM) terms, in which we are interested, are different, since now the individual frequency components are chosen to satisfy the above frequency condition. To find an equation for a_3 we need the applicable part of the Hamiltonian, H_3. Comparing with the procedure for the OPA in Section 9.3, we find that the SPM and XPM parts remain unchanged but that the all-important FWM terms are given by

$$H_{1234}^{WE} = 24\rho a_1' a_2 a_3' a_4 \exp[i(\omega_1 - \omega_2 + \omega_3 - \omega_4)t - i(\beta_1 - \beta_2 + \beta_3 - \beta_4)z]$$
$$= 24\rho a_1' a_2 a_3' a_4 e^{-i\Delta\beta z}, \qquad (9.32)$$

where $\Delta\beta = \beta_1 - \beta_2 + \beta_3 - \beta_4$.

The combination of operators in Eq. (9.32) shows that, during an interaction, pump 1 gains one photon while pump 2 loses one photon and the signal gains one photon while the idler loses one photon. Hence photons are exchanged between signal and idler (as well as between the pumps). Thus, rather than wavelength exchanger (WE), a better name for this device might be "photon exchanger." Note that because the photons do not have the same energies there is no conservation of energy between signal and idler, but of course energy is conserved when considering all four waves.

The Manley–Rowe relations again apply, albeit with different labels since we have relabeled the frequencies.

Calculating $[a_3, H_3]$, we now find that

$$i\frac{\hbar c}{12\rho}\frac{da_3}{dz} = [a_3, H_3]$$
$$= (N_3 + 1)a_3 + [2(N_1 + N_2 + N_4) + 3]a_3 + 2a_4 a_1' a_2 \exp^{(-i\Delta\beta z)}, \qquad (9.33)$$

where ρ is a coefficient that can be calculated from Eq. (9.16).

We see that when we have two strong pumps at ω_1 and ω_2 the equations become identical to the classical equations that we derived in subsection 3.5.5, except that they now apply to operators instead of classical fields. The integration of these equations then proceeds as in subsection 3.5.5.

Noise figure

With the notation of Appendix 1, the relations between the initial and final field operators are given by

$$a_3(z) = m_{33}a_3 + m_{34}a_4,$$
$$a_4(z) = -m_{34}a_3 + m_{33}a_4, \qquad (9.34)$$

where $m_{33} = \cos\theta$, $m_{34} = \sin\theta$, and θ is real. We note that the form of the transfer matrix M is exactly the same as that for a lossless, passive, beamsplitter (BS) [6, 7]. One difference between a WE and a BS is that in a WE the inputs (and outputs) are distinguished by their wavelengths whereas for a BS they are separated by their physical location. The transfer matrix M is a rotation matrix, which is unitary, an indication that in a WE, as in a BS, photon number is conserved.

Another similarity between a WE and a BS is that in a BS incident photons go either way in a random fashion. In a WE the same is true: an input photon changes wavelength, or not, on a random basis. We therefore expect that the same probabilities will apply in both cases, resulting in the same photon statistics, etc.

Some important consequences of the similarity of a WE and a BS are as follows.

1. There is no ASE in a WE.
2. Total photon number is conserved (and, approximately, the total power of signal and idler if the wavelengths are very close).
3. The NF for a WE is the same as for a BS, which itself acts as an attenuator in each path between the signal input and the outputs. So the NF in each case is simply the reciprocal of the transmittance. Hence we have the following NFs for the signal and the idler:

$$F_s = \frac{1}{\cos^2\theta}, \quad F_i = \frac{1}{\sin^2\theta}. \tag{9.35}$$

   In particular, when $\theta = \pi/2$ the input signal is entirely converted to the idler wavelength, with $F_i = 1$, i.e. with no deterioration of the signal quality. (This may be viewed as a consequence of the fact that there is no ASE associated with wavelength exchange.)
4. To our knowledge, wavelength exchange is the only wavelength-converting mechanism that can convert a signal with 100% efficiency and no degradation in signal quality. In comparison, an OPA-based wavelength converter with 100% conversion efficiency ($G_i = 1$) has $F_i = 3$ (5 dB).
5. When $\theta = \pi/2$, we have $a_4(z) = -a_3$. Hence the output idler annihilation operator is equal (within a trivial minus sign) to the input signal annihilation operator. This implies that a moment calculation performed on the output idler will yield the same result as for the input signal. Hence the output idler must have the same wavefunction as the input signal. In other words, when $\theta = \pi/2$, wavelength exchange performs a perfect translation of the input signal quantum state to the idler wavelength. To our knowledge, this is the first optical device reported to have this interesting property.
6. One can use wavelength exchange to perform the same type of interference experiments as with a BS, but the inputs are at two different wavelengths.

A WE does something that no other device can do. In principle, it can take the power of two different lasers and add them. Since it is obviously a phase-sensitive device, this may be hard to demonstrate with independent sources. Some optical (phase-locked loop (PLL) or equivalent device would be required. Also, by changing the relative phase of the waves one can switch the output from one wavelength to another.

9.5 Noise figure of degenerate fiber OPAs

In Section 9.3 we showed that a non-degenerate fiber OPA can have an NF approaching 3 dB under standard conditions. One way to interpret this result is to say that even when there is no finite idler at the input a vacuum state is still present at the input at the idler frequency, with fluctuations that are not correlated with the fluctuations of the CS at the signal frequency. The parametric amplification process amplifies both types of fluctuations and also mixes them. As a result, at the output the signal has fluctuations resulting both from signal amplification and from wavelength conversion of idler-frequency input fluctuations. In effect this results in a doubling of the output fluctuations in comparison with what would occur if there were no wavelength conversion of idler fluctuations; this is the origin of the 3 dB NF.

In a degenerate fiber OPA the signal and idler are indistinguishable, since the frequencies of all the waves coincide. Thus the fluctuations that are present in the signal at the input are the only fluctuations that can give rise to output fluctuations. These input fluctuations are amplified along with the signal, and with the same gain. There is no doubling of the output fluctuations owing to an idler contribution, as the idler is non-existent here. As a result, since the input signal and fluctuations experience exactly the same gain, it is clear that the NF is 1 (0 dB). This is in agreement with the general statement that phase-sensitive amplifiers (PSAs) can have a zero-decibel NF [2–4].

While noiseless amplification is a potentially interesting phenomenon, the fact that it is necessarily narrowband makes it unattractive for most optical communication systems, which are designed to carry very broad optical spectra. However there might be other applications, such as in sensing or special-purpose narrowband optical communication, where a zero-decibel NF could be of potential interest.

9.6 Effect of Raman gain on OPA noise figure

In subsection 3.6.4 we showed that in a well-phase-matched fiber OPA the Raman gain has little influence on the parametric gain spectrum. It has been shown, however, that this does not necessarily mean that the NF for this nonlinear amplifier is the same as that for an ideal OPA with no Raman gain [8–10]. In fact the NF can be several decibels higher than the 3 dB expected for an ideal amplifier.

To show this, it is necessary to modify the quantum field equations that we introduced earlier for a Raman-free OPA. Here, because of the asymmetry between signal and idler we restrict the signal (subscript 3) to be on the low-frequency side of the pump (the Stokes wave) while the idler (subscript 4) is on the other side (the anti-Stokes wave). It can be shown that Eq. (9.21) now becomes

$$a_3(z) = m_{33}a_3 + m_{34}a'_4 + c_{31}\hat{t}_1 + c_{32}\hat{t}_2,$$
$$a'_4(z) = m^*_{34}a_3 + m^*_{33}a'_4 + c^*_4\hat{t}_1, \tag{9.36}$$

where m_{33} and m_{34} correspond to an OPA with Raman gain, subsection 3.6.4 and $\hat{t}_1$ and $\hat{t}_2$ are thermal field operators. They represent the interaction between the photons and the phonons that is the origin of the Raman gain. The phonons are in thermal equilibrium at temperature T, with a mean phonon occupation number

$$n_{\text{th}} = \frac{1}{\exp \hbar\Omega/K_B T - 1}, \tag{9.37}$$

where Ω is the phonon frequency and K_B is Boltzmann's constant. The coefficients of these operators are given by

$$c_{31} = Kc_3, \quad c_{32} = \sqrt{1 - |K|^2}c_3, \quad K = |m_{34}|^2 - |m_{33}|^2$$
$$c_3c_4, \quad c_3 = (1 - |m_{33}|^2 + |m_{34}|^2)^{1/2}, \quad c_4 = (1 - |m_{34}|^2 + |m_{33}|^2)^{1/2}.$$

9.6 Effect of Raman gain on OPA noise figure

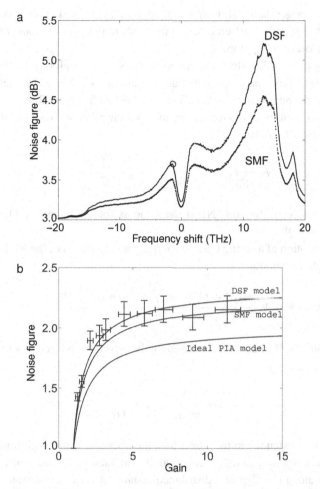

Fig. 9.1 (a) Theoretical plots of fiber OPA NF versus the pump-signal detuning. (The OPA is phase matched at each detuning.) Upper line, dispersion-shifted filter (DSF); dotted line, single-mode filter (SMF); circle, experimental data point from [8]. (b) Noise factor versus gain for a phase-matched ideal phase-insensitive amplifier (PIA), a PIA made with SMF-28, and a PIA made with DSF at 1.38 THz pump-signal detuning. The crosses represent data points for DSF from [8]. (After [9]).

Under the assumption of a strong CS at the input, the calculation of the NF for the signal gain yields

$$F_s = 1 + \frac{|m_{34}|^2 + (1 + 2n_{th})||-1+|m_{33}|^2 - |m_{34}|^2|}{|m_{33}|^2}. \tag{9.38}$$

By calculating the Raman contributions to the various coefficients from experimental data, one can then calculate the NF as a function of wavelength. The results for a typical fiber OPA are plotted in Fig. 9.1(a) in the limit of high gain (Fig. 9.1(b)) as would be the case if the OPA were phase-matched for maximum gain at each detuning. The noise figure is larger than 5 dB in the vicinity of the Raman-gain peak and drops as the wavelength decreases. It is still larger than 3 dB near the pump, where the Raman gain is very small. The NF does not approach 3 dB until the signal is more than one Raman shift on the short-wavelength side of the pump.

It is interesting to consider what happens when the signal is well beyond the Raman shift, on either side of the pump. High (non-Raman) gain can be achieved for such signals, by suitably tuning the pump (Section 5.3). Then the OPA is free from Raman influence, and photon conservation in four-wave mixing imposes the relation $|m_{33}|^2 - |m_{34}|^2 = 1$. Substituting into Eq. (9.38) then yields

$$F_3 \equiv F_s = 2 - \frac{1}{|m_{33}|^2} = 2 - \frac{1}{G_s}, \qquad (9.39)$$

regardless of n_{th}. As expected, this NF is the same as for a Raman-free OPA, and it approaches 3 dB at high gain.

Under the assumption of a strong CS at the input, the calculation of the NF for signal-to-idler wavelength conversion yields

$$F_4 \equiv F_i = 1 + \frac{|m_{33}|^2 + (1 + 2n_{\text{th}})\left|-1 + |m_{33}|^2 - |m_{34}|^2\right|}{|m_{34}|^2}. \qquad (9.40)$$

For a signal beyond the Raman shift the wavelength-conversion NF reduces to that of a Raman-free OPA, i.e.

$$F_i = 2 + \frac{1}{|m_{34}|^2} = 2 + \frac{1}{G_i}. \qquad (9.41)$$

This theory has been extended to the case where the fiber loss is not negligible [10]. In that case it is necessary to take into account additional vacuum fluctuations due to the loss, which enter along the fiber in a distributed fashion. A series expansion is used to calculate the gain in that case. The closed-form solution of subsection 3.6.5 could be used to obtain an alternate formulation.

The conclusion of these studies is that Raman gain can add several decibels to the NF of a fiber OPA for signals that are within a Raman shift from the pump. However, for signals that are outside that range, the 3 dB NF limit is still achievable.

Another way to avoid NF degradation due to Raman gain is to use orthogonally polarized signal and pump, i.e. the YXY architecture (see Chapter 4). Since the Raman gain essentially vanishes (at detunings larger than 10 nm) for light orthogonal to the pump [11], the signal and idler are not affected by the Raman gain. The Raman ASE associated with the pump is polarized parallel to it and does not interfere with signal or idler. In this case it should be possible to approach the 3 dB NF limit at signal wavelengths beyond 10 nm.

The influence of Raman gain on co-polarized phase-sensitive non-degenerate fiber OPAs has also been investigated [12]. It was shown that the minimum NF is about 0.4 dB, higher that the 0 dB expected in the absence of Raman gain but substantially lower than for a phase-insensitive OPA.

9.7 Conclusion

We have investigated the noise figure of fiber OPAs, starting from the appropriate quantum-mechanical field equations. We have shown that for an ideal OPA, in the limit of high gain and a coherent state at the input, the NF approaches 3 dB and for a frequency-degenerate phase-sensitive amplifier (PSA) the limit is 0 dB. Even though Raman gain can add a few decibels to these numbers, it remains the case that the quantum-mechanical NFs of fiber OPAs are sufficiently attractive to be of interest for optical communication. In the next chapter we will see that high-quality pumps are required for the OPA NFs to approach the quantum limit.

The fact that fiber OPAs emit signal and idler photons in pairs leads to the emission of correlated photons. This can be used for a number of applications, which will be described in Chapter 12.

For readers interested in learning more about OPA NFs by reading the literature on the subject [12, 14–20], it should be pointed out that the NF is a complex subject, which should be approached with care. For example, one should realize that there are several alternate definitions of the NF, which prove useful under different circumstances [12]. The definition used in this chapter is a valid one, although the actual measurement of this NF poses some difficulties.

References

1. *Quantum Electronics*, Yariv, A. Wiley, New York; 1989; p. 384.
2. "Quantum limits on noise in linear amplifiers," Caves, C. M. *Phys. Rev. D*; 1982; vol. 26, pp. 1817–39.
3. "Noise in amplifiers," Yamamoto, Y., Inoue, K. *J. Lightwave Technol.*; 2003; vol. 21, pp. 2895–915.
4. "Fundamentals of optical amplifiers," Yamamoto, Y., Mukai, T. *Optical and Quantum Electron.*, 1989; vol. 21, pp. S1–14.
5. "Noise characteristics of fiber-based optical phase conjugators," Hedekvist, P. O., Andrekson, P. A. *J. Lightwave Technol.*; 1999; vol. 17, pp. 74–79.
6. "Quantum physics of simple optical instruments," Leonhardt, U. *Rep. Prog. Phys.*; 2003; vol. 66, pp. 1207–49.
7. "Nonclassical fields in a linear directional coupler," Lai, W. K., Buek, V., Knight, P. L. *Phys. Rev. A*; 1991; vol. 43, pp. 6323–36.
8. "Measurement of the photon statistics and the noise figure of a fiber-optic parametric amplifier," Voss, P. L., Tang, R., Kumar, P. *Opt. Lett.*; 2003; vol. 28, pp. 549–51. "Measurement of the photon statistics and the noise figure of a fiber-optic parametric amplifier: erratum," *Opt. Lett.*; 2004; vol. 29, pp. 2815.
9. "Raman-noise-induced noise-figure limit for $\chi^{(3)}$ parametric amplifiers," Voss, P. L., Kumar, P., *Opt. Lett.*; 2004; vol. 29, pp. 445–7.
10. "Noise-figure limit of fiber-optical parametric amplifiers and wavelength converters: experimental investigation," Tang, R., Voss, P. L., Lasri, J., Devgan, P., Kumar, P. *Opt. Lett.*; 2004; vol. 29, pp. 2372–4.
11. "Measurement of the Raman gain spectrum of optical fibers," Dougherty, D. J., Kartner, F. X., Haus, H. A., Ippen, E. P. *Opt. Lett.*; 1995; vol. 20, pp. 31–3.

12. "Raman-noise-induced quantum limits for $\chi^{(3)}$ nondegenerate phase-sensitive amplification and quadrature squeezing," Voss, P. L., Köprülü, K. G., Kumar, P. *J. Opt. Soc. Amer. B*; 2006; vol. 23, pp. 598–610.
13. "Noise figure definition valid from RF to optical frequencies," Haus, HA. *IEEE J. Select. Topics Quantum Electron.*; 2000; vol. 6, pp. 240–7. "The proper definition of noise figure of optical amplifiers" Haus, H. A. In *Proc. Conf. on Optical Amplifiers and Their Applications*, Stresa, Italy, 2001.
14. "Intuitive classical approach of intensity noise in linear phase-insensitive optical amplifiers," Dallot, V; Gallion, P. In *Conf. on Lasers and Electro-Optics*, Baltimore MD, May 2001; *Technical Digest*, pp. 297–8.
15. "Optical noise figure: theory and measurement," Tucker, R.S., Baney, D. M.; In *Proc. Optical Fiber Communication Conf.*, Anaheim CA, March 2001; vol. 3, pp. WI1-1-3.
16. "Quantum and nonlinearity limitations of the optical communication channel," Desurvire, E. *C. R. Physique*; 2003; vol. 4, pp. 11–28.
17. "Quantum noise properties of parametric amplifiers driven by two pump waves," McKinstrie, C. J., Radic, S., Raymer, M. G. *Optics Express*; 2004; vol. 12, pp. 5037–66.
18. "Semiclassical model for noise propagation in depleted-pump optical amplifiers," Annovazzi-Lodi, V., Merlo, S. *IEEE J. Quantum Electron.*; 1998; vol. 34, pp. 1823–9.
19. "On the physical origin of the 3-dB noise figure limit in laser and parametric optical amplifiers," Desurvire, E. *Optical Fiber Technology: Materials, Devices and Systems*; 1999; vol. 5, pp. 40–6
20. "Quantum optics of traveling-wave attenuators and amplifiers," Jeffers, J. R., Imoto, N., Loudon, R. *Phys. Rev. A*; 1993; vol. 47, pp. 3346–59.

10 Pump requirements

10.1 Introduction

The pumping requirements for fiber OPAs are substantially more demanding than those for other types of optical amplifier. The difficulties are related to: (i) the relatively high power levels required; (ii) the polarization requirements; (iii) the requirement of phase matching, which imposes restrictions on the pump frequencies; (iv) the ease with which stimulated Brillouin scattering (SBS) can be excited, which may require that the pump spectrum be broadened by phase or frequency modulation; (v) the nearly instantaneous nature of the basic nonlinearity, which provides little or no smoothing of gain fluctuations resulting from the pump fluctuations.

The latter point is the most critical for communication applications, because it means that any pump fluctuations (of either amplitude or frequency) can affect the gain and therefore result in modulation of the signal and idler, which would lead to a deterioration of the signal-to-noise ratio (SNR) and noise figure (NF).

In this chapter we discuss these issues, with particular emphasis on the last one as it is of considerable importance for determining the levels of pump fluctuations that can be tolerated if fiber OPAs are to be viable in communication applications.

10.2 Pump power requirements

Maximum gain and gain bandwidth are the primary considerations determining the required pump power for an OPA.

10.2.1 Gain considerations

The phase-matched gain for an OPA with large gain is well approximated by $G_s \approx 8.6(\gamma P_0 L) - 6$, in dB. This implies that to have 20 dB of gain we need $\gamma P_0 L \approx 3$. For currently available highly nonlinear fibers (HNLFs), with $\gamma \approx 20$ W^{-1} km^{-1} we need $P_0 L \approx 0.15$ W km. We can then trade P_0 and L to obtain the desired gain. Very long fibers, say longer than a few kilometers, are not desirable as they can be expensive, lead to bulky packages, exhibit large zero-dispersion wavelength (ZDW) fluctuations, and introduce losses that make performance deteriorate. Very short fibers would be desirable but would lead to very high P_0 values: for example, for $L = 10$ m we would need

$P_0 = 15$ W, which would not be acceptable for communication applications but could be of interest for high-power wavelength conversion. Recent work with dispersion-shifted fibers DSFs or highly nonlinear fibers HNLFs has used fiber lengths ranging from tens of meters to a few kilometers and pump powers ranging from tens of milliwatts to several watts. To be competitive in communication applications, pump powers should probably not exceed those currently used for pumping Raman amplifiers; hence operating with P_0 values of a few watts or less would be desirable. This would correspond to a few hundred meters of HNLF.

10.2.2 Gain bandwidth considerations

As shown in Chapter 5, the gain bandwidth of fiber OPAs is an increasing function of the pump power P_0. Depending upon the location of the pump wavelength(s) with respect to λ_0, the wavevector mismatch $\Delta\beta$ will be dominated by either $\beta^{(2)}$ or $\beta^{(4)}$. The gain bandwidth, defined as the wavelength region to one side of λ_c where $g^2 > 0$, is of the order of

$$\Delta\lambda_{2m} \approx \frac{\lambda_c^2}{2\pi c}\left[-\frac{2(2m)!\gamma P_0}{\beta^{(2m)}}\right]^{1/2m}, \qquad m = 1, 2. \tag{10.1}$$

Hence, for a given fiber, $\Delta\lambda_{2m}$ scales as $(P_0)^{1/2}$ or $(P_0)^{1/4}$, i.e. sublinearly with respect to P_0. Thus obtaining large increases in bandwidth may require increasing P_0 by many orders of magnitude, and at some point P_0 might become too large to be practical.

We note that $\Delta\lambda_{2m}$ is a function only of $\gamma P_0/\beta^{(2m)}$. Thus, when we compare different types of fiber, if we find two that have similar $\beta^{(2m)}$ values but different γ values then the fiber with the larger γ will have a larger bandwidth than the other for the same P_0. This indicates that having fibers with a large γ may be beneficial for obtaining large gain bandwidths. This is indeed the case when one compares DSF and HNL-DSF, as they have similar dispersion properties but γ values differing by about one order of magnitude: then using HNL-DSF is clearly a way of increasing the gain bandwidth.

Using a high-γ fiber will not necessarily lead to a larger bandwidth, however because $\beta^{(2m)}$ may be very large for that particular fiber. This turns out to be the case for several new types of high-γ fibers, such as holey silica fibers or tellurite fibers, which have very high values of $\beta^{(4)}$ and consequently a relatively small OPA gain bandwidth [1]. Hence, to exploit fully the potential increase in bandwidth associated with γ and P_0, the design and manufacture of fibers with low $\beta^{(4)}$ values will be necessary.

State-of-the-art HNL-DSF, such as that manufactured by Sumitomo Electric, has $\beta^{(4)} \approx -5 \times 10^{-56}$ s^4 m^{-1} and $\gamma = 20$ W^{-1} km^{-1}. For $\lambda_c = 1550$ nm and $P_0 = 1$ W, the theory of [2] yields a gain bandwidth $\Delta\lambda_4 \approx 84$ nm. This number is about twice as large as the bandwidth of a C-band erbuim-doped fiber amplifier (EDFA) and shows that fiber OPAs have the potential for making wideband fiber amplifiers of interest for optical communication systems.

By going to a pump power of 16 W (pulsed) this bandwidth can be doubled, yielding an amplifier with a full bandwidth $2\Delta\lambda_4 \approx 336$ nm. This has recently been experimentally verified: an OPA operating under similar conditions exhibited a bandwidth of the order of 400 nm, a record bandwidth for any fiber amplifier [3].

10.3 Polarization considerations

The theory presented in Chapter 3 assumes that all waves are in same linear SOP; this ensures that the OPA will have maximum gain compared with that arising from other choices. In Chapter 4 we saw that alternative choices of SOPs can be made, leading to OPAs with interesting properties such as polarization independence. Hence in general it is desirable to control the SOPs of the pumps. One way to do this in practice is to use linearly polarized lasers with polarization-maintaining fiber (PMF) pigtails and polarization-maintaining couplers to combine them into a PM HNLF. This approach guarantees the long-term stability of the system once it is constructed.

For laboratory experiments, one can use laser pigtails and couplers that are not made of PMF, together with a PM HNLF (or a non-PM HNLF; these are more commonly available). In that case one has to place a number of polarization controllers in the system, which have to be adjusted to make sure that all the SOPs are correctly aligned.

10.4 Pump amplitude fluctuations (pump RIN)

If the pump intensity does not remain constant but varies in time in either a deterministic or random manner then this will introduce both intensity and phase modulation in the signal and the idler. Clearly the effect of pump relative intensity noise (RIN) is critical for optical communication systems, and as such it has been the subject of a number of detailed investigations [4–10].

In this section we analyze in detail how the pump intensity fluctuations evolve along the OPA and are transferred to the signal and idler.

10.4.1 Evolution of pump amplitude modulation in fiber OPAs

In this section we first study how the pump intensity fluctuations themselves evolve along the fiber. This is important because if these fluctuations changed, this would complicate the calculation of their impact. However, we will establish that, under realistic assumptions, they actually do not change along the fiber length. This then allows us readily to calculate their impact on the signal and idler co-propagating with the pump(s).

(a) One-pump OPA

To simplify the analysis, we will assume that the initial pump intensity modulation (IM) is not accompanied by frequency modulation (FM); in that case IM is equivalent to amplitude modulation (AM). We thus consider an OPA with a pump that at the input has sinusoidal AM at a frequency f_m. This can be viewed as the superposition of three waves: a carrier with real amplitude A_{p0} (average power $P_0 = (A_{p0})^2$); a signal sideband with real amplitude A_{s0} at a distance f_m from the carrier (signal power $P_{s0} = (A_{s0})^2$); an idler sideband with real amplitude $A_{i0} = A_{s0}$ at a distance $-f_m$ from the carrier. In the small-signal limit the carrier power remains unchanged throughout the OPA and the carrier thus acts as a constant pump, providing parametric gain for the two sidebands. The output amplitudes of the signal and idler are then given by the OPA output field expressions (see Chapter 3):

$$A_s(L) = \left[A_{s0}\cosh(gL) + \frac{i}{g}\left(\frac{\kappa}{2}A_{s0} + rA_{i0}^*\right)\sinh(gL)\right]\exp\left[-i\left(\frac{\kappa}{2} - 2\gamma P_0\right)L\right],$$

$$A_i(L) = \left[A_{i0}\cosh(gL) + \frac{i}{g}\left(\frac{\kappa}{2}A_{i0} + rA_{s0}^*\right)\sinh(gL)\right]\exp\left[-i\left(\frac{\kappa}{2} - 2\gamma P_0\right)L\right],$$

(10.2)

where, for a one-pump OPA, $g^2 = r^2 - (\kappa/2)^2$, $r = \gamma P_0$, and $\kappa = \Delta\beta + 2\gamma P_0$. Because of the conditions on A_{s0} and A_{i0}, these two equations yield the same result, i.e.

$$A_s(L) = A_i(L) = A_{s0}\left[\cosh(gL) + \frac{i}{g}\left(\frac{\kappa}{2} + r\right)\sinh(gL)\right]\exp\left[-i\left(\frac{\kappa}{2} - 2\gamma P_0\right)L\right].$$

(10.3)

Assuming relatively small modulation frequences, we have $\Delta\beta \approx 0$, $g \approx 0$, $\kappa \approx 2\gamma P_0$, and hence $A_s(L) = A_i(L) \approx A_{s0}(1 + 2i\Phi)\exp^{(i\Phi)}$, where $\Phi = \gamma P_0 L$. This shows that the signal and idler amplitudes grow linearly with distance. On that basis we might expect the AM to grow at the same rate. This, however, is not true, as shown by the following argument.

Since we are assuming that $\Delta\beta \approx 0$, we can use the exact solution of the nonlinear Schrödinger equation for the total field (see Section 6.4), i.e.

$$A(L, t) = A(0, t')\exp\left[i\gamma L|A(0, t')|^2\right]$$

(10.4)

where $t' = t - L/v$, v is the speed of propagation of light in the fiber, and

$$A(0, t') = A_{p0} + A_{s0}e^{i\omega_m t'} + A_{i0}e^{-i\omega_m t'} = A_{p0} + 2A_{s0}\cos(\omega_m t')$$

(10.5)

is the total input field, which is real. Since the exponential in Eq. (10.4) is purely a phase term, we see that the instantaneous amplitude of the total output field is exactly the same as that of the total input field (at the corresponding time). Therefore, there is no growth in pump AM along the fiber in spite of the growth of the individual modulation sidebands. The fact that the pump IM does not grow in spite of the finite parametric gain of the individual sidebands is due to the fact that the growth of the latter corresponds to self-phase modulation (SPM), not to IM.

(b) Two-pump OPA

We now consider a two-pump OPA, with pump frequencies ω_1 and ω_2; pump 1 is assumed to have initial IM, as above. Its modulation sidebands are then coupled to sidebands of the other pump, because non-degenerate four-wave mixing (FWM) involving the two pumps is well phase matched. However, we may assume that the FWM interactions generating new frequency components at or near $2\omega_1 - \omega_2$ and $2\omega_2 - \omega_1$ are poorly phase matched, and so we neglect them. We thus consider the interactions between six frequencies, those of the two pumps and the two sidebands for each pump. This situation was previously investigated in subsection 3.5.6. It is known that, in the limit of very low frequency modulation, g vanishes and that the OPA exhibits little gain if a single sideband is present at the input. However, since IM involves the initial presence of two sidebands for one pump, we need to examine this situation closely to ascertain whether the IM grows, as we did for the one-pump case.

We consider only phase shifts due to pump SPM and XPM, and we neglect Raman gain. The equations for the six waves are as in subsection 3.5.6 and lead to (see Eq. (3.109))

$$\frac{d}{dz}(P_1 + P_3 + P_5) = \frac{d}{dz}(P_2 + P_4 + P_6) = 0.$$

As a result, the total field of pump 1 at L must be of the form $A_1(L,t) = A_1(0,t') \exp^{i\varphi_1(L,t)}$, where $\varphi_1(L,t)$ is a real phase. Thus, if $A_1(0,t')$ has IM with a form similar to Eq. (10.5), we conclude as in subsection 10.4.1(a) that $A_1(L,t)$ has exactly the same IM. Therefore there is no growth in the IM of pump 1. Repeating the reasoning for pump 2, which originally has no IM, we find that it does not acquire any IM due to pump 1. (A more detailed proof of these properties can be found in subsection 3.5.6.)

In summary, we have shown that in a two-pump OPA the initial IM of each pump remains unchanged over the amplifier length, as in the one-pump OPA case.

10.4.2 Impact of pump IM

Since the basic nonlinearity leading to parametric gain is very fast, the gain will follow the pump intensity fluctuations almost exactly and this will result in IM of the signal and idler. The generation of signal IM by pump relative initensity noise (RIN) is referred to as RIN transfer as pump RIN induces signal RIN, with a magnification factor ρ. In addition pump IM induces XPM of the signal and idler. In this section we present analyses to quantify the impact of pump IM on signal and idler.

(a) Transfer of pump IM to signal IM

If a pump has IM, the variations in instantaneous power will modulate the OPA gain and this will produce undesirable modulation of the amplitude of amplified signals. We will calculate the impact of this effect in the small-signal-gain regime, by making use of the gain expressions obtained in Chapter 3.

To simplify the notation, we use a quasi-steady-state approach, assuming that the pump varies so slowly that we can apply the usual steady-state expressions to calculate the gain. This allows us to suppress any explicit time dependence and to simply calculate the changes in gain by means of derivatives with respect to pump power. Thus we now write the pump power in the form $P_p = P_0(1 + m_p)$, where P_0 is the average power and m_p is the IM index. Similarly, we write the signal output power as

$$P_{s,out} = G_s P_{s,in} = G_{s0}(1 + m_s) P_{s,in}, \qquad (10.6)$$

where $P_{s,in}$ is the signal input power, free of modulation, G_s is the signal power gain, modulated by the pump power, G_{s0} is the reference power gain, i.e. G_s for $m_p = 0$, and m_s is the signal output IM index, which is the same as the signal power gain modulation index. The relationship between m_p and m_s is obtained by expressing the fact that G_s depends on m_p through P_p, as follows:

$$G_s = G_{s0} + m_p \frac{dG_s}{dm_p} = G_{s0} + m_p P_0 \frac{dG_s}{dP_p} = G_{s0}(1 + m_s). \qquad (10.7)$$

Hence we find that the RIN magnification factor is

$$\rho \equiv \frac{m_s}{m_p} = \frac{P_0}{G_{s0}} \frac{dG_s}{dP_p} = \frac{P_0}{1 + G_{i0}} \frac{dG_i}{dP_p}, \qquad (10.8)$$

where we have used the fact that in a lossless fiber OPA $G_s = 1 + G_i$; here G_i is the signal-to-idler conversion gain, given by

$$\sqrt{G_i} = \frac{\gamma P_p}{g} \sinh(gL). \qquad (10.9)$$

To calculate ρ it is convenient to calculate the derivative of Eq. (10.9) with respect to P_p. This leads to

$$\frac{1}{2\sqrt{G_i}} \frac{dG_i}{dP_p} = \frac{\sqrt{G_i}}{P_p} + \sqrt{G_i} [gL \coth(gL) - 1] \frac{1}{g} \frac{dg}{dP_p}, \qquad (10.10)$$

or

$$\frac{P_p}{2G_i} \frac{dG_i}{dP_p} = 1 + [gL \coth(gL) - 1] \frac{P_p}{g} \frac{dg}{dP_p}. \qquad (10.11)$$

Since $g^2 = (\gamma P_p)^2 - (\Delta\beta + u\gamma P_p)^2/4$, where $u = 1$ (2) for a two-pump (one-pump) OPA, we have

$$4g \frac{dg}{dP_p} = \gamma^2 P_p (4 - u^2) - u\gamma \Delta\beta \qquad (10.12)$$

and

$$\frac{P_p}{g} \frac{dg}{dP_p} = \frac{4 - u(x + u)}{4 - (x + u)^2} \qquad (10.13)$$

where $x = \Delta\beta/(\gamma P_0)$ is the ratio of the linear and the nonlinear phase shifts. Finally,

$$\rho = \frac{P_p}{G_s} \frac{dG_i}{dP_p} = \frac{G_i}{G_s} \frac{P_p}{G_i} \frac{dG_i}{dP_p} = 2 \frac{G_i}{G_s} \left\{ 1 + [gL \coth(gL) - 1] \frac{P_p}{g} \frac{dg}{dP_p} \right\}$$

$$= \frac{2}{1 + [g'/\sinh(g'\Phi)]^2} \left\{ 1 + [g'\Phi \coth(g'\Phi) - 1] \frac{4 - u(x + u)}{4 - (x + u)^2} \right\}, \qquad (10.14)$$

10.4 Pump amplitude fluctuations (pump RIN)

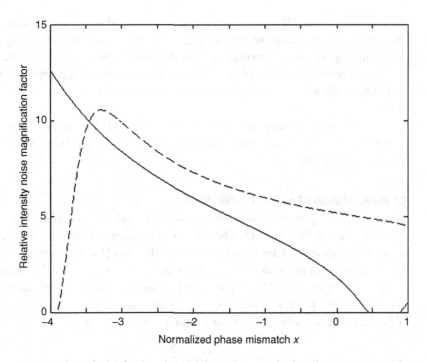

Fig. 10.1 Plots of $\rho(x)$ for $\Phi = 3$: solid line, one pump; broken line, two pumps. The maximum gain corresponds to $x = -2$ for the one-pump case and $x = -1$ for the two-pump case.

where $g' = g/(\gamma P_0) = \sqrt{1 - (x+u)^2/4}$ is the parametric gain coefficient normalized to its maximum value and $\Phi = \gamma P_0 L$ is the pump nonlinear self-phase shift.

Maximum parametric gain ($g' = 1$) is obtained for $x = -u$. For a typical OPA, $\Phi \approx 3$ and so $\rho \approx 2\Phi = 6$. This is a non-negligible value and shows that in a typical fiber OPA the pump-to-signal RIN transfer can indeed be significant near maximum gain, which is the region of main interest.

Figure 10.1 shows theoretical plots of the RIN magnification factor $\rho(x)$ for $\Phi = 3$. We see that ρ can be considerably larger than 6 away from maximum gain: for a one-pump OPA, ρ varies from 2 to 13 across the gain bandwidth and for a two-pump OPA it varies from 4 to 10. This strong wavelength dependence may have implications for the design of practical fiber OPAs.

(b) Signal and idler phase modulation

In a one-pump OPA, if the pump has IM then it will induce phase modulation of both the signal and idler, by simple XPM. This time-varying XPM is itself equivalent to FM, and so the signal and the idler spectra will be broadened.

As an example, consider a pump of average power P_0 and intensity modulated at frequency f_m, with modulation index m. At the OPA output the maximum frequency excursion due to pump XPM (in Hz) is $(\delta f)_{max} = 2 f_m \gamma P_0 L m$. If we have $\gamma P_0 L \approx 3$

then $(\delta f)_{\max} = 6 m f_m$. If $m = 1$ then $(\delta f)_{\max} = 6 f_m$, and if f_m is of the same order as the signal data rate then we see that $(\delta f)_{\max}$ would exceed the data rate and significantly distort the signal optical spectrum; this could cause difficulties in narrowband filters and lead to performance degradation. This would occur when the pump is intentionally modulated with $m = 1$ to exploit the benefits associated with the increased peak power [11].

For a good-quality pump, however, the r.m.s. amplitude fluctuations are likely to be well below 1% in the signal bandwidth; then spectral broadening by a single OPA stage should be negligible.

(c) Impact of pump IM on the OPA noise figure

Since NF is an important performance measure for amplifiers, it is of interest to find out the effect of pump RIN on the OPA NF. Because of the importance of this subject, several in-depth investigations have been performed in this area [5–13]. In practice it is found that the transfer of pump IM to signal IM is the main cause of OPA NF deterioration, except for very low signal input powers [5]. Assuming that the signal fluctuations are dominated by the pump RIN transfer, we can make use of ρ to express the electrical NF, NF_e. We adopt the definition of F_e as a ratio of SNRs, namely

$$F_e = \frac{SNR_{in}}{SNR_{out}}, \quad (10.15)$$

where SNR_{in} is the ratio of the square of the signal photocurrent and its noise variance due to shot noise. It can be shown that $SNR_{in} = P_{s,in}/(2h\nu \Delta \nu)$, where $P_{s,in}$ is the signal input power and $\Delta \nu$ is the receiver bandwidth [11]; SNR_{out} can be understood in purely classical terms when pump-induced fluctuations dominate the output noise. Then we have

$$SNR_{out} = \frac{(P_{s,out})^2}{(m_s P_{s,out})^2} = \frac{1}{m_s^2} = \frac{1}{(m_p \rho)^2}, \quad (10.16)$$

and finally

$$F_e = \frac{P_{s,in}}{2h\nu \Delta \nu}(m_p \rho)^2. \quad (10.17)$$

The form of this result is in agreement with that published in the literature [6, 7]. Equation (10.17) then says that F_e scales linearly with $P_{s,in}$ and quadratically with ρ and m_p. The dependences on ρ and m_p are in agreement with intuition: we expect the quality of the output signal to deteriorate as RIN transfer increases (larger ρ) and/or the optical signal-to-noise ratio (OSNR) of the pump gets worse (larger modulation index m_p). Also, since we have seen that ρ increases with gain, it follows that the NF will deteriorate as the gain increases.

The dependence on $P_{s,in}$, however, is not as intuitive. Equation (10.17) says that as the signal input power increases, F_e increases. Since we regard the NF as an important measure of performance, we might think that we should therefore try to operate with a low value of $P_{s,in}$. This reasoning, however, is misleading because, from a system performance standpoint, what really counts is SNR_{out}. But in this regime we always have $SNR_{out} = 1/(m_p\rho)^2$, which shows that $SNR_{s,out}$ is actually independent of $P_{s,in}$. Therefore, in this regime it is actually not necessary to reduce $P_{s,in}$ in order to improve system performance.

It is also important to realize that the NF limitation due to RIN transfer is important in current OPA experiments because most of these use erbium-doped fiber amplifiers (EDFAs) to generate high-power pumps. High-power EDFAs generate a great deal of amplified spontaneous emission (ASE) at and near the pump wavelength, which is essentially impossible to filter out completely with the kinds of filter that are available today. This ASE then introduces pump RIN, which degrades the amplified signals by RIN transfer. One should realize, however, that this method for generating high-power pumps is not the only one. It is in fact entirely possible that other methods will be developed that can generate low-RIN high-power pumps, because this would not violate any fundamental principle of physics. A step in that direction might be to use a so-called ASE-free laser, made with a Sagnac ring interferometer within an external cavity [14].

10.5 Pump phase or frequency fluctuations

There are two types of frequency variation that can affect a pump. The first is the natural linewidth of the pump laser itself; if necessary, this linewidth can be reduced to very low values (in the kilohertz range). The second comes from the pump-phase modulation, or frequency modulation, required for stimulated Brillouin scattering (SBS) suppression; a frequency excursion of several gigahertz is typically required (see subsection 2.7.2) and so this contribution is generally more important than the first one.

If the frequency of the pump in a one-pump OPA (or the center frequency in a two-pump OPA) varies in time by an amount δf from some reference value, two potentially detrimental effects can take place: idler frequency modulation and gain modulation.

In this section we will show that pump FM remains essentially unchanged along the fiber when dispersion effects can be neglected, for both one- and two-pump OPAs. On this basis we will calculate the impact of pump FM on OPA performance.

10.5.1 Evolution of pump frequency modulation in fiber OPAs

In the case of a one-pump fiber OPA, we can use the same starting point as that used in subsection 10.4.1. We now take it that the signal and idler sidebands are imaginary and have opposite amplitudes. Equations (10.2)–(10.4) remain the same, but Eq. (10.5) is replaced by

$$A(0, t') = A_{p0} + A_{s0}e^{i\omega_m t'} + A_{i0}^* e^{-i\omega_m t'} = A_{p0} + 2i|A_{s0}|\sin(\omega_m t'). \quad (10.18)$$

As a result,

$$A(L,t) = A(0,t')\exp[i\gamma L|A(0,t')|^2]$$
$$= A(0,t')\exp[i\gamma L(A_{p0}^2 + 2|A_{s0}|^2 \sin^2(\omega_m t'))]. \quad (10.19)$$

The exponential in Eq. (10.19) contains terms that might lead to phase modulation. The first term shows that all three waves acquire phase modulation due to pump SPM. The second term is of second order in $|A_{s0}|$ and so under most practical circumstances is negligible compared to the first. There is no first-order term in $|A_{s0}|$ and, as a result, no additional PM at ω_m. So we conclude that the input pump FM remains unchanged through the entire length. This conclusion is in agreement with the work of Bar-Joseph et al. [15].

In the case of a two-pump OPA we can use results obtained in subsection 3.5.6. Specifically, Eqs. (3.121)–(3.124) are valid in the absence of dispersion. Assuming that both pumps only have FM at the input, we have $I_{10} = I_{20} = 0$, which implies that $B_k = A_{k0}$, $k = 3$–6. Therefore there is no change in the initial pump sidebands, and the FM of each pump remains the same along the entire fiber. It should be noted that if dispersion effects cannot be neglected then the preceding conclusions may be invalid. In the case of a one-pump OPA, the pump and its modulation sidebands are generally near the zero-dispersion wavelength (ZDW), and therefore neglecting dispersion effects is likely to be a good approximation under most circumstances.

In the case of a two-pump OPA, however, it is possible to imagine circumstances where dispersion would not be negligible. The reason for this is that for broadband two-pump OPAs the pumps can be far from the ZDW, perhaps as much as 50 or 100 nm away. In this case, if the fiber is long there may be enough group-velocity dispersion (GVD) to transform the FM of each pump into a significant amount of IM, which will then have effects similar to those studied in Section 10.4. Such a situation was investigated in [16], where it was shown that this phenomenon can indeed lead to additional performance degradation.

10.5.2 Impact of pump FM

Having established that pump FM remains unchanged along the fiber, for either type of OPA, we now investigate the impact that pump FM has on OPA characteristics.

(a) Idler frequency modulation

If the signal frequency is fixed and the pump frequency is changed by δf then the idler frequency itself will vary by $2\delta f$. This can be a problem in wavelength conversion if δf is larger than the linewidth desired for the idler. If this idler spectrum broadening must be avoided, one has to resort to techniques such as the use of a two-pump OPA with the two pumps modulated 180° out of phase. However, the signal frequency is not affected and therefore this effect is not a problem if one is only interested in signal amplification.

(b) Gain modulation

If δf is large enough the shape of the gain spectrum may be altered instantaneously, leading to detrimental gain modulation for the signals being amplified. Because of its important practical significance, this effect has been the object of several in-depth investigations [16–20]. (In Chapter 5 we saw that the shape of the OPA gain spectrum can be very sensitive to the location of the pump frequency with respect to λ_0, particularly for OPAs designed to have a very wide gain spectrum.)

This effect can be analyzed by taking the gain expression for an OPA and modulating its pump frequency(ies). We will handle one- and two-pump OPAs simultaneously. For a one-pump OPA, the modulated pump frequency has the form $\omega_1 = \omega_{10} + \delta\omega_p$, where ω_{10} is the average pump frequency and $\delta\omega_p$ is a small fluctuation. For a two-pump OPA, we assume that for the second pump $\omega_2 = \omega_{20} + \varepsilon\delta\omega_p$, where $\varepsilon = 1\,(-1)$ for two pumps with identical (opposite) FM. An amplified signal will acquire IM, because the signal gain G_s depends on $\delta\omega_p$ through $\Delta\beta$. For either type of OPA, the relative signal-intensity modulation is found to be

$$m_{\text{IM}} = \frac{\delta\omega_p}{G_s} \frac{dG_s}{d(\delta\omega_p)}$$

$$= \frac{1-G_s}{2(g')^2 G_s}[g'\Phi\coth(g'\Phi) - 1](x+u)\frac{dx}{d(\delta\omega_p)}\delta\omega_p, \qquad (10.20)$$

where

$$\gamma P_0 \frac{dx}{d(\delta\omega_p)} = \frac{\varepsilon+1}{2}\beta^{(3)}\left[(\Delta\omega_{s0})^2 - (\Delta\omega_{p0})^2\right] - \beta^{(2)}\left[\Delta\omega_{s0}(\varepsilon+1) + \Delta\omega_{p0}(\varepsilon-1)\right]$$

$$- \frac{\beta^{(4)}}{6}\left[(\Delta\omega_{s0})^3(\varepsilon+1) + (\Delta\omega_{p0})^3(\varepsilon-1)\right]; \qquad (10.21)$$

$G_s = 1 + [\sinh(g'\Phi)/g']^2$ is the signal power gain, $g' = \sqrt{1-(x+u)^2/4}$ is the normalized parametric gain coefficient, $u = 1\,(2)$ for a two-pump (one-pump) OPA ($\varepsilon = 1$ for a single pump), $x = \Delta\beta/(\gamma P_0)$, $\Phi = \gamma P_0 L$, $\Delta\omega_{p0} = (\omega_{20} - \omega_{10})/2$, $\Delta\omega_{s0} = \omega_{s0} - \omega_{c0}$, and $\beta^{(m)} = d^m\beta/d\omega^m$ evaluated at the center frequency $\omega_{c0} = (\omega_{10} + \omega_{20})/2$. Other notation is as in subsection 10.4.2. The modulation index m_{IM} vanishes for $x = -u$ (maximum gain), but it can reach large values elsewhere.

Figure 10.2 (left) shows the gain spectra to the right of λ_0 for one-pump and two-pump OPAs. Figure 10.2 (right) shows the corresponding graphs of the signal IM modulation index m_{IM} across the OPA gain bandwidths. We have assumed a maximum gain of 20 dB, with $\Phi \approx 3$, $\delta\omega_p/2\pi = 3$ GHz, $\gamma = 17$ km^{-1} W^{-1}, $P_0 = 0.33$ W, $L = 0.5$ km, $D_\lambda = 0.02$ ps nm^{-2} km^{-1}, and $\lambda_0 = 1550$ nm. Very large values of m_{IM} are obtained at the edge of the gain spectrum. For the one-pump OPA, m_{IM} reaches 14% within the useful part of the gain spectrum, which is of a considerable size. For the two-pump OPA with $\varepsilon = 1$, m_{IM} reaches 8%. For $\varepsilon = -1$ (shown with a magnification of 100), m_{IM} remains below 1% over the entire gain bandwidth. This shows that one way greatly to reduce this effect is to use an OPA with two pumps having opposite FM.

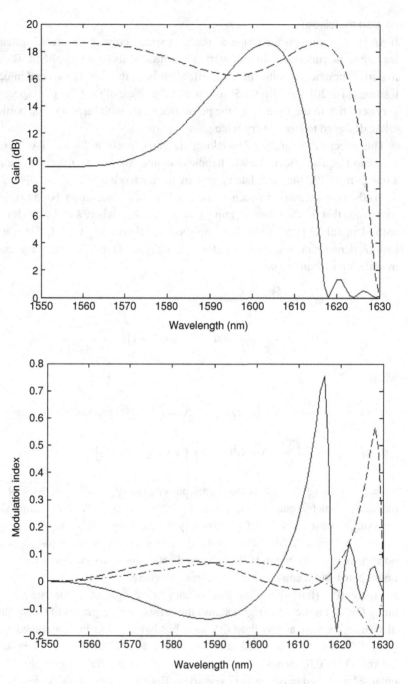

Fig. 10.2 Plots for one- and two-pump OPAs. Upper figure, gain spectra: solid line, one pump; broken line, two pumps. Lower figure, plots of m_{IM}: solid line, one pump, broken line, two pumps, co-dithered; broken-and-dotted line, two pumps, counter-dithered, shown at 100 magnification.

This theory is valid for fibers with a fixed ZDW. In long realistic fibers, longitudinal variations of ZDW are always present and may mitigate the effect of pump FM.

10.6 Conclusion

The choice of the pump(s) is critical for the operation of fiber OPAs in applications such as communication or remote sensing, for which it is desirable to avoid detrimental intensity or frequency modulation of the signal or the idler caused by similar pump fluctuations due to the mechanisms described above.

Ideally one would therefore like to have pumps that have phase and amplitude fluctuations due only to the inevitable quantum fluctuations. It is not possible to employ such pumps in practice, and therefore one must deal with additional pump fluctuations due to a number of sources. Most prominent are: the frequency modulation used to suppress stimulated Brillouin scattering (SBS); intensity and phase noise due to amplified spontaneous emission (ASE) added by amplification by EDFAs; intensity and phase noise associated with lasers with a relatively large linewidth.

As shown in the preceding discussion, these pump fluctuations will lead to corresponding signal and idler fluctuations. The latter will have to be evaluated to determine whether a particular type of pump is suitable for a given application.

The considerations in this chapter show that it is generally not a good idea to consider using a laser with a relatively large bandwidth to pump an OPA for communication applications. Consider for example a semiconductor Fabry–Pérot laser, with a bandwidth of a few nm and a longitudinal mode spacing of 0.1 nm. The output of this laser consists of a superposition of different frequencies, and thus the total intensity varies chaotically in time. This intensity variation then gets transferred to the signal (and the idler) by means of the instantaneously varying OPA gain. As a result, if the input signal is monochromatic then the output signal has a spectrum that is broadened in the same manner as the pump laser, because the signal intensity variations follow those of the pump. Other types of laser would lead to similar results; this would be the case for EDFLs or Raman lasers, which typically have output spectral widths determined by fiber Bragg gratings that are typically of the order of 1 nm. Such lasers would not be suitable for communication applications or sensing applications requiring narrow linewidths.

For such applications, it is necessary to use pumps with narrow linewidths and low RIN and to keep ASE addition by amplification to a minimum by using narrowband filters. These requirements, combined with the need for high power, make the design of pumps for fiber OPAs a challenging matter.

References

1. "Design of highly-nonlinear tellurite fibers with zero dispersion near 1550 nm," Hu, E. S., Hsueh, Y. L., Marhic, M. E., Kazovsky, L. G. In *Proc. European Conf. on Optical Communication*, Copenhagen, September 2002; vol. 2, paper 3.2.3.

2. "Broadband fiber optical parametric amplifiers," Marhic, M. E., Kagi, N., Chiang, T. K., Kazovsky, L. G. 1996; *Opt. Lett.*, vol. 21, pp. 573–5.
3. "Wideband tuning of the gain spectra of one-pump fiber optical parametric amplifiers," Marhic, M. E., Wong, K. K. Y., Kazovsky, L. G. *IEEE J. Select. Topics in Quantum Electron.*; 2004; vol. 10, pp. 1133–41.
4. "Pump to signal transfer of intensity modulation in fiber optical parametric amplifiers," Marhic, M. E., Kalogerakis, G., Wong, K. K. Y., Kazovsky, L. G. *J. Lightwave Technol.*; 2005; vol. 23, pp. 1049–55.
5. "Impact of pump OSNR on noise figure for fiber-optical parametric amplifiers," Durecu-Legrand, A., Simonneau, C., Bayart, D., Mussot, A., Sylvestre, T., Lantz, E., Maillotte, H. *IEEE Photon. Technol. Lett.*; 2005; vol. 17, pp. 1178–80.
6. "Noise characteristics of fiber optical parametric amplifiers," Kylemark, P., Hedekvist, P. O., Sunnerud, H., Karlsson, M., Andrekson, P. A. *J. Lightwave Technol.*; 2004; vol. 22, pp. 409–16.
7. "Correction to 'Noise characteristics of fiber optical parametric amplifiers'," Kylemark, P., Hedekvist, P. O., Sunnerud, H., Karlsson, M., Andrekson, P. A. *J. Lightwave Technol.*; 2005; vol. 23, p. 2192.
8. "Noise characteristics of dual-pump fiber-optic parametric amplifiers," Bogris, A., Syvridis, D., Kylemark, P., Andrekson, P. A. *J. Lightwave Technol.*; 2005; vol. 23, pp. 2788–95.
9. "Gain and wavelength dependence of the noise-figure in fiber optical parametric amplification," Kylemark, P., Karlsson, M., Andrekson, P. A. *IEEE Photon. Technol. Lett.*; 2006; vol. 18, pp. 1255–7.
10. "Pump-noise transfer in dual-pump fiber-optic parametric amplifiers: walk-off effects," Yaman, F., Lin, Q., Agrawal, G. P., Radic, S. *Opt. Lett.*; 2005; vol. 30, pp. 1048–50.
11. "High-repetition-rate pulsed-pump fiber OPA for amplification of communication signals," Kalogerakis, G., Shimizu, K., Marhic, M. E., Wong, K. K. Y., Uesaka, K., Kazovsky, L. G. *J. Lightwave Technol.*; 2006; vol. 24: pp. 3021–7.
12. "Noise statistics of fiber optical parametric amplifiers," Kylemark, P., Karlsson, M., Torounidis, T., Andrekson, P. A. In *Proc. Optical Fiber Communications Conf.*, Anaheim CA, March 2006; paper OWT5. "Noise statistics in fiber optical parametric amplifiers," Kylemark, P., Karlsson, M., Andrekson, P. A. *J. Lightwave Technol.*, in press.
13. "Signal-to-noise considerations in fiber links with periodic or distributed optical amplification," Yariv, A. *Opt. Lett.*; 1990; vol. 15, pp. 1064–6.
14. "High power ASE-free tunable laser using a Sagnac ring interferometer within the external cavity," Fulop, L., Souhaite, G., Moulinet, X., Graindorge, P., Lefevre, H. C. In *Proc. Optical Fiber Communication Conf.*, Anaheim CA, March 2001: vol. 2, pp. TuJ6-1–3.
15. "Parametric interaction of a modulated wave in a single-mode fiber," Bar-Joseph, I., Friesem, A. A., Waarts, R. G., Yaffe, H. H. *Opt. Lett.*; 1986; vol. 11, pp. 534–6.
16. "Impact of pump-phase modulation on dual-pump fiber-optic parametric amplifiers and wavelength converters," Yaman, F., Lin, Q., Radic, S., Agrawal, G. P. *IEEE Photon. Technol. Lett.*; 2005; vol. 17, pp. 2053–5.
17. "Pump FM to signal IM conversion in fiber OPAs," Marhic, M. E., Kalogerakis, G., Kazovsky, L. G. In *Proc. OECC/COIN* 2004, Yokohama, Japan, July 2004; paper 13P-87.
18. "Impact of pump phase modulation on the gain of fiber optical parametric amplifier," Mussot, A., Durecu-Legrand, A., Lantz, E., Simonneau, C., Bayart, D., Maillotte, H., Sylvestre, T. *IEEE Photon. Technol. Lett.*; 2004; vol. 16, pp. 1289–91.

19. "Impact of pump phase modulation on system performance of fibre-optical parametric amplifiers," Durecu-Legrand, A., Mussot, A., Simonneau, C., Bayart, D., Sylvestre, T., Lantz, E., Maillotte, H. *Electron. Lett.*; 2005; vol. 41, pp. 350–2.
20. "Performance of fiber parametric-processing devices using binary-phase-shift-keyed pump modulation," Radic, S., McKinstrie, C. J., Jopson, R. M., Gnauck, A. H., Centanni, J. C., Chraplyvy, A. R. *IEEE Photon. Technol. Lett.*; 2004; vol. 16, pp. 548–50.

11 Performance results

11.1 Introduction

In this chapter we present the best results that have been obtained with fiber OPAs, in terms of verifying their basic characteristics. We begin with pulsed devices, for which it is easy to obtain the high pump powers that enable wide bandwidths and high gains. We then cover continuous-wave (CW) devices, whose performance is important for applications such as optical communications.

11.2 Pulsed devices

11.2.1 Pulsed OPAs

(a) One-pump OPA with 400 nm gain bandwidth and 70 dB gain

A fiber OPA with a gain bandwidth of 400 nm was reported in [1]. With such a large bandwidth, it was actually difficult to find enough signal lasers at different wavelengths to perform gain measurements. For that reason, the OPA amplified spontaneous emission (ASE) spectrum was used as a substitute for the gain spectrum (since the two are closely related); as we have seen, the ASE power in a certain optical frequency interval is proportional to $2G_s - 1$. Hence, for an OPA with a relatively large gain, the ASE power is almost proportional to the gain and can therefore be used for estimating the gain. This measurement is valid provided that only vacuum fluctuations are present at the input. In a typical fiber OPA setup the pump is provided by a low-power distributed-fiber Bragg (DFB) laser, whose output is amplified by a cascade of two or three erbuim-doped fiber amplifiers (EDFAs). These EDFAs generate a considerable amount of ASE, which must be attenuated in order that at the signal wavelength it is negligible compared with the vacuum level. This can be accomplished by cascading several thin-film optical filters or by using a single fiber Bragg grating (FBG) with a high rejection ratio. The latter approach was used in [1], thereby eliminating the influence of EDFA ASE from the measurements (except close to the pump where the filter cannot remove all the pump ASE).

An important consequence of this approach is that, since signal lasers at various different wavelengths are not needed, the measurements can proceed very quickly: one only needs to record the OPA ASE spectrum, which can be done in a short time with

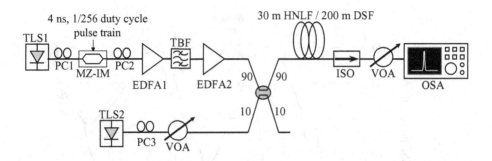

Fig. 11.1 Experimental setup for measuring OPA ASE or gain spectra of a pulsed fiber OPA. TLS, tunable laser system; MZ, Mach–Zehnder; VOA, variable optical attenuator; ISO, isolator. (After [1], © 2004 IEEE.)

an optical spectrum analyzer (OSA). One can then use a tunable pump laser and rapidly optimize the gain spectrum by monitoring its shape as λ_p is varied. In this manner, one can easily maximize the gain bandwidth.

The experimental setup is shown in Fig. 11.1. The highly nonlinear fiber (HNLF), manufactured by Furukawa Electric, had the following parameters: $L = 30$ m, $\gamma = 18$ W^{-1} km^{-1}, $\lambda_0 = 1556$ nm, $\beta_3 = 0.49 \times 10^{-40}$ s^3 m^{-1} ($D_\lambda = 0.031$ ps nm^{-2} km^{-1}), and $\beta^{(4)} = -5.8 \times 10^{-56}$ s^4 m^{-1}. The pump wavelength was $\lambda_p = 1555.54$ nm. The peak pump power in the HNLF was 20 W. The FBG and the circulator removed EDFA ASE from the pump.

Measured OPA ASE spectra for this HNL-DSF are shown in Fig. 11.2. The widest one-piece spectrum was obtained when the lower edge of the gain region reached about 1380 nm. Trying to obtain wider spectra by tuning the pump led to breakup of the spectrum and the formation of a narrow peak moving toward shorter wavelengths (see the next part of the subsection).

The average OPA gain was also measured, with the OSA, over a smaller spectral region and then converted to a peak OPA gain. The similarity of the details of the graphs of the ASE and the gain in the 1400–1500 nm region [1] confirms that the ASE spectrum and gain spectrum are proportional at high gain.

The total ASE spectral width exceeded the range of the OSA. Indirect means were used to estimate the bandwidth, relying on the fact that the gain spectrum is symmetric in the frequency domain (see Chapter 5). Since the gain drops to 20 dB at 1380 nm on the short-wavelength side, it will also drop to 20 dB on the long-wavelength side, at 1782 nm ($1/\lambda_4 = 2/\lambda_1 - 1/\lambda_3 = 2/1555.4 - 1/1380 = 1/1782$). This corresponds to a 20 dB gain bandwidth that is slightly over 400 nm. This is in good agreement with the theoretical value calculated using the experimental parameters. A maximum gain of 70 dB was also measured with the same experimental setup. Such a high gain is relatively easy to obtain with a pulsed pump, as compared with a CW one, since the need for

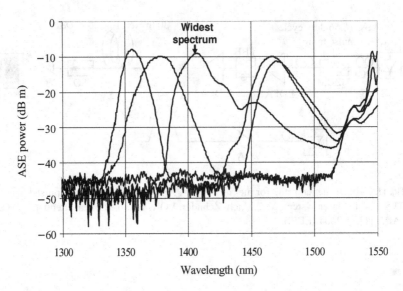

Fig. 11.2 Measured broadband OPA ASE spectrum of an HNL-DSF pulsed fiber OPA. The widest one-piece spectrum is obtained when the short-wavelength edge is about 1380 nm; this corresponds to a full width of about 400 nm. The pump was near 1550 nm. (After [1], © 2004 IEEE.)

stimulated Brillouin scattering (SBS) suppression is greatly reduced by using a low duty cycle.

(b) Tunable narrowband one-pump OPA

As shown in Section 5.3, with a one-pump OPA it is also possible to obtain narrow gain regions away from the pump, by tuning the pump away from λ_0. This was observed experimentally with the same setup as that used for obtaining a 400 nm bandwidth, described in the preceding section. The only difference was the change in pump wavelength, which led to dramatically different gain spectra. Here again the existence of gain was inferred from the presence of OPA ASE.

Experiments were performed with both dispersion-shifted fiber (DSF) and HNL-DSF. In both cases, narrow gain peaks were obtained as low as 1350 nm, i.e. about 200 nm away from the pump. The location of the peaks could readily be varied by tuning the pump wavelength. As expected from the theory, it was observed that the peaks became narrower as they moved further from the pump. The peaks were also narrower for the DSF than for the HNL-DSF, owing to the smaller γ value of the former. The results for HNL-DSF are shown in Fig. 11.2 and those for DSF in Fig. 11.3.

Similar experiments have been performed with photonic crystal fibers [2–4]. In [4] over 185 nm of tunability was reported, using a 1.6 W quasi-CW pump. Also, similar experiments have been performed in the context of modulation instability (MI) [5]. Finally, the generation of picosecond pulses, tunable over 100 THz, was obtained in a 7-cm-long HNLF; tunability was achieved by changing the changing the chirp of the pump pulses instead of the pump frequency [6].

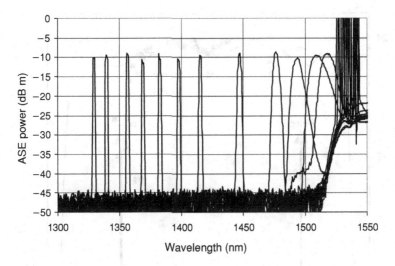

Fig. 11.3 Measured OPA ASE spectra of a DSF pulsed fiber OPA. The various spectra were obtained by tuning the pump near 1550 nm. (After [1], © 2004 IEEE.)

(c) Two-pump fiber OPA with 95 nm low-ripple bandwidth

While the 400-nm-wide gain spectrum obtained with a one-pump OPA is impressive, it is far from being flat. It is thus desirable to investigate the feasibility of making fiber OPAs that display wide regions with constant gain, so that they can be used directly without resorting to gain-flattening filters.

It is well known that two-pump fiber OPAs have inherently flatter gain spectra than one-pump OPAs, and so they present a promising approach for developing flat-gain OPAs. In recent work, a 95 nm bandwidth with less than 2 dB of ripple was obtained with a two-pump OPA [7]. The fiber was a 160-m-long HNLF, with $\gamma = 7\,\text{W}^{-1}\,\text{km}^{-1}$. It had zero-dispersion wavelength (ZDW) variations smaller than 0.15 nm. A gain of 23.5 ± 1 dB was obtained over a 95 nm bandwidth, with total pump power 4.2 W; see Fig. 11.4. This work is important because the bandwidth achieved is similar to the entire EDFA range, and it is achieved in spite of fiber imperfections, namely ZDW variations and random birefringence [8].

Another interesting aspect of the work is that it used pumps at 1499.6 nm and 1607.7 nm. The long-wavelength pump was obtained by amplification of the output of a laser diode (LD) by an L-band EDFA, but the short wavelength fell below the range of C-band EDFAs and had to be obtained by different means. This was done by amplifying the output of an LD at 1499.6 nm in an auxiliary one-pump fiber OPA. By using 11 W peak power for the pump at 1560 nm, power up to 3 W was readily obtained at 1499.6 nm, which was quite sufficient for pumping the two-pump OPA. This demonstrates the possibility of using different OPAs for fulfilling different roles in a single system.

By using a fiber OPA to generate the short-wavelength pump, the resulting pump quality is similar to that of the initial pump and signal LDs, which can be quite good

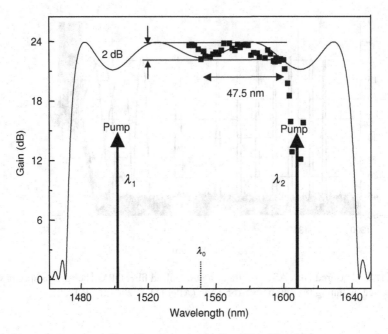

Fig. 11.4 Spectrum of broadband two-pump fiber OPA. The gain was measured only on one side of the center frequency, and the total bandwidth was calculated by assuming a symmetric spectrum. (After [7].)

(with a linewidth below 1 MHz). This is in contrast with what would be obtained with a Raman fiber laser, which typically has a much larger linewidth, of the order of 1 nm. This approach would therefore be suitable in principle for communication applications, where it is important to minimize spectral broadening.

(d) Phase-insensitive OPA with 3.7 dB NF

The NF of a pulsed fiber OPA was recently measured by a photon-counting technique [9]. This was a non-degenerate OPA, operated in a phase-insensitive mode by injecting only the pump and signal at the input of a nonlinear fiber. Launching weak signal pulses into the OPA ensured that they were very close to being in a coherent state (CS), the standard assumption for the derivation of the NF. To model a digital communication system, "0" and "1" pulses were launched. For a "0" pulse there is no light at the input, and therefore what is measured at the output corresponds to light generated within the OPA, i.e. OPA ASE; the photons are then Bose–Einstein distributed. For a "1", a finite signal is launched into the OPA and the output is a combination of amplified signal and ASE; the photons are then Laguerre distributed. The probability density functions for photon numbers at the OPA output were measured for both cases and were found to be in excellent agreement with theoretical predictions.

The OPA noise figure was calculated from

$$F_s = [1 + 2n_{sp}(G_3 - 1)]/G_s,$$

with $n_{sp} = 1.17$; n_{sp} is the excess spontaneous-emission factor. The plot asymptotically approaches a value corresponding to about 3.7 dB for large gain. This is larger than the 3 dB standard quantum limit (Section 9.3). The difference is attributed to the effect of Raman gain [10, 11]. This experiment confirms that fiber OPAs can have NFs comparable with those of EDFAs and thus are possible candidates for optical communication networks.

(e) Degenerate OPA with 1.8 dB NF

Degenerate OPAs are phase sensitive, and therefore their NFs can in principle approach the ideal limit of 0 dB. Experiments were performed with a nonlinear Sagnac interferometer to verify this [12]. The fiber was a 3-km-long HNLF with $\gamma = 15.8$ W^{-1} km^{-1} and a loss of 0.7 dB km^{-1}. Square pump and signal pulses, 160 ps long, were derived from a frequency-stabilized erbuim-doped fiber laser (EDFL). The pump pulses were then amplified to a peak power of 0.4 W by means of a low-noise EDFA. The resulting gain was 16 dB, sufficient to put the phase-sensitive amplifier (PSA) in the high-gain regime. Under these conditions, the theory for a lossy-fiber PSA predicts an NF of 0.3 dB [13]. After detection of the output pulses, the NF was measured by means of a calibrated electrical spectrum analyzer (ESA). As expected, the NF around 1 GHz was not good, because of low-frequency fluctuations introduced by guided acoustic-wave Brillouin scattering (GAWBS) [14]. However, the effect of GAWBS was negligible at higher frequencies, and an NF of 1.8 dB was measured near 16 GHz.

While this result clearly verifies the ability of fiber-based PSAs to beat the 3 dB NF limit that applies to phase-insensitive amplifiers (PIAs), the NF value achieved is still significantly above the ideal 0 dB limit. Several factors contribute to the discrepancy, including fiber loss (2 dB), pump RIN introduced by the ASE of the EDFA used for boosting the pump, and possibly a contribution due to the Raman gain, which has been found to be significant for non-degenerate OPAs (see the discussion in the previous part of the subsection). Thus it is possible that, with improvements in some of these areas, lower NFs will eventually be obtained.

11.2.2 Pulsed wavelength conversion

As described in Chapter 3, in parametric amplification one can in principle obtain 100% pump depletion, which implies a conversion efficiency of pump power into idler power of 50% (if the signal input power is included in the initial power, a somewhat lower efficiency is calculated). Maximum conversion efficiency occurs for a signal wavelength different from that corresponding to the maximum small-signal gain.

Experimental results showing high conversion efficiency by a pulsed fiber OPA were first presented in [15]. Orthogonally polarized nanosecond signal and pump pulses were injected into a 7-m-long ultralow-birefringence fiber, with $\gamma = 23.3$ W^{-1} km^{-1}. The weak linear birefringence was introduced by winding the fiber onto a drum with diameter 14.5 cm, which resulted in a beat length of 5.8 m. The signal was obtained by frequency doubling the output of an Nd–YAG laser and shifting it into a CO_2 Raman cell; its wavelength was 574 nm.

Conversion efficiencies of the order of 20% were obtained over a bandwidth of about 0.5 THz (the full width at half maximum of the conversion spectrum), i.e. about 0.55 nm. Good agreement with the results of numerical simulations was obtained. A quasi-CW model was used, assuming that the exact Jacobian elliptic solutions could be employed with instantaneous values of the parameters. Since the measured conversion efficiencies corresponded to an average over the pulse duration, it is clear that higher efficiencies were obtained at one or more instants during the pulse.

Similar results have been obtained with a single pump equally divided along the two axes of a highly birefringent fiber [16].

11.2.3 Pulsed OPOs

Since high gain and large bandwidth can easily be obtained with pulsed OPAs, the demonstration of pulsed fiber OPOs is relatively simple. One difficulty, however, is that the pump and signal pulses must be synchronized within the cavity, since the OPA gain does not exist outside the pump pulse. On the basis of this idea, a number of non-degenerate pulsed fiber OPOs have been demonstrated [17–19].

The fiber OPO of [19] was tunable over a 40 nm range by adjusting the angle of a ruled diffraction grating. The gain medium was a 105-m-long DSF with $\lambda_0 \approx 1538$ nm and $\gamma = 2.4$ W^{-1} km^{-1}. Together with a 50% coupler it formed a Sagnac loop, which acted as a mirror keeping the pump on one side and the signal and idler on the other side (see Fig. 8.1(d)). The signal round-trip transmittance of the cavity with the pump off was $T = 0.3$ (5 dB loss), indicating that oscillation could be obtained with a gain of 5 dB or higher. The pump was provided by a color-center laser with $\lambda_1 \approx 1539$ nm.

Threshold was obtained for a pump power of 1.43 W. A pump depletion of 15% was obtained for $P_1 = 2$ W, indicating fairly efficient internal conversion of pump to signal and idler. A fraction of the signal and idler powers was tapped by means of a 10/90 coupler, and the output optical spectrum was observed on an OSA. The OPO output was tunable over a 40 nm range.

More recently, a pulsed fiber OPO based on a flattened-dispersion fiber has been demonstrated [20]; it was tunable in the ranges 1415–1505 nm or 1600–1750 nm. Microstructure-fiber-based OPOs have also been demonstrated [21–23]; the OPA in [23] could be tuned over 200 nm.

If a fiber OPO were operated with the fiber and pump used for obtaining an OPA with a 400 nm gain bandwidth [1], one could make a source tunable over such a wavelength range. This type of source could be of interest for investigating ultrawideband optical communication systems.

A degenerate fiber OPO has also been reported [24]. It employed a nonlinear-fiber Sagnac interferometer as a parametric amplifier. Synchronous pumping with 3.9 ps pulses at 1544 nm yielded 0.83 ps pulses. The wide bandwidth of the fiber parametric amplifier caused the oscillator to act as a pulse compressor. The output signal pulses exhibited improved spectral symmetry and a reduced time–bandwidth product in comparison with the pump pulses.

11.3 CW devices

11.3.1 CW fiber OPAs

(a) High-gain OPAs

Obtaining high gain with continuous-wave (CW) OPAs is fairly easy in theory. All that is needed is a large value for $\Phi = \gamma P_0 L$. For example, to obtain 70 dB of gain, $\Phi \approx 8.8$ is required. Hence, for a given type of fiber and pump power, one needs only to use a sufficiently long fiber. However, large values of Φ also imply that the pump power is well above the stimulated Brillouin scattering (SBS) threshold, and that very strong SBS-suppression measures must be used.

In practice, it is found that if only frequency dithering is used, with a single RF tone, the required RF driving power for an external phase modulator is beyond the capabilities of common laboratory equipment for high values of Φ. Hence more advanced SBS-suppression techniques must be used for obtaining high CW gains. In the following we describe two approaches with which record high gains were obtained.

CW OPA with 70 dB gain

In [25] the approach taken for SBS suppression was to use two phase modulators (PMs) in series to dither the pump frequency. The first PM was driven by a 10 GHz wave, while the second was driven by the sum of four different RF tones (105, 325, 1000, and 3110 MHz). These five frequencies yielded $3^5 = 243$ new frequencies, covering a 25 GHz range and providing a theoretical stimulated Brillouin scattering (SBS) increase of $10 \log_{10}(243) = 24$ dB.

The gain medium consisted of three pieces of highly nonlinear fiber (HNLF) spliced together, with a total length of 500 m and γ value 11.4 W^{-1} km^{-1}. A maximum fiber gain of 70 dB was measured for 1.9 W of pump power in the fiber. The gain was limited by SBS on the amplified signal and by saturation due to ASE.

CW OPA with 60 dB gain

In [26], an isolator was used in combination with pump phase modulation for SBS suppression. The phase modulation was implemented by means of two cascaded phase modulators: the first was driven by a sine wave at 2.5 GHz and the second by a pseudo-random bit sequence (PRBS). The fiber was a 1-km-long HNLF, with $\gamma = 17$ W^{-1} km^{-1}. With phase modulation only, a maximum gain of about 30 dB was obtained. To improve SBS suppression the fiber was cut in half and an isolator was placed in the middle. It is well known that such an arrangement stops the light reflected by SBS, so that the SBS threshold power was then equal to that of one fiber segment, i.e. it was increased by a factor 2. This improvement can be very beneficial when used in conjunction with phase modulation, because the gain in decibels can itself be increased by nearly a factor 2: hence an OPA with 40 dB gain without isolator has the potential for providing nearly 80 dB of gain with an isolator. Measurements confirmed this: after adding the isolator, a 60 dB maximum gain was measured. The gain was actually no longer limited by SBS but by saturation of the gain by EDFA ASE.

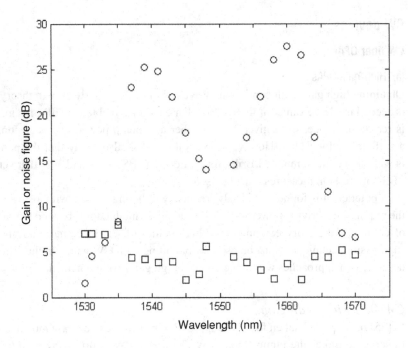

Fig. 11.5 EDFA ASE spectra, with (squares) and without (circles) the FBG.

(b) Non-degenerate OPAs with low optical noise figure

While non-degenerate (and phase-insentive) OPAs in principle have the potential for approaching the 3 dB NF quantum limit, a number of practical difficulties have to be overcome for this limit to be achieved (see Sections 9.3 and 9.6). A major difficulty is that currently OPAs operating in the telecommunication window around 1550 nm all use as pumps DFB lasers amplified by EDFAs, which generate a large amount of ASE. In a typical setup designed to generate a 1 W pump, the ASE level may be as high as -30 dB m nm^{-1}. This EDFA ASE is then combined with the signal by means of a coupler and amplified in the OPA. If the ASE has not been reduced by optical filtering prior to this combining process, it is then amplified to a very high level: for 30 dB of gain, the OPA output ASE would be 0 dB m nm^{-1}, which would lead to a very high NF. To avoid this, the pump EDFA ASE must be greatly reduced before it is combined with the signal. One can use a thin-film filter and achieve a reduction of about 30 dB, but an insertion loss of about 3 dB is incurred. By cascading two such filters one could achieve 60 dB reduction, but the resulting cumulative loss would reduce the pump power so much that it would no longer be sufficient for the OPA. If, however, one uses a fiber Bragg grating (FBG) and a circulator to inject the pump into the OPA, one can obtain a reduction in the ASE by up to 30 dB, with an insertion loss of only about 1.5 dB.

This idea was first implemented in [27]. By using an FBG, the EDFA ASE was reduced to a level essentially equal to that of vacuum fluctuations near 1560 nm, the peak of the EDFA gain; see Fig. 11.5. From the considerations of Chapter 9, it would appear that such a level of input noise is still fairly large and could lead to an NF of the order of 5 dB. Noise factor measurements over the OPA gain bandwidth gave a fairly constant NF value, with an average of 4.2 dB. These values were the lowest reported at the time and attracted attention to the field. This experiment, as well as the NF measurement of a pulsed OPA described in subsection 11.2.1(d), yielded NF values around 4 dB. Such NFs are comparable with those of commercial EDFAs and indicate that, with appropriate attention to design and manufacturing, fiber OPAs should exhibit NFs sufficient for use in optical communication systems.

It should be emphasized that the measurements just described were based on experimental values of signal and optical power and thus constituted a measurement of the *optical* noise figure (F_o). While the optical NF is a useful performance measure for fiber OPAs, the quantity that is eventually of most interest in optical communication systems is the *electrical* noise figure F_e. The latter is obtained by using a modulated electrical signal, detecting it, and comparing the values of the electrical SNR obtained with and without the device under test. The electrical noise figure F_e is a more complete way of measuring the amplifier noise figure than F_o, because the electrical noise contains some fluctuations that are not captured by ASE measurements. In particular, the signal in a fiber OPA may deteriorate because it experiences relative intensity noise (RIN) transfer from the pump, a fact that cannot be seen by ASE measurements.

Measurements of F_e have also been performed [28, 29]. It has been found that a low value of F_e, in the range 4.4–5.0 dB, can be achieved, with a gain in excess of 20 dB, for low signal input powers [28, 29]. When the pump optical SNR was 52 dB, which is known to be too low to avoid RIN transfer [28], F_e increased linearly with the signal input power, as shown by Eq. (10.17).

(c) One-pump OPA with 100 nm bandwidth

We saw in subsection 11.2.1(a) that a gain bandwidth of the order of 400 nm can be obtained with a fiber OPA if about 10 W of pump power is used. At the present time such high powers can only be obtained on a pulsed basis. However, it is possible to pass several watts through a fiber on a continuous basis and this can lead to fairly large bandwidths, particularly if a fiber with suitable dispersion properties is used.

Recently a 100 nm bandwidth was obtained with a single-pump OPA [30]; see Fig. 11.6. The fiber used was 100 m long and was similar to that reported in [31]. It had $\gamma = 25$ W^{-1} km^{-1} and $\beta^{(4)} = 1 \times 10^{-56}$ s^4 m^{-1}. The CW pump power was 4 W. The high value γ, the low value of $\beta^{(4)}$, with the high pump power combined to yield a large gain bandwidth. It was found that the gain had an average value 11.5 dB with a 1.5 dB excursion, over a 100 nm bandwidth. However, the product $\Phi = \gamma PL = 10$ leads to a theoretical maximum gain of 80 dB, much larger than that observed. From the 11 dB experimental gain value near the pump, which is theoretically equal to $1 + \Phi^2$, we infer an experimental value of $\Phi = 3$, in which case the maximum gain would have been reduced to 20 dB. That the maximum gain observed was even lower than this may be due to the fact that the spectrum was optimized for maximum bandwidth and not necessarily

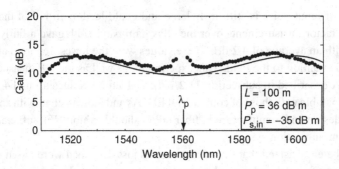

Fig. 11.6 100-nm-wide gain spectrum of a one-pump fiber OPA. (After [30], © 2007 IEEE.)

for maximum gain. One possible reason for the discrepancy in Φ is that the actual value of γ appropriate for OPA calculations is probably smaller than that reported in [31], as that value was obtained by measuring cross-phase modulation (XPM), not parametric gain or four-wave mixing (FWM). Also, there may have been some uncertainty in the pump power. In addition the presence of zero-dispersion wavelength (ZDW) variations tends to reduce the peak gain, and flatten the spectrum.

Regardless of the mediocre agreement with theory, the characteristics of this optical amplifier are impressive. In particular, the 100 nm bandwidth is very close to the 105 nm bandwidth obtained by combining a C-band and an L-band EDFA in parallel [32].

Since this amplifier is CW, it can amplify standard telecommunication signals with arbitrary formats. The fact that a high pump power is used should lead to a significant reduction in nonlinear crosstalk of wavelength-division multiplexing (WDM) signals passing through such an OPA (see Chapter 13 for a detailed discussion of this important topic). This work therefore represents a significant step forward in the development of practical fiber OPAs for optical communication systems.

11.3.2 CW wavelength converters

(a) CW wavelength converters with low optical noise figure

The NF of an OPA used as a wavelength converter (WC) was investigated in [33]. From the theory of Section 9.3, we would expect that, at high gain, the OPA would exhibit almost the same NF whether used as an amplifier or as a WC. It was observed that the WC exhibited a minimum NF as low as 6.8 dB for a conversion gain of +40 dB (this number is 3 dB higher than that quoted in [33], because the formula used for calculating the NF in that paper did not account for the fact that the ASE of a one-pump OPA is not randomly polarized). It should be noted that all other types of WCs (based on semiconductor or $\chi^{(2)}$ devices) are generally lossy, i.e. have negative conversion gains and, as a result, very high noise figures, generally in excess of 10 dB.

The fact that fiber WCs exhibit low NFs implies that they can be used in systems in much the same way as EDFAs. For example, they could in principle be cascaded many times to form long-haul systems. This represents a major improvement in performance for WCs of any kind and could have an impact on how network designers think about WCs. This could then bring about the development of networks exploiting the properties of WCs to a much greater extent than has been heretofore possible.

(b) CW wavelength converters with narrow idler spectrum

A difficulty with CW fiber OPAs is that they require strong SBS suppression, which is generally implemented by means of frequency or phase modulation (phase modulation is equivalent to frequency modulation for continuous phase variations). For a one-pump OPA, $\omega_3 + \omega_4 = 2\omega_1$ is applicable on an instantaneous basis. If we let $\omega_k = \omega_{k0} + \delta\omega_k$, where $\omega_k = \omega_{k0}$ is the mean frequency of the kth wave and $\delta\omega_k$ represents modulation about the mean, this leads to

$$2\delta\omega_1 = \delta\omega_3 + \delta\omega_4. \tag{11.1}$$

Equation (11.1) shows that if the signal frequency is fixed ($\delta\omega_3 = 0$) then the idler frequency is modulated twice as much as that of the pump ($\delta\omega_4 = 2\delta\omega_1$). This leads to idler spectrum broadening, which could be unacceptable in optical communication systems. One way to avoid this is to modulate the signal frequency with twice the amplitude of the pump ($2\delta\omega_1 = \delta\omega_3$); then the idler does not experience any frequency modulation [34].

Another possibility is to use binary phase modulation for the pump, by arranging for its phase to alternate between $\pi/2$ and $-\pi/2$. In the absence of signal phase modulation, the idler's phase then alternates between π and $-\pi$; its total phase excursion is 2π, which is equivalent to 0 radians or no phase modulation. This has been experimentally demonstrated for both one-pump and two-pump OPAs [35]. A problem with this approach is that it is difficult to obtain very abrupt transitions at high modulation rates. During finite-time transitions the ideal phase condition does not hold and idler frequency modulation results. This has been found to be a substantial problem in practice [36].

For two-pump OPAs, Eq. (11.1) is replaced by

$$\delta\omega_1 + \delta\omega_2 = \delta\omega_3 + \delta\omega_4. \tag{11.2}$$

To avoid idler frequency modulation, we must have $\delta\omega_1 + \delta\omega_2 = \delta\omega_3$, which offers several possibilities. In particular, we can have $\delta\omega_3 = 0$ and $\delta\omega_1 = -\delta\omega_2$, i.e. no signal modulation and equal but opposite modulation for the two pumps [37, 38]. This approach is attractive because the signal is left intact, which could be useful if it were desired to use the OPA as an amplifier as well as a WC. Appropriate modulation of the pumps can be obtained by means of electro-optical phase modulators or by direct modulation of the pump laser diodes; efficient reduction in idler spectrum broadening has been demonstrated by both of these techniques [38]. Figure 11.7 shows the considerable reduction in idler spectrum broadening that can be obtained: the idler spectral width was reduced by almost two orders of magnitude in this fashion [39].

Pumps with opposite modulation can also be derived from a single modulated pump by means of FWM; this automatically generates the correct amplitudes and phases for

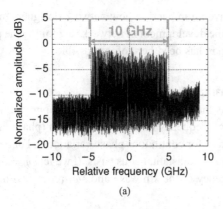 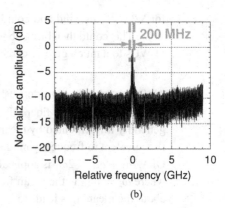

Fig. 11.7 Idler radiofrequency spectrum for pump FMs that are: (a) not controlled; (b) equal in amplitude and 180° out of phase. (After [38], © 2002 IEEE.)

the two modulations, which might be difficult to accomplish by electrical means. This method has been experimentally demonstrated and shown to be effective for reducing idler broadening [39].

(c) CW wavelength converters with strong pump depletion and high pump conversion efficiency

For WCs it is desirable to obtain a large conversion efficiency of pump power into idler power or, equivalently, a strong depletion of pump power. This has been demonstrated for both one- and two-pump fiber OPAs.

One-pump OPA

This experiment was performed with an 11-km-long DSF, with $\alpha = 0.26$ dB km^{-1}, $\gamma = 2.4$ W^{-1} km^{-1}, and a nominal $\lambda_0 = 1560$ nm [40]. The pump wavelength was set at $\lambda_1 = 1560.9$ nm to optimize the gain spectrum; its power was set at 19.54 dB m at the fiber output. The signal wavelength was set at $\lambda_3 = 1557.6$ nm to provide a $\Delta\beta$ value close to that required by theory. The signal power (measured at the fiber output) was initially set at more than 10 dB m; it was then reduced by means of a variable attenuator and the pump output power was measured during the process.

Figure 11.8 shows the experimental results, as well as a numerical simulation corresponding to the experimental conditions. It can be seen that they are in excellent agreement. At best 92% pump depletion was observed. Some spurious FWM terms were observed, but they were typically at least 10 dB below the signal and idler levels. Hence most of the power extracted from the pump was indeed converted to the signal and idler. Thus in this case large pump depletion is essentially synonymous with high conversion efficiency.

Two-pump OPA

This experiment was performed with an 11-km-long DSF (a Corning LS fiber) with $\alpha = 0.2$ dB km^1, $\gamma = 2.1$ W^{-1} km^{-1}, $D_\lambda = 0.07$ ps nm^{-2} km^{-1}, and an average

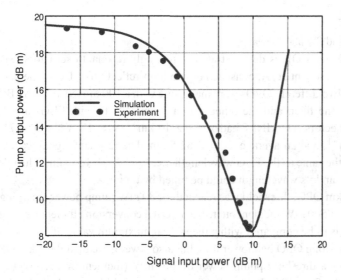

Fig. 11.8 Pump output power versus signal input power, showing a maximum pump depletion of 92%. (After [40].)

$\lambda_0 = 1568.6$ nm [41]. The pump wavelengths were set at $\lambda_1 = 1556.0$ nm and $\lambda_2 = 1581.2$ nm. The signal wavelength was set at $\lambda_3 = 1557.6$ nm. All three input waves had the same state of polarization (SOP). Here the performance measure was taken to be the conversion efficiency

$$\eta = \frac{P_3(L) + P_4(L) - P_3(0)\,e^{-\alpha L}}{P_1(L) + P_2(L) + P_3(L) + P_4(L) - P_3(0)\,e^{-\alpha L}}. \tag{11.3}$$

This definition coincides with that for pump depletion used in the preceding experiment when spurious FWM and ASE are negligible. In a typical output spectrum, spurious FWM terms were found to be at least 30 dB down from the signal and idler levels, which indicates that their effect on the results should be minimal. Values of η in excess of 83% were achieved, the highest recorded value being 88%.

An important feature of two-pump OPAs is that they can be made to exhibit a nearly constant $\Delta\beta$ over a wider range than one-pump OPAs (see Section 5.3). As a result, one can maintain the phase-matching conditions required for high conversion efficiency over a wide wavelength range, which could be beneficial in some applications. The authors report flat efficiency, within a ±3% range, over a 25 nm wavelength range (the signal power was adjusted at each wavelength to maximize η).

These two experiments confirm that high pump conversion efficiencies can indeed be obtained in practice, with values close to those predicted by theory.

11.3.3 CW fiber OPOs

(a) HNLF-based CW fiber OPOs

The first CW fiber OPOs demonstrated were singly resonant oscillators (SROs) [42]. The cavities, resonant at ω_3, consisted of two high-reflectivity FBGs. The gain medium was HNLF. Two different OPOs were made, one with a 100 m HNLF, with FBGs written directly into the fiber, and the other with a 1 km HNLF, with FBGs attached at the ends by connectors. The HNLF had $\alpha = 0.2$ dB km^{-1}, $\gamma = 17$ W^{-1} km^{-1}, $D_\lambda = 0.019$ ps nm^{-2} km^{-1}, and an average $\lambda_0 = 1562.5$ nm. The resonant signal wavelength was fixed, being the same as the FBG wavelength, i.e. $\lambda_g = \lambda_3 = 1560$ nm. The idler was the OPO output, and its wavelength could be tuned by tuning λ_1.

For the 1 km OPO, threshold was reached for 80 mW pump power emerging from the OPO, or about 500 mW incident on it. The internal conversion efficiency reached almost 30%, which is to be compared with a theoretical maximum value of 50%.

The 100-m-long OPO had lower loss, but also lower gain, than the 1 km OPO. It was found to have a threshold pump power of 240 mW (incident on the OPO), i.e. about half that of the first OPO. It was also found that it could be tuned over a wider range: oscillation was obtained over an 80 nm range. The output powers were low because of the short fiber length, which was insufficient for high gain and substantial pump depletion to be achieved.

More recently a CW OPO using a unidirectional ring cavity has been reported [30]. Up to 1.8 W of pump power was used; it was injected into the ring by means of a 50/50 coupler. A tunable filter with 2 nm bandwidth was placed in the cavity; it could be tuned from 1510 nm to 1600 nm to select the signal wavelength. The pump and the idler did not pass through the filter and were removed after each pass. The signal was extracted by means of a 10/90 coupler. The cavity loss was 7 dB and the oscillation threshold was reached for 1 W of pump power. Output signal powers between 1 and 20 mW were obtained over the tuning range.

(b) Holey-fiber-based CW fiber OPOs

Continuous-wave fiber OPOs made with microstructured fibers have been reported [4, 43]. In [43] the source reached threshold for a pump power of 1.28 W and saturated for pump powers in excess of ~ 1.6 W. In [4], over 30 nm of sideband tunability was demonstrated using a 300 mW pump.

11.3.4 CW wavelength exchange

(a) Demonstration of efficient wavelength exchange

In subsection 3.5.5 we described a variation in parametric amplification obtained by having two pumps located asymmetrically with respect to the center frequency ω_c in a non-degenerate FWM interaction. We showed that under certain circumstances this arrangement is predicted to perform an essentially complete exchange of photons between two signals symmetric to the two pumps with respect to ω_c.

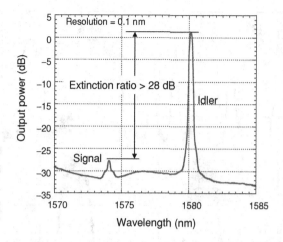

Fig. 11.9 Output spectrum of a WE. At the input the signal was at $\lambda_3 = 1573.44$ nm and there was no idler at $\lambda_4 = 1579.88$ nm. At the output, most light appeared at $\lambda_4 = 1579.88$ nm and there was very little light left at $\lambda_3 = 1573.44$ nm. (After [44], © 2002 IEEE.)

Experiments have been performed to verify this property [44]. An HNL-DSF had a nominal λ_0 value of 1562.5 nm, $dD/d\lambda = 0.03$ ps nm^{-2} km^{-1}, and $\gamma = 17$ W^{-1} km^{-1}. Two tunable laser sources set at $\lambda_1 = 1544.94$ nm and $\lambda_2 = 1551.06$ nm provided the pumps, which were amplified to a maximum output power of 27 dB m. Two other tunable laser sources set at $\lambda_3 = 1573.44$ nm and $\lambda_4 = 1579.88$ nm provided the first and second signal, respectively. The first signal was intensity-modulated by a 2.5 Gb/s 2^7-1 PRBS, while the second signal was intensity-modulated by a 10 Gb/s 2^7-1 PRBS. The input power of each signal to the HNL-DSF was $+3$ dB m. The SOP of each pump and signal was adjusted by polarization controllers.

Partial wavelength exchange with only one input signal was first investigated in order to verify the wavelength-exchange transfer characteristics. Figure 11.9 shows the output spectrum in the vicinity of the signal and the idler: an extinction ratio of more than 28 dB was obtained, confirming the ability of wavelength exchange to switch photons from one wavelength to another with nearly 100% efficiency.

Full wavelength exchange was then demonstrated by feeding both signals into the HNL-DSF. However, since both SOPs could not be optimized simultaneously, the extinction ratios of the first and the second signal were only 20 dB and 18.5 dB, respectively. Figure 11.10 shows the output waveforms of the monitor receiver at λ_3 (1573.44 nm) and λ_4 (1579.88 nm) respectively. The bottom trace in each panel represents the output with only one pump, i.e. without wavelength exchange, while the top trace corresponds to the output with two pumps, i.e. with wavelength exchange. We see that the original 2.5 Gb/s signal was successfully replaced by the 10 Gb/s signal at λ_3, and vice versa at λ_4.

It is important to note that because the exchange is not complete (approximately 1% of the power of each signal is left behind by the exchange), there is crosstalk with the output signal. Furthermore, since the exchanged and residual signals have virtually the same frequencies, the crosstalk is quasi-coherent and can therefore degrade system

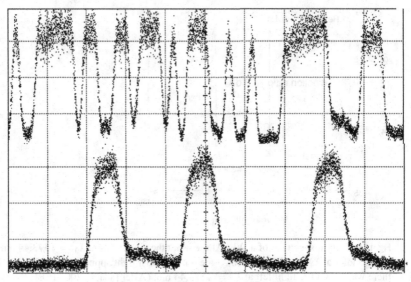

(a) Output waveform at λ_3 = 1573.44 nm

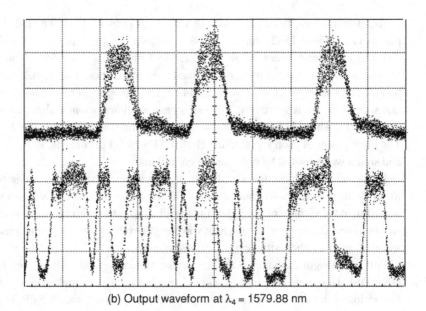

(b) Output waveform at λ_4 = 1579.88 nm

Fig. 11.10 Demonstration of full wavelength exchange. In each panel the top (bottom) trace corresponds to the pump off (on). (After Ref. [44], © 2002 IEEE.)

performance. This was investigated by measuring bit-error rates (BERs) for the output signals with the pumps on and off. It was found that wavelength exchange introduced a power penalty of the order of 3 dB for a wavelength spacing of 0.01 nm between signal and crosstalk. In order to reduce this penalty, the crosstalk needs to be reduced further.

(b) Demonstration of low-NF wavelength exchange

In Section 9.4 we discussed the quantum noise properties of wavelength exchange. We showed that since a wavelength exchanger (WE) does not exhibit gain, there is no ASE associated with it. As a result, its noise properties are the same as those of a lossless beamsplitter. In particular, when the conversion efficiency is 100% the NF is 0 dB (this does not include possible noise contributions due to the Raman gain or to coupling and fiber losses).

Experiments have been performed to verify that low NFs can be achieved with a WE [45]. The setup was similar to that for the preceding experiments. An external conversion efficiency of 90% was measured. Taking coupling losses into account, a 99% internal conversion efficiency was inferred.

A key point is that the residual noise from the pumps in the signal wavelength region (1570–1580 nm) was well below −55 dB m, which corresponds to the vacuum noise level (i.e. $2h\nu\Delta\nu$ for 0.1 nm optical spectrum analyzer (OSA) resolution bandwidth). This is shown in Fig. 11.11: when the pump is on, the optical noise that it contributes between the signal and idler frequencies is about 10 dB below the vacuum level. This noise is due to EDFA amplified spontaneous emission (ASE), attenuated by optical filtering. The fact that the noise density is constant and does not show any increase at the signal and idler wavelengths confirms that the WE does not add any ASE at these wavelengths; this is the basis for the very low NF of the device.

Under these circumstances, and assuming that the signal was in a coherent state at both input and output, a 2.1% external NF was calculated. Taking coupling losses into account, an internal NF of 0.06 dB was inferred. This is indeed very close to the predicted 0 dB NF for ideal wavelength exchange. It should be possible to improve the external NF further by reducing the insertion loss of the WDM coupler and the fiber splicing loss, thereby approaching the sub-decibel value of the internal NF.

The NF calculated above is optical and does not include noise sources that could affect measurements of electrical NF. In particular, the electrical NF would be affected by pump RIN as in the case of an OPA. However, there are reasons why the impact of pump RIN on wavelength-exchange electrical NF should be smaller than for an OPA. These are: (i) there is no large gain, only unit conversion efficiency; (ii) the magnification factor ρ contains the derivative of the gain with respect to the pump power, which should vanish at maximum conversion efficiency. Hence the electrical NF may also turn out to be very low.

(c) Wavelength exchange with 180 nm shift

In recent work [46], wavelength conversion by wavelength exchange with a shift as large as 180 nm was demonstrated. One pump was near 1620 nm and was obtained from an amplified tunable L-band laser. The other pump was obtained from a tunable Raman

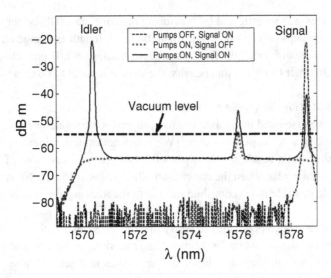

Fig. 11.11 Output optical spectrum of an HNLF for several conditions. (After [45], © 2006 SEE.)

laser (RL) and was in the 1420–1450 nm range. These wavelength assignments, together with a zero-dispersion wavelength (ZDW) of 1487.5 nm, permitted the conversion of a signal in the 1515–1550 nm range to the upper range of the O-band (1260–1360 nm), i.e. with a wavelength shift of the order of 180 nm, significantly larger than had been obtained by wavelength exchange previously.

The fiber used was a 1.35-km-long large-effective-area fiber (LEAF), which has a high SBS threshold but a correspondingly low value of γ. This fiber has the advantage of having low water-peak attenuation, which permits the use of wavelengths in the vicinity of that peak. The pump power was not given but the conversion efficiency achieved was only 1%, indicating that the pump power was probably low.

Another reason cited for the low conversion efficiency is the low coherence of the RL. Raman lasers have a linewidth, determined by the FBGs making up the Fabry–Pérot cavity, that is typically of the order of 1 nm. In addition to causing a low conversion efficiency, the large pump linewidth is transferred to the converted signal (the idler). This is undesirable for communication applications, because the idler then has amplitude and/or frequency fluctuations that corrupt the desired information modulated onto the carrier. It is thus apparent that, for this type of large-shift wavelength conversion to be attractive for communication applications, progress will have to be made with high-power narrow-linewidth pumps.

11.3.5 CW phase-sensitive OPA with two pumps

The phase-sensitive amplification of communication signals by a single pump having the same frequency as the signal (see subsection 11.2.1(e)). It requires the use of an interferometric arrangement to combine and then separate the signal and the pump.

One way to avoid the above frequency degeneracy is to use a two-pump approach, wherein the signal frequency is placed exactly halfway between the pump frequencies. In this manner the signal and idler are exactly the same, which leads to phase-sensitive amplification of the signal without requiring a separate idler. The advantage of this approach is that the signal can be combined with the pumps by means of WDM couplers, without resorting to an interferometric arrangement.

While this may sound much simpler than the all-degenerate approach, one should realize that phase-sensitive operation can be achieved only if all the waves involved are in a well-defined frequency and phase relationship, which generally requires some sort of careful stabilization (much more so than with a phase-insensitive amplifier (PIA)).

An experiment has been performed to demonstrate this approach [47]. In order to obtain the proper frequencies and phases, the three frequencies were derived from a single laser source. Its output was phase-modulated at 40 GHz by a lithium niobate modulator, and an FBG selected modulation sidebands located at $+80$ GHz and -80 GHz from the carrier; these were then amplified to form the pump while the center frequency served as the signal. Next, a low-frequency phase modulator was used to modulate the signal phase. Phase-sensitive operation was verified by looking at the signal output variations, whose power was a function of the signal phase, as expected. The gain varied from -4.5 dB to 5 dB as the phase was varied.

It is also interesting to note that pump SBS was suppressed in a novel manner: the 40 GHz RF was frequency-modulated by a 500 kHz sawtooth. This broadened the spectrum of each pump with opposite FM, in such a way that the signal output spectrum was not affected. This is an interesting method for simultaneously broadening the spectra of two pumps spaced a few nanometers apart.

References

1. "Wide-band tuning of the gain spectra of one-pump fiber optical parametric amplifiers," Marhic, M. E., Wong, K. K. Y., Kazovsky, L. G. *IEEE J. Select. Topics in Quantum Electron.*; 2004; vol. 10, pp. 1133–41.
2. "Scalar modulation instability in the normal dispersion regime by use of a photonic crystal fiber," Harvey, J. D., Leonhardt, R., Coen, S., Wong, G. K. L., Knight, J. C., Wadsworth, W. J., Russell, P. St J. *Opt. Lett.*; 2003; vol. 28, pp. 2225–7.
3. "Widely tunable optical parametric generation in a photonic crystal fiber," Chen, A. Y. H., Wong, G. K. L., Murdoch, S. G., Leonhardt, R., Harvey, J. D., Knight, J. C., Wadsworth, W. J., Russell, P. St J. *Opt. Lett.*; 2005; vol. 30, pp. 762–4.
4. "Continuous-wave tunable optical parametric generation in a photonic-crystal fiber," Wong, G. K. L., Chen, A. Y. H., Murdoch, S. G., Leonhardt, R., Harvey, J. D., Joly, N. Y., Knight, J. C., Wadsworth, W. J., Russell, P. St J. *J. Opt. Soc. Amer*; 2005; vol. 22, pp. 2505–11.
5. "Experimental observation of new modulation instability spectral window induced by fourth-order dispersion in a normally dispersive single-mode optical fiber," Pitois, S., Millot, G. *Opt. Comm.*; 2003; vol. 226, pp. 415–22.
6. "Widely tunable sub-30-fs pulses from a compact erbium-doped fiber source," Tauser, F., Adler, F., Leitenstorfer, A. *Opt. Lett.*; 2004; vol. 29, pp. 516–8.

7. "Nearly 100 nm bandwidth of flat gain with a double-pumped fiber optic parametric amplifier," Marconi, J. D., Chavez Boggio, J. M., Fragnito, H. L. In *Proc. Optical Fiber Communication Conf.*, Anaheim, CA, March 2007; paper OWB1.
8. "Impact of dispersion fluctuations on dual-pump fiber-optic parametric amplifiers," Yaman, E., Lin, Q., Radic, S., Agrawal, G. P. *IEEE Photon. Technol. Lett.*; 2004; vol. 16, pp. 1292–4.
9. "Measurement of the photon statistics and the noise figure of a fiber-optic parametric amplifier," Voss, P. L., Tang, R., Kumar, P. *Opt. Lett.*; 2003; vol. 28, pp. 549–51. "Measurement of the photon statistics and the noise figure of a fiber-optic parametric amplifier: erratum," Voss, P. L., Tang, R., Kumar, P. *Opt. Lett.*; 2004; vol. 29, pp. 2815.
10. "Raman-noise-induced noise-figure limit for $\chi^{(3)}$ parametric amplifiers," Voss, P. L., Kumar, P. *Opt. Lett.*; 2004; vol. 29, pp. 445–7.
11. "Noise-figure limit of fiber-optical parametric amplifiers and wavelength converters: experimental investigation," Tang, R., Voss, P. L., Lasri, J., Devgan, P., Kumar, P. *Opt. Lett.*; 2004; vol. 29, pp. 2372–4.
12. "Low-noise amplification under the 3 dB noise figure in high-gain phase-sensitive fibre amplifier," Imajuku, W., Takada, A., Yamabayashi, Y. *Electron. Lett.*; 1999; vol. 35, pp. 1954–5.
13. "Noise figure of phase-sensitive parametric amplifier using a Mach–Zehnder interferometer with lossy Kerr media and noisy pump," Imajuku, W., Takada, A. *IEEE J. Quantum Electron.*; 2003; vol. 39; pp. 799–812.
14. "Generation and detection of squeezed states of light by nondegenerate four-wave mixing in an optical fiber," Levenson, M. D., Shelby, R. M., Perlmutter, S. H. *Phys. Rev. A*; 1985; vol. 32, pp. 1550–62.
15. "Failure of phase-matching concept in large-signal parametric frequency conversion," Trillo, S., Millot, G., Seve, E., Wabnitz, S. *Appl. Phys. Lett.*; 1998; vol. 72, pp. 150–2.
16. "Strong four-photon conversion regime of cross-phase-modulation-induced modulational instability," Seve, E., Millot, G., Trillo, S. *Phys. Rev. E*; 2000; vol. 61, pp. 3139–50.
17. "Four-photon fiber laser," Margulis, W., Osterberg, U. *Opt. Lett.*; 1987; vol. 12; pp. 519–21.
18. "Parametric soliton laser," Suzuki, K., Nakazawa, M., Haus, H. A. *Opt. Lett.*; 1989; vol. 14; pp. 320–2.
19. "Tunable fiber-optic parametric oscillator," Serkland, D. K., Kumar, P. *Opt. Lett.*; 1999; vol. 24, pp. 92–4.
20. "Dispersion-flattened-fibre optical parametric oscillator for wideband wavelength-tunable ps pulse generation," Saito, S., Kishi, M., Tsuchiya, M. *Electron. Lett.*; 2003; vol. 39, pp. 86–8.
21. "A microstructure-fiber-based 10-GHz synchronized tunable optical parametric oscillator in the 1550-nm regime," Lasri, J., Devgan, P., Tang, R., Sharping, J. E., Kumar, P. *IEEE Photon. Technol. Lett.*; 2001; vol. 15, pp. 1058–60.
22. "Optical parametric oscillator based on four-wave mixing in microstructure fiber," Sharping, J. E., Fiorentino, M., Kumar, P., Windeler, R. S. *Opt. Lett.*; 2002; vol. 27, pp. 1675–7.
23. "Broadly tunable femtosecond parametric oscillator using a photonic crystal fiber," Deng, Y., Lin, Q., Lu, F., Agrawal, G. P., Knox, W. H. *Opt. Lett.*; 2005; vol. 30, pp. 1234–6.
24. "Pulsed degenerate optical parametric oscillator based on a nonlinear-fiber Sagnac interferometer," Serkland, D. K., Bartolini, G. D., Agarwal, A., Kumar, P., Kath, W. L. *Opt. Lett.*; 1998; vol. 23, pp. 795–7.
25. "Fiber-optical parametric amplifier with 70-dB gain," Torounidis, T., Andrekson, P. A., Olsson, B. A. *IEEE Photon. Technol. Lett.*; 2006; vol. 18, pp. 1194–6.

26. "Continuous-wave fiber OPA with 60 dB gain using a novel two-segment design," Wong, K. K. Y., Shimizu, K., Uesaka, K., Kalogerakis, G., Marhic, M. E., Kazovsky, L. G. *IEEE Photon. Technol. Lett.*; 2003; vol. 15, pp. 1707–9.
27. "Low-noise-figure optical parametric amplifier with a continuous-wave frequency-modulated pump," Blows, J. L., French, S. E. *Opt. Lett.*; 2002; vol. 27, pp. 491–3.
28. "Impact of pump OSNR on noise figure for fiber-optical parametric amplifiers," Durecu-Legrand, A., Simonneau, C., Bayart, D., Mussot, A., Sylvestre, T., Lantz, E., Maillotte, H. *IEEE Photon. Technol. Lett.*; 2005; vol. 17, pp. 1178–80.
29. "Gain and wavelength dependence of the noise-figure in fiber optical parametric amplification," Kylemark, P., Karlsson, M., Andrekson, P. A. *IEEE Photon. Technol. Lett.*; 2006; vol. 18, pp. 1255–7.
30. "Broadband single-pumped fiber-optic parametric amplifiers," Torounidis, T., Andrekson, P. *IEEE Photon. Technol. Lett.*; 2007; vol. 19, in press.
31. "Broadband wavelength conversion over 193-nm by HNL-DSF improving higher-order dispersion performance," Hirano, M., Nakanishi, T., Okuno, T., Onishi, M. In *Proc. 31st European Conf. on Optical Communication*, September 2005, Glasgow, vol. 6, pp. 43–4.
32. "A 105-nm ultrawide-band gain-flattened amplifier combining C- and L-band dual-core EDFAs in a parallel configuration," Yu, Y. B., Chu, P. L., Alphones, A., Shum, P. *IEEE Photon. Technol. Lett*; 2004; vol. 16, pp. 1640–2.
33. "Continuous-wave fiber optical parametric wavelength converter with +40-dB conversion efficiency and 3.8-dB noise figure," Wong, K. K. Y., Shimizu, K., Marhic, M. E., Uesaka, K., Kalogerakis, G. and Kazovsky, L. G. 2003; *Opt. Lett.*, 2003; vol. 28, pp. 692–4.
34. "Cancellation of spectral spread in highly-efficient optical fibre wavelength converters," Yamashita, S., Torii, K. *IEEE Electron. Lett.*; 2000; vol. 36, pp. 1997–8.
35. "Suppression of idler broadening in highly-efficient fiber four-wave mixing by binary-phase-shift-keying modulation of pump wave," Tanemura, Lim, H. C., Kikuchi, K. *IEEE Photon. Technol. Lett.*; 2001; vol. 13, pp. 1328–30.
36. "System impact of pump phase modulation for fiber optical parametric amplifiers," Legrand, A., Lanne, S., Simonneau, C., Bayart, D., Mussot, A., Sylvestre, T., Lantz, R., Maillotte, H. In *Proc. Optical Fiber Communication Conf.*, Los Angeles CA, February 2004; paper TuK2.
37. "Demonstration of two-pump fiber optical parametric amplification," Yang, F. S., Ho, M. C., Marhic, M. E., Kazovsky, L. G. *Electron. Lett.*; 1997; vol. 33, pp. 1812–3.
38. "Narrow linewidth idler generation in fiber four-wave mixing and parametric amplification by dithering two pumps in opposition of phase," Ho, M. C., Marhic, M. E., Wong, K. K. Y., Kazovsky, L. G. *J. Lightwave Technol.*; 2002; vol. 20, pp. 469–76.
39. "Phase-conjugate pump dithering for high-quality idler generation in a fiber optical parametric amplifier," Wong, K. K. Y., Marhic, M. E., Kazovsky, L. G. *IEEE Photon. Technol. Lett.*; 2003; vol., pp. 33–5.
40. "92% pump depletion in a continuous-wave one-pump fiber optical parametric amplifier," Marhic, M. E., Wong, K. K. Y., Ho, M. C., Kazovsky, L. G. *Opt. Lett.*; 2001; vol. 26, pp. 620–2.
41. "Broad-band 88% efficient two-pump fiber optical parametric amplifier," Chavez Boggio, J. M., Dainese, P., Karlsson, F., Fragnito, H. L. *IEEE Photon. Technol. Lett.*; 2003; vol. 15; pp. 1528–30.
42. "Continuous-wave fiber optical parametric oscillator," Marhic, M. E., Wong, K. K. Y., Kazovsky, L. G., Tsai, T. E. *Opt. Lett.*, 2002; vol. 27, pp. 1439–41.

43. "Continuous-wave, totally fiber integrated optical parametric oscillator using holey fiber," de Matos, C. J. S., Taylor, J. R., Hansen, K. P. *Opt. Lett.*; 2004; vol. 29, pp. 983–5.
44. "Wavelength exchange in a highly-nonlinear dispersion-shifted fiber: theory and experiments," Uesaka, K., Wong, K. K. Y., Marhic, M. E., Kazovsky, L. G. *IEEE J. Selected Topics in Quantum Electron.*; 2002; vol. 8, pp. 560–8.
45. "Low noise figure efficient wavelength exchange in an optical fiber," Kalogerakis, G., Marhic, M. E., Kazovsky, L. G. In *Proc. European Conf. on Optical Communication*, Cannes, September 2006; vol. 4, paper Th1.3.1, pp. 13–4.
46. "Experimental demonstration of a 180 nm wavelength conversion based on a potentially noise-free Bragg scattering process," Méchin, D., Harvey, J. D., McKinstrie, C. J. In *Proc. European Conf. on Optical Communication*, Cannes, September 2006; vol. 6, paper Tu1.1.4, pp. 7–8.
47. "Phase-sensitive amplifier based on two-pump four-wave mixing in an optical fiber," Takano, K., Tanemura, T., Kikuchi, K. In *Proc. European Conf. on Optical Communication*, Cannes, September 2006; vol. 4, paper Tu1.3.6, pp. 23–4.

12 Potential applications of fiber OPAs and OPOs

12.1 Introduction

The potential for fiber OPAs and OPOs is based on some characteristic features that are not present in other types of optical amplifiers and oscillators. Specifically, the features that can be exploited are: (i) the availability of gain at essentially arbitrary wavelengths, limited only by the availability of the necessary pumps and of fibers with suitable nonlinearity, loss, and dispersion; (ii) the availability of an idler; (iii) the nearly instantaneous response of the gain to pump power variations; (iv) the all-fiber structure, which is capable of withstanding high powers.

In this chapter we will discuss the main applications that can be envisioned and describe the experiments that have been performed in some areas. For OPAs, the main areas of applications that have been envisioned so far are in optical communication and high-power wavelength conversion. We now discuss each of these areas in some depth.

12.2 OPAs in optical communication

This is the most advanced area, in part because of the availability of convenient fibers and pumps, developed for conventional communication systems. These have been exploited in recent years to demonstrate several applications of OPAs, naturally at wavelengths of interest for optical communication.

12.2.1 Amplification

The realization that fiber OPAs could exhibit gain bandwidths as large as or larger than those of erbium-doped fiber amplifiers (EDFAs), by the use of commercially available fibers and the watt-level pump powers that are now considered reasonable, sparked interest in their possible use in communication systems. The potential for obtaining gain in any band within the 1200–1700 nm transparency window of silica fibers is attractive for providing gain in regions not well served by EDFAs, such as between 1300 and 1500 nm or beyond 1600 nm. The other attractive feature is that it is in principle possible for a single CW fiber OPA to cover a very wide band, perhaps in excess of 200 nm, if fiber with suitable dispersion properties can be obtained.

In this subsection we discuss modifications of the basic OPA architecture.

(a) Polarization-insensitive OPAs

Fiber OPAs face some limitations that need to be addressed before they can become practical. A severe difficulty is that in their simplest forms, described in Chapter 3, fiber OPAs exhibit gains that are a function of the SOP of the incident signal. This is undesirable in optical communication systems as, over their length, transmission fibers exhibit a random birefringence that also varies in time: as a result, the input SOPs of the optical carriers drift randomly in time and would not remain aligned with the pump(s) of a fiber OPA. The result would be that the amplified signal would experience strong random intensity variations, which would be unacceptable in a communication system.

Fiber OPAs are not alone in exhibiting polarization sensitivity: SOAs and Raman amplifiers also exhibit it to a certain extent, and so do wavelength conversion schemes based on four-wave mixing (FWM) in fibers. Similarly, coherent optical receivers are basically sensitive to the signal SOP. A variety of means have been used to overcome polarization sensitivity in these devices, and variations on these have been investigated for fiber OPAs. Good results have been obtained with two methods, which we describe below.

Polarization diversity

The basic idea is similar to one employed first for FWM [1, 2]. The incident signal SOP is broken up into a horizontal and a vertical component, and these components are amplified separately in such a way that they experience the same gain. Thus, when they are recombined the total signal gain is always the same, regardless of the ratio of the polarization components, i.e. of the input SOP. This can in principle be applied to both one-pump and two-pump OPAs.

This was first demonstrated for a one-pump OPA [3–5]. The pump was linearly polarized at 45° with respect to the vertical and was incident on a polarizing beamsplitter (PBS), which split it into horizontal and vertical components with equal power. These two pump components were coupled into the opposite ends of a highly nonlinear fiber (HNLF), which constituted the OPA. A polarization controller in the loop was adjusted in such a way that the counter-propagating pump SOPs were orthogonal at the HNLF input, and hence everywhere throughout the fiber. The PBS also decomposed the input signal into its horizontal and vertical components and sent them toward the HNLF together with the corresponding pump components. The HNLF then acted as an OPA in each direction, simultaneously amplifying the two signal components with the same gain since the two pump components had equal power. The waves were then recombined by the PBS. The SOPs were such that all the waves emerged from the fourth port of the PBS, and nothing was coupled back into the input fiber.

An obvious trade-off associated with this arrangement is that if the total pump power is P_0 then only $P_0/2$ is available for amplification in each direction. As a result, both the gain bandwidth and the maximum gain are reduced in comparison with those of a unidirectional OPA having all waves in the same linear SOP. The maximum gain can in principle be made the same by doubling the fiber length, but the bandwidth remains unchanged.

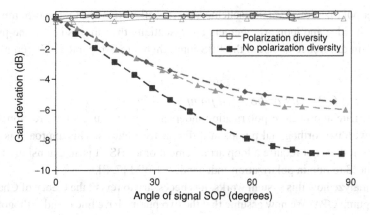

Fig. 12.1 One-pump OPA gain versus angle of the linear SOP of the input signal. (After [4], © 2002 IEEE.) Open symbols, polarization diversity; closed symbols, no polarization diversity. Squares, $\lambda_p = 1571.54$ nm; triangles, $\lambda_p = 1569.13$ nm; diamonds, $\lambda_p = 1567.50$ nm.

By suitable adjustment of the SOPs, it is possible to obtain good performance with this arrangement. This is illustrated in Fig. 12.1, where the signal gain is plotted versus the angle of a linearly polarized signal; the angle is changed by rotating a half-wave plate. As a reference, the gain for a conventional unidirectional OPA is also shown; it can be seen that there is a very strong dependence on polarization angle, as the gain drops dramatically when the signal is perpendicular to the pump. By contrast, with the bidirectional setup the gain remains virtually constant over the whole range of angles. This good performance was obtained for several signal wavelengths, which indicates that this approach is suitable for wideband fiber OPAs.

The fact that this arrangement relies on the bidirectional propagation of light in the loop raises some concerns that are not present in conventional unidirectional OPAs. One concern is that waves traveling in one direction in the loop experience cross-phase modulation (XPM) or self-phase modulation (SPM) due to the pump traveling in the other direction. We discussed a similar situation occurring in linear-cavity OPOs in subsection 8.2.2. There we showed that in practice these contributions should all be equal XPM contributions when pump dithering is used. As a result there should be no modification of the gain or gain spectrum in comparison with a unidirectional OPA. Other possible concerns are related to reflections. Any reflection of light traveling in one direction will generate waves that will be superimposed on those traveling in the opposite direction, potentially leading to performance degradation. Reflections may be due to imperfections in the optical components, i.e. the PBS, splices, connectors, etc. Another source is Rayleigh scattering, which is inevitable and increases with fiber length. Also, stimulated Brillouin scattering (SBS) may occur even though pump-frequency modulation is used to suppress it; suppression may not be complete for a moderate level of modulation. All these effects will in general co-exist in such a system, and their relative influences will have to be quantified and kept at a low level before these systems can be used in practice.

This polarization-diversity scheme was recently extended to a two-pump OPA [6]. The basic idea and loop arrangement are essentially the same as in the one-pump case. The main difference is that at the PBS input there are two parallel LP pumps, inclined at 45°.

Two-pump OPA with orthogonal pumps

An alternate approach to polarization independence is to use a unidirectional arrangement, with two orthogonal pumps at different frequencies. This approach is appealing because it does not require a loop arrangement or a PBS. It is an extension of a method introduced to obtain polarization-independent FWM [7, 8].

To analyze how this system works, it is necessary to revisit the theory of Chapter 4 for a two-pump OPA. We now assume that the two pumps have linear and orthogonal SOPs, aligned with $\hat{x}$ and $\hat{y}$. An incident signal with an arbitrary SOP can be decomposed into $\hat{x}$ and $\hat{y}$ components, and each component can be studied independently. The $\hat{x}$ ($\hat{y}$) signal generates an idler polarized along $\hat{y}$ ($\hat{x}$). The main modification of the basic propagation equations is that the coefficient of the FWM term needs to be changed from $\chi^{(3)}_{xxxx}$ to either $\chi^{(3)}_{xyxy} = \chi^{(3)}_{xxxx}/3$ or $\chi^{(3)}_{yyxx} = \chi^{(3)}_{xxxx}/3$. Since the maximum parametric gain is proportional to this coefficient, we see that it is the same for both signal polarization components, and is one-third that for the two-pump OPA with parallel pumps. The SPM and XPM terms in the basic equations are affected, and this modifies $\Delta\beta_{NL}$; as a result the shape of the gain spectrum is also affected.

The main point, however, is that the gain spectra are exactly the same for both signal polarization components, and this implies that the amplifier gain is insensitive to the SOP of the incident signal. Of course, since the applicable element of the tensor $\chi^{(3)}$ is only one-third as large as that for a conventional OPA, one would need three times the pump power, or the fiber length, to obtain the same gain in both cases. This is a trade-off similar to that encountered with the preceding scheme.

Because of its relative simplicity, this arrangement has received a great deal of attention and experiments have been performed by several groups [9–12]. In the experiment described in [9], polarization controllers were used to ensure that the two pumps had orthogonal SOPs. Less than 0.5 dB variation was observed for a gain of about 15 dB. As expected, this was a major improvement compared with the case of parallel pumps. However, subsequent theoretical and experimental studies have shown that there are difficulties associated with this approach. One aspect is that if the pumps are aligned with the fiber PSPs then fiber birefringence introduces a tilt in the gain spectrum, which increases with pump separation [13]. This is a problem when one is trying to make a wideband amplifier, which requires a large pump spacing.

Also, in fibers with random birefringence, significant gain tilt and polarization-dependent gain (PDG) can occur even if the pumps are not aligned with the principal states of polarization (PSPs) [14]. Again, this is more pronounced for large pump spacings. In addition the Raman gain associated with the pumps can also introduce gain tilt and PDG [15].

For all these reasons it now appears that the two-pump polarization-diversity scheme may have a better chance of being practical than the orthogonal-pumps scheme.

(b) Phase-sensitive amplification

In the preceding chapters we have been mainly concerned with OPAs with no idler at the input. In the non-degenerate case we saw that this resulted in a gain that is independent of the phase of the input signal. In such cases, the OPA is said to be operated as a phase-insensitive amplifier (PIA).

We did encounter one case where the gain depended on the phase of the input signal, the degenerate case of a one-pump OPA in which the signal and the pump have exactly the same frequency. This arrangement is then referred to as a phase-sensitive amplifier (PSA). We saw that this PSA has some very interesting properties, including squeezing and the possibility of obtaining a noise figure approaching 0 dB. Also, such PSAs can in principle be used to combat the effect of dispersion in fiber transmission lines [16–18]. The analysis of pulse propagation in such a fiber line shows that, owing to attenuation in the quadrature orthogonal to the amplified quadrature, it is possible for a pulse to propagate without significant broadening over lengths much greater than the usual dispersion length of the fiber. Similar work has been done for soliton propagation [19]. Degenerate OPAs can also be used for the regeneration of differential phase-shift keying (DPSK) signals [20] and for compensating the Raman self-frequency shift of solitons in transmission lines [21].

One should not conclude from the preceding discussion that the only PSAs are degenerate OPAs. In fact, any non-degenerate OPA is in general a PSA provided that both finite signal and idler are present at the input. A classic demonstration of this utilized the sidebands of a modulated carrier to provide simultaneously a signal and an idler with a well-defined input phase relationship [22].

The physical reason for this phase sensitivity is simple to understand when the pumps are not depleted. Then the signal and idler fields at the OPA output are related to their input counterparts by linear relationships (a 2×2 matrix). Thus the output fields can be viewed as the sum of contributions separately due to the input signal and idler. This superposition of fields constitutes interference, and thus clearly the output fields depend on the phases of the input fields.

These non-degenerate PSAs have properties similar to those of the degenerate PSAs. In particular the noise figure (NF) of such an OPA, suitably defined, can also approach 0 dB. This offers the prospect of noiseless amplification without the severe practical constraints associated with degenerate PSAs.

Because of these attractive properties, non-degenerate PSAs have been the object of increasing attention in recent years. Novel configurations have been investigated experimentally and theoretically [23–25].

In an experimental demonstration a non-degenerate OPA in the telecom band phase-sensitively amplified an input signal, and the measured gain response agreed well with the theory [26]. Efficient phase-sensitive amplification was achieved even after transmitting an optical double-sideband signal over 25 km of single-mode fiber.

(c) Quasi-CW signal amplification in a pulsed-pump OPA

The highly nonlinear fibers available at the present time have a zero-dispersion wavelength (ZDW) λ_0 that can vary by several nanometers over a few hundred meters. This

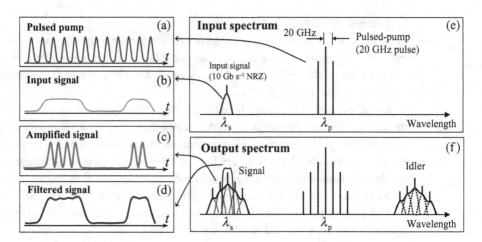

Fig. 12.2 Principle of a one-pump OPA with a pump pulsed at a high repetition rate. (a) pump waveform; (b) input signal waveform; (c) output signal waveform; (d) waveform of the filtered output signal; (e) signal and pump input spectra; (f) output signal spectrum. (After [27], © 2006 IEEE.)

prevents the phase-matching condition from being maintained at its optimum value along the fiber. As a result, to date continuous-wave (CW) fiber OPAs have not exhibited gain spectra as wide as is predicted by the standard OPA theory, which assumes a fiber with constant λ_0. In contrast, by using a pulsed pump consisting of high-peak-power pulses and a relatively short fiber, one can obtain very wide (>200 nm) gain spectra, with a shape very close to that predicted by theory.

Unfortunately, the use of a pulsed pump means that amplified signals cannot in general maintain their original temporal format, since the gain will not be constant in time but will occur only during the pump pulses. The resulting amplitude modulation of the signal appears to make this approach unsuitable for arbitrary signal formats.

Nevertheless, one method makes it possible to use a pulsed-pump OPA for signals with arbitrary modulation formats, namely, the use of a high-repetition-rate pulsed pump OPA followed by a narrowband optical filter [27]. (A similar idea was investigated, wherein a gain increase was obtained by using a pump with a broadened spectrum [28].) As a result, this arrangement works as a CW OPA but with a larger gain and gain bandwidth that are determined by the peak power used rather than by the average power. A 90 nm bandwidth and low-penalty amplification of 10 Gb/s non-return-to-zero (NRZ) signals have been demonstrated using a 20 GHz pulsed-pump OPA. Asynchronous pulses can also be utilized, as is essentially required for the amplification of signals in practical systems.

The principle of operation is shown in Fig. 12.2. The pulsed pump at λ_1 is intensity-modulated at 20 GHz, Fig. 12.2(a). The input signal at λ_3 is NRZ-modulated at 10 Gb/s, Fig. 12.2(b). The amplified signal, Fig. 12.2(c), is the product of the input signal and the periodic gain, which is a waveform with the same period as the pump. After optical filtering the waveform of the output signal becomes similar to that of the input signal, Fig. 12.2(d). The signal and pump input spectra are shown in Fig. 12.2(e). Since the pump

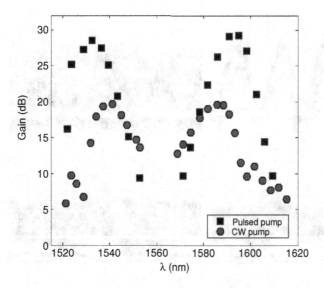

Fig. 12.3 Experimental OPA gain spectra for both the CW and pulsed-pump cases. The average pump power at input was 28 dB m, the signal input power was −30 dB m, the center pump wavelength was 1561.96 nm, and the fiber characteristics were $\alpha = 0.5$ dB km^{-1}, $\gamma = 11$ km^{-1}W^{-1}, $\lambda_0 = 1561.14$ nm, $D_\lambda = 0.03$ ps nm^{-2} km^{-1}, and $\beta^{(4)} = -5.8 \times 10^{-56}$ ps^4 km^{-1}. (After [27], © 2006 IEEE.)

is periodic in the time domain, its spectrum consists of evenly spaced delta functions. The spectrum of the output signal is the convolution of the spectrum of the input signal and that of the periodic gain and therefore consists of equally spaced replicas of the input signal spectrum, Fig. 12.2(f). The narrowband optical filter then selects the central lobe of that spectrum, resulting in the waveform of Fig. 12.2(d); it closely resembles that of Fig. 12.2(b), indicating that the pulsed-pump OPA is suitable for the amplification of modulated communication signals.

The calculation of the gain then proceeds as in Chapter 7. Experiments have been performed to compare the CW regime and the pulsed regime. In the pulsed case, two electro-absorption modulators were used to generate a 20 Gb/s pulse train, with a duty factor of 20%. Figure 12.3 shows the signal gain spectrum measured for the two cases, with a total average pump power of 28 dB m (631 mW). As expected the gain for the pulsed case is significantly higher, and the gain bandwidth is also larger.

The input signal was then modulated by a 10 Gb/s NRZ sequence. The performance of the pulsed-pump OPA is shown in Fig. 12.4. The "eye" patterns with synchronous pump, where the 20 GHz pulsed pump was synchronized with the 10 Gb/s signal, are shown on the left. Figure 12.4(a) is the waveform before the fiber Bragg grating (FBG) filter. It shows that the input 10 Gb/s NRZ signal was sampled at 20 GHz by the pulsed OPA. After passage through the FBG the 20 GHz spectral component was rejected and the 10 Gb/s NRZ signal was recovered, as shown in Fig. 12.4(b).

In order for this OPA to amplify input signals transparently, a narrowband optical filter is necessary. However, if a pulsed-pump OPA is used as the pre-amplifier for an optical

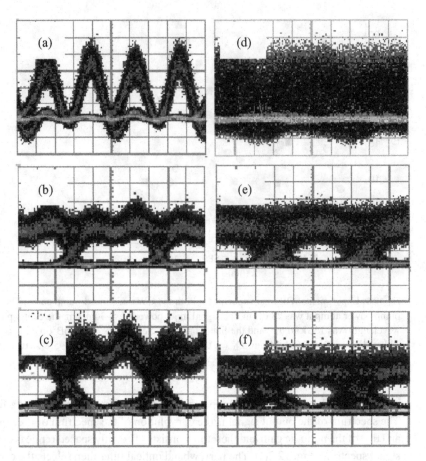

Fig. 12.4 Eye patterns for the pulsed-pump OPA. On the left, synchronous pump; On the right, asynchronous pump. (After [27], © 2006 IEEE.)

receiver, an electrical low-pass filter can be a good alternative to the narrowband optical filter. Figure 12.4(c) shows the received signal waveform obtained by using an electrical low-pass filter instead of the narrowband optical filter.

In Figs. 12.4(b) and 12.4(c) the residual 20 GHz ripple is due to the imperfect characteristics of the available filters; it could be reduced by using better filters.

Bit-error-rate plots for the output signal, corresponding to Fig. 12.4(b), are shown in Fig. 12.5. Although a penalty of 1.5 dB was observed, error-free operation was confirmed, which proves the feasibility of using the pulsed-pump OPA for amplifying NRZ signals. The penalty is thought to result mainly from the imperfections of the optical filter characteristics, i.e. its bandwidth, dispersion, and dispersion ripple.

Asynchronous pumping was also investigated, where the repetition rate of the pulsed pump was 2 Hz higher than twice the bit-rate of the input NRZ signal. The output eye pattern from the asynchronous pulsed-pump OPA, triggered by the signal clock, is shown

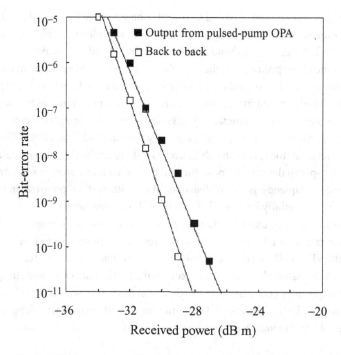

Fig. 12.5 Bit-error-rate performance of a 10 Gb s^{-1} NRZ signal amplified by a 20 GHz pulsed-pump OPA followed by a narrowband optical filter. (After [27], © 2006 IEEE.)

in Fig. 12.4(d). The eye opening for the signal recovered by the optical FBG filter is shown in Fig. 12.4(e). Figure 12.4(f) shows the eye opening when an electrical low-pass filter was used. Asynchronous pumping is not expected to induce any additional penalty, since the only limitation of the proposed technique is that the repetition rate of the pulsed pump should be higher than twice the bit rate of the input signal. A sufficiently narrowband optical filter at the output can recover the original (amplified) signal irrespective of any pump and signal synchronization mismatch.

12.2.2 Wavelength conversion

(a) Fixed-pump wavelength conversion

In an OPA where only a signal and the pump(s) are fed into the input, the generation of the idler at the output can be interpreted as conversion of the energy of the input waves to the idler wavelength. In optical communication, if the pump has constant intensity and the signal is modulated then the idler itself is modulated, and it is natural to regard this as a conversion (or transfer) of modulation from the signal to the idler. This wavelength conversion of a signal could find a number of practical applications. For example, it could be used for routing signals according to their wavelengths, by passing them through dispersive elements such as prisms, arrayed waveguide gratings, etc. It could also be used for switching received signals to wavelengths that are optimized for long-haul transmission (e.g. conversion from the 1300 nm band to the C-band). Thus, in the context of communication, it would be useful to be able to switch signals, individually or in bands, from one wavelength range to another. For example one might want to convert from C-band to L-band, from C-band to S-band, from S-band to L-band, etc.

For communication applications, the use of a CW pump would be necessary to accommodate signals with typical modulation formats. Hence, a knowledge of the performance of CW amplifiers, described in Section 11.3, is useful in understanding the potential of wavelength conversion in communication. Basically, the wavelength coverage for such wavelength converters can be obtained from the gain bandwidth of a fiber OPA used as an amplifier. Hence all the results in subsection 11.3.1 are applicable. In particular we see that one- or two-pump devices are possible and that conversion over tens of nanometers about the center frequency is easily feasible; also, with further progress in nonlinear fibers this could potentially be extended to hundreds of nanometers.

It is important to note that fiber OPAs are unique among wavelength converters in having high conversion efficiencies as well as low noise figures (NFs). A wavelength converter with NF 6.8 dB together with +40 dB of conversion gain, over tens of nanometers, has been demonstrated [29]. This far exceeds the performance of competing devices, such as semiconductor optical amplipfiers (SOAs) or devices based on periodically poled lithium niobate (PPLN). Conversion for numerous closely spaced wavelength-division multiplexing (WDM) signals has been investigated [5].

(b) Tunable-pump wavelength conversion

Another desirable type of wavelength conversion is the so-called any-to-any conversion, meaning that a signal at an arbitrary wavelength can be converted to an idler with another arbitrary wavelength. Clearly the only way that this can be accomplished with an OPA is to have the freedom of adjusting the pump wavelength at will while maintaining a high conversion efficiency.

From what we have seen in Chapter 3, we know that the conversion efficiency depends on the wavevector mismatch $\Delta \beta$, which itself depends strongly on the location of the center frequency with respect to the ZDW of the fiber. As a result, it is clear that for a given fiber we will not be able to choose the signal and idler wavelengths completely arbitrarily, but that they will have to be confined to certain ranges over which phase-matching conditions hold relatively well.

One way to obtain this type of any-to-any conversion is to start from an OPA designed with a broad gain spectrum, and to tune the pump about its initial location [30]. One can then define a tuning range for the pump, which might for example correspond to a drop in idler power by 3 dB. This approach was considered in [31], where it was shown that, for typical OPA parameters, an idler tuning range of the order of 10–20 nm can be obtained. Using a standard HNLF, a conversion bandwidth of 61 nm, together with a pump tuning range of 24 nm, was obtained [32]. By using a fiber with a large ZDW variation, achieved by applying controlled stress distribution, an idler tuning range of 70 nm has been achieved [33].

It is important to note that this is actually not the only way to obtain tunable conversion with a fiber OPA. One can also employ the phase-matching condition used to obtain a widely tunable narrowband OPA [34]. By using a different part of the graph of $\Delta \beta$ versus

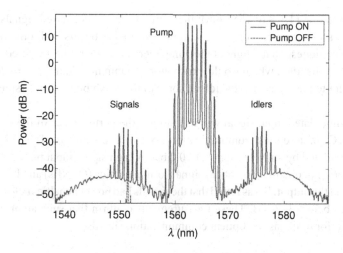

Fig. 12.6 Spectrum at the output of a fiber OPA when the pump is either on or off. The signal input power was −30.7 dB m. (After [27], © 2006 IEEE.)

$\Delta\omega$, one can obtain phase matching over a narrow region, far from the pump. Then, by tuning the pump, one can move this gain region over a wide range, of the order of 100 nm. Since the signal and idler must remain symmetric with respect to the pump this type of wavelength conversion does not fit into the any-to-any category. Nevertheless, this type of conversion is attractive for obtaining conversion between a tunable signal and an idler that are widely separated.

(c) High-repetition-rate pulsed-OPA wavelength conversion

A very different approach to wavelength conversion is obtained by using a periodically pulsed pump. As we saw in subsections 7.5.3 and 12.2.1(a), when such a pump is used and a modulated signal is fed into the input, the output spectrum consists of a number of peaks that all carry a replica of the signal modulation. As a result, if we select any of these peaks by means of a narrowband optical filter, we can recover the original modulation in the electrical domain after detection. In other words, this arrangement provides multicasting of the original signal to a number of output wavelengths. It should be noted that both the signals and the idlers carry the modulations, and so they can all be used as signal replicas.

Experiments have been performed to demonstrate this mode of operation [35]. The pump was intensity-modulated by combining two tunable lasers at 1563.13 nm and 1563.93 nm. Their 0.80 nm wavelength separation provided sinusoidal intensity modulation of the input pump power at about 100 GHz. The average pump power at the input of the highly nonlinear dispersion-shifted fiber (HNL-DSF) were 23 dB m per pump. A low signal input power was used in order to avoid gain saturation.

The optical spectrum at the output of the HNL-DSF is shown in Fig. 12.6. The central lobe (1558 to 1568 nm) corresponds to the pump spectrum, which consists of 12 main peaks; the increase in the number of peaks, from two at the fiber input, is due to pump SPM. The original signal and the multiple converted signals are located in the lobe to the left of the pump (1548 to 1555 nm); there are nine main converted signals. Several of these have a net conversion gain, in contrast with what occurs with other multiple-conversion techniques. To the right of the pump there is another lobe, centered about the frequency ω_i of the idler, which would exist without pump modulation; $\omega_i = 2\omega_p - \omega_s$. Hence we refer to these peaks as the idlers; some of them exhibit net conversion gains as well.

An NRZ-modulated input signal was used to test the performance of the wavelength converter (WC) in an optical communication system. The signal (at 1551.32 nm) was intensity-modulated by an NRZ signal at 10 Gb/s. A high signal input power, 5.6 dB m, was used in order to obtain high optical signal-to-noise ratios (OSNRs) and low bit-error rates (BERs) at the output. It was found that the signals had power penalties below 1.24 dB, and the idlers below 3.66 dB. These measurements confirm that such an arrangement could be used for multicasting optical communication signals.

(d) Multiband wavelength conversion

Yet another approach to making a WC is to use two pumps that are spaced close enough for FWM terms other than the usual signal and idler to be well phased matched and to grow to significant levels [36]. Under these circumstances, one finds that the following types of wave need to be taken into account:

1. strong waves obtained by FWM between the pumps, which may be considered as additional pumps;
2. weaker waves, which can be considered as new signals or idlers depending on their frequency relationship with the original signal and idler. The new signals (idlers) have the same spectrum as the original signal (idler), and so they constitute frequency-shifted replicas of the original signal or idler. The presence of these multiple replicas indicates that this arrangement can be used for multicasting, an important feature in optical networking.

A useful feature of this arrangement is that the broadening of the new waves due to phase modulation of the two pumps for stimulated Brillouin scattering (SBS) suppression can be controlled. In particular, if the two pumps are modulated in phase then the original idler has strong phase modulation but none of the new signals do [37]. Because it requires only a single phase modulator, this method is potentially attractive compared with opposite-phase modulation of the two pumps, which requires two modulators.

It is interesting to note that high-repetition-rate pulsed-OPA wavelength conversion, described in (c) and multiband wavelength conversion, described here, are fundamentally part of the same phenomenon: in both cases the initial waves are two pumps and one signal, and in both cases FWM between the two pumps generates a number of additional pumps. The new signals and idlers can then be viewed as arising from FWM between the original signal and the new pumps.

What distinguishes these two wavelength conversion mechanisms is the relative position of the signal with respect to the pumps. For the first we have $\Delta\lambda_s \gg \Delta\lambda_p$, whereas for the second we have $\Delta\lambda_s \approx \Delta\lambda_p$. These frequency relationships explain why the new signals and idlers appear where they do in the two cases.

12.2.3 Phase conjugation and spectral inversion

Since the idler A_4 arises from the FWM term in the basic propagation equations (see Eq. (3.34)), we see that $A_4 \propto (A_3)^*$, i.e. that the idler is proportional to the complex conjugate of the signal. Complex conjugation has two related consequences, phase conjugation and spectral inversion.

Phase conjugation applies to monochromatic signals. If a signal has phase modulation $\phi(t)$ then the corresponding idler will have phase modulation $-\phi(t)$. This property can potentially be used for canceling undesirable phase modulation due to SPM or XPM, which accumulates along transmission lines. The idea is to use phase conjugation in the middle of a transmission link, so that the accumulated phase modulation at the end has the same amplitude but opposite signs for the two halves, thereby leading to cancellation.

Spectral inversion applies to polychromatic signals. For such a signal, frequency components initially lower (higher) than the carrier frequency become higher (lower) than the new carrier frequency after FWM; hence the name spectral inversion. An important application of spectral inversion is that mid-span spectral inversion (MSSI) can be used for canceling the effect of the chromatic dispersion D [38, 39]. This can be understood as follows. A part of the spectrum located $\Delta\lambda$ away from the carrier experiences a differential group delay (DGD) $\Delta\tau = D\Delta\lambda L$, where L is the fiber length. After MSSI, that part of the spectrum is now $-\Delta\lambda$ away from the new carrier and thus experiences a second DGD equal to $-\Delta\tau$, in a second fiber opposite to the first. We thus see that this part of the spectrum, and by extension the entire spectrum, experiences no net DGD as it passes through this fiber span of length $2L$, with MSSI.

Cancellation of DGD by MSSI was demonstrated with $\chi^{(2)}$-based devices a few years ago. A limitation of these devices, however, is that they are lossy and therefore have a high noise figure. They must be supplemented by conventional optical amplifiers such as erbium-doped fiber amplifiers (EDFAs) to restore signal amplitude. By contrast, fiber OPAs used as wavelength converters can provide high gain and a low noise figure [29] and thus in principle do not require the use of auxiliary EDFAs. Differential group delay cancellation by a fiber OPA used for MSSI has been demonstrated [40–43]. Numerical simulations have shown that periodic phase conjugation by OPAs in long-haul networks could be used to reduce the timing jitter of solitons [44].

12.2.4 Optical delay lines

A desirable function in optical networks would be the ability to delay incoming signals in the optical domain. If the time delay were large enough this could allow for some processing to take place in the node before sending the signal on to its next destination. Delaying in the optical domain is also referred to as optical buffering.

One area where optical buffering is always desirable is for avoiding packet collisions at a node. For example, if two packets arrive on different fibers in such a way that they would collide at a node, one way to avoid a collision would be to deflect one packet by means of an optical switch into an optical delay element and to release it only after the first packet has passed.

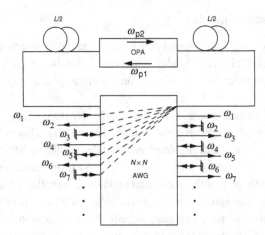

Fig. 12.7 Schematic of a delay line based on a highly nonlinear fiber (HNLF) and an arranged waveguide grating (AWG) used bidirectionally to implement two different OPAs. (The arrangement of the output frequencies may not correspond to that of an actual AWG, although they could be rearranged by routing waveguides.)

A natural way of obtaining an optical delay is simply to use the propagation delay in a long fiber. For example a 10-km-long fiber would provide a delay of about 50 μs, which is sufficient for many applications. One difficulty with this approach, however, is that, in order to accommodate a number of different delays, one must have a number of fibers of different lengths and optical space switches to switch between them. These switches introduce losses and also they can be expensive if they have to be fast.

An interesting alternative to space switches has been proposed, based on wavelength conversion by fiber OPAs [45, 46]. The idea is to use a single length of fiber to obtain a number of possible different delays, in the following manner. If a packet with frequency ω_1 enters the delay fiber, it passes through an OPA, which converts it to $\omega_1 + \Delta\omega$. A filter then separates the new frequency from the original one and redirects it through the fiber. This process is then repeated N times. In the end, one has available at the filter outputs N wavelengths with N different delays. One can then choose the desired delay by selecting a particular wavelength with a wavelength switch.

One advantage of this method is that all these different delays require only a single fiber length, so that long time delays can be obtained with a relatively short fiber. This can be a significant cost advantage for very long delays. In addition, the OPAs can provide gain to compensate for the losses that would accumulate for long delays.

Figure 12.7 shows a possible implementation [46]. A single HNLF is placed in the middle of a delay line of length L. It is pumped in each direction by a single pump: the clockwise (counter-clockwise) pump is at ω_{p1} (ω_{p2}). An arrayed waveguide grating (AWG) is used bidirectionally as a wavelength multiplexer and demultiplexer. The system works as illustrated in Fig. 12.7. The initial signal at ω_1 enters the AWG and is coupled into the first half of the delay line in the counter-clockwise direction. When it passes through the OPA it generates a new frequency $\omega_2 = \omega_1 + \Delta\omega$ (Fig. 12.8); ω_1 and ω_2 then pass through the second half of the delay line and reach the AWG's input port. The AWG sends

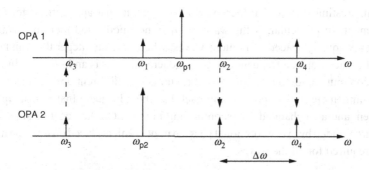

Fig. 12.8 Diagram showing the consecutive conversion of frequencies taking place in the two OPAs. The upper (lower) diagram corresponds to the clockwise (counter-clockwise) direction.

ω_1 and ω_2 to two different output ports on the right: ω_2 is reflected by a mirror and retraces its path to the HNLF, where the OPA then generates $\omega_3 = \omega_2 - 2\Delta\omega = \omega_1 - \Delta\omega$; ω_2 and ω_3 then pass through the second half of the delay line and reach the AWG's right input port. The AWG sends ω_2 and ω_3 to two different output ports on the left: ω_3 is reflected by a mirror and then undergoes the same evolution as ω_1, except for the difference in frequencies. The process is then repeated, and the sequence is eventually terminated by omitting a reflecting mirror. The new frequencies form two series, namely $\omega_{2m} = \omega_1 + m\Delta\omega$ and $\omega_{2m+1} = \omega_1 - m\Delta\omega$, $m = 1, 2, 3, \ldots$

Figure 12.7 shows the use of a single bidirectional $N \times N$ AWG. Alternatively, one could use two $1 \times N$ AWGs, each employed bidirectionally. Also, the converted frequencies shown in Fig. 12.8 are obtained by assuming that $\omega_{p1} - \omega_{p2} = \Delta\omega/2$. Other distributions would be obtained by choosing ω_{p1}, ω_{p2}, and ω_1 differently [45], but that would not change the basic principle of operation.

This arrangement has some features that distinguish it from other types of delay line. Since there is no switching in time, there is no limit on the temporal duration of the input signal and one can have streaming data of arbitrary duration entering the delay line. A potentially interesting aspect is that, if desired, one can simultaneously extract from the system all the outputs at $\omega_1, \omega_2, \omega_3, \ldots$, i.e. versions of the input signal that are delayed by different amounts. Space switches could be used for selecting the desired version(s).

Alternative architectures based on the same principle are described in [45]. A major difference from Fig. 12.7 is that propagation in the loop is unidirectional; as a result, two different HNLFs are required to implement the two frequency conversions. There is a single output, extracted by means of a fixed dropping filter; to obtain different delays, the frequency of the input signal is shifted by means of an additional wavelength converter.

In any implementation, the practical number of different time delays that can be achieved is limited by signal deterioration due to a number of causes: (i) amplified spontaneous emission (ASE) accumulation after multiple passes; (ii) nonlinear interaction (FWM, cross-gain modulation (XGM)) between the various wavelengths. In [45] it was shown experimentally that up to seven passes could be made around the loop; also simulations for an optimized system indicated that up to 100 passes could be made around a 1-km-long loop, providing up to 0.5 ms of delay in 5 µs increments.

It is interesting to note that there may also be additional applications for this basic arrangement. In particular, if the wave at ω_1 is not modulated then the arrangement generates N equally spaced continuous waves, where N is an integer that can be chosen at will. In other words, a frequency comb is generated. This is an important function for optical communication applications, and a number of different approaches have been used to implement it. The present approach has the advantage that in principle N can be chosen, and also changed, by reconfiguring mirrors. On the negative side, since the OSNR is degraded by every additional pass, an upper limit is placed on N by the minimum quality required for all the waves.

12.2.5 Slow and fast light

Another desirable optical processing function would be the ability to vary the propagation time of light in a continuous manner. Clearly this could be used for synchronizing optical signals, etc. In recent years a variety of physical mechanisms have been proposed to increase and/or decrease the propagation time of light in guiding structures. The terms fast and slow light are often used to refer to this active field of research.

Recently fiber OPAs have been used for this purpose. In one approach, changes in pump power are used to affect the shape of the gain spectrum, which is also accompanied by a change in propagation delay [47]. The maximum delay and the tuning range can be optimized with respect to each other by considering saturation effects in long fibers. The proposed scheme offers tunable delay in the presence of gain, with a bandwidth that is sufficiently wide to process digital data streams at a rate of tens of gigabits per second as well as picosecond pulses.

In another approach the wavelength of the incident pulses is converted by an OPA to an idler wavelength that can be varied by tuning the pump wavelength [48]. As the idler pulses are propagated in a dispersive fiber they acquire a time delay that depends on their wavelength. Finally, the idler pulses are reconverted to the original wavelength by a second OPA. So the overall system can impart a variable delay on the input pulses while conserving their wavelength. A major advantage of this approach is that the delay is continuously variable. The demonstration in [48] used a pulsed pump for convenience. Experimental conditions limited the maximum delay to about 800 ps. Simulations with a CW pump showed that, with an optimized system, delays of tens of nanoseconds could in principle be obtained.

Further progress in this area was made by using a two-pump fiber OPA with CW pumps [49]. A maximum delay of 12.47 ns was demonstrated. Signals at 10 Gb/s were delayed in an error-free manner. Because of the CW nature of this system, it could in principle be used to delay signals with arbitrary modulation formats.

12.2.6 Storage of optical pulse trains

Another way to obtain long time delays is to store optical pulses in a fiber loop. If the losses are compensated for by an amplifier inserted in the loop then the pulses can in principle circulate forever around the loop. In reality there are several problems that make the pulses deteriorate in a finite amount of time. The first is the ASE of the amplifier, which will degrade the OSNR over time. In addition, the same problems as those occuring in long transmission lines, namely dispersion and nonlinear effects, will also cause a loss in pulse quality.

In order to obtain long confinement times in such a loop, it is thus necessary to add some means for regenerating the quality of the signals. For on–off signaling this can be achieved by a variety of means. One possibility is to use a fiber OPA with a pump modulated in synchronism with the data. This approach was implemented in [50], where a degenerate fiber OPA which exploited a nonlinear Sagnac loop made of standard polarization-maintaining fiber (PMF) was used. More recently, non-degenerate OPAs were used [51, 52].

An important issue with a storage loop is how to load the information into it. A variety of means involving switches have been used. In an arrangement based on four-wave mixing, a 20 Gb/s-packet pattern was loaded and stored for up to several seconds, corresponding to millions of circulations of the packet within the buffer [51]. In the latest version, the signal injection was accomplished by means of XPM [52].

12.2.7 High-repetition-rate source of short pulses

The nearly instantaneous nature of the fiber nonlinearity implies that its gain can be modulated as fast as one can modulate the pump. Since modulation techniques have been developed to modulate communication signals at up to 160 Gb/s, fiber OPAs can therefore also be modulated at such speeds.

In addition to this high-modulation-rate capability, fiber OPAs can generate output pulses that are considerably shorter than the pump pulses. The reason for this is the nearly exponential dependence of the gain on pump power: for a Gaussian-shaped pump pulse, the gain will be highest near the maximum and will drop rapidly near the edges. The result is output pulses that can be several times shorter than the pump pulses. These properties can therefore be exploited for making a source of very short pulses, emitted at a very high rate.

These features have been experimentally demonstrated [53–58]. Short return-to-zero pulses ($\sim$2 ps) were generated at bit rates of 40, 80, and 160 Gb/s using a fiber OPA. The performance of the parametric pulse source was evaluated both back-to-back and in a 110 km transmission link. A receiver sensitivity of -33 dB m back-to-back was achieved after demultiplexing from 160 to 10 Gb/s. The power penalty at 160 Gb/s due to 110 km transmission was less than 2 dB. Very short pulses (0.5 ps) were also achieved when the parametric amplifier was used as a compressor.

In an alternative approach, a fiber OPO was used for generating dual-wavelength picosecond pulses at a rate of 10 GHz simultaneously in the C- and L-bands [59]. A self-starting regeneratively mode-locked fiber OPO utilizing parametric gain in an HNLF, together with intracavity soliton formation, yielded picosecond pulses. An electroabsorption-modulator-based optoelectronic oscillator was used as an ultralow-jitter pump source, enabling, by means of direct feedback injection locking, a significant reduction in the timing jitter of the pulses.

12.2.8 Optical clock recovery and multiplication

An important function in optical communication is to derive a stable clock from the incoming data stream itself. Numerous all-optical techniques have been developed for doing so. In [60] a wavelength-tunable clock recovery scheme for all-optical 3R (three error) regeneration and simultaneous wavelength conversion was reported. The scheme was based on a fiber OPO. The 2.9 ps clock pulses were recovered from the incident pseudo-random data stream; the clock wavelength was tunable over the whole EDFA bandwidth. Radiofrequency spectrum measurements showed that the recovered clock had less timing jitter than the incident data stream. Two-channel clock-recovery experiments demonstrated the feasibility of the scheme at a data rate of 100 Gb/s.

In addition, soliton trains can be parametrically generated in fiber loops when an incoming soliton train interacts with a CW pump in the nonlinear fiber [61]. The idler may either reproduce or multiply the incoming repetition rate, depending on whether the corresponding cavity harmonic is exactly matched or differs by an amount equal to an integer fraction of the longitudinal mode spacing.

12.2.9 Optical signal processing

(a) Pulsed-pump OPAs

The applications considered above rely on the properties of CW fiber OPAs. We now turn to applications using pulsed fiber OPAs. In a pulsed OPA the pump consists of a train of pulses. As a result, the gain is strongly modulated in time since it is a function of the instantaneous pump power (recall that the maximum gain of a fiber OPA is approximately $G = 8.6\gamma P_0 L - 6$ in decibels). The output signal and idler will then be essentially turned on and off by the OPA, which can be thought of as acting as a type of switch. Several applications make use of this feature.

TDM demultiplexing

A common problem in optical communication is the demultiplexing of time-division multiplexed (TDM) signals. In a TDM system, the modulated pulses associated with a number N_c of channels are interlaced in the time domain and grouped in frames of N_c pulses. For example, the nth pulse in a frame can be associated with the nth channel. The role of the demultiplexer is then to pick out the nth pulse in each frame, in order to recover the pulses associated with the nth channel and hence the data carried by that channel. The demultiplexing can be destructive if the pulses associated with the other channels are lost in the process. However, it can in principle be non-destructive if all the other pulses are preserved by the process; in this case, further stages of demultiplexing can extract the other channels. In this case one could insert new pulses in the now empty nth slot in each frame, thereby realizing an "add or drop" operation.

To make a TDM demultiplexer from a pulsed-pump OPA, one needs to have only one pump pulse per frame and its duration should be at most that of a signal bit. The pump pulses must be periodic and synchronized with the bits to be extracted. Under these conditions, at the output of the OPA the signal pulses coinciding with the pump pulses will be amplified compared with the rest of the signal pulses. If the gain is large enough, say 20 to 30 dB, most energy in a frame might correspond to the desired channel. Then one could ignore the contributions from the other channels and use this output to extract the desired channel by detection and thresholding. However, if the gain is not large enough or the frames are very long then the unamplified pulses of the undesired channels might contribute a significant amount of energy per frame and, hence, noise that could cause errors in detecting the desired channel. In that case, it might be preferable to use the idler as the output of the demultiplexer (DMUX). The advantage of the idler is that only the desired pulses are present, which leads to a better SNR. With the idler, one could in principle use relatively low gain and still obtain a good SNR.

The demultiplexing of TDM signals using the idler was first demonstrated for 20 Gb/s signals with a gain of 19 dB and a 1.4 dB power penalty [62]; TDM demultiplexing using the signal was demonstrated for 40 Gb/s signals with a gain of 40 dB [63] and a signal that could be tuned over a 39 nm range.

We note that using the idler also leads to a non-destructive DMUX, since the pulses of the other channels at the signal wavelength are essentially undisturbed and could potentially be demultiplexed by further OPA stages.

Multiband bit-level switching

In related work, a pump modulated with a 10 Gb/s bit sequence was used for switching bits from a signal synchronously modulated with another 10 Gb/s bit sequence [36]. This work is interesting in that it was based on a two-pump OPA and the pump spacing was small enough that multiple converted frequencies were created at the output. Multiple-band bit-level optical switching was demonstrated over a bandwidth greater than 50 nm, with a maximal conversion efficiency of 24.2 dB.

2R or 3R regeneration

In long-haul communication systems, signals can become distorted over long distances because of the combined effects of fiber nonlinear effects, dispersion, and ASE added by optical amplifiers. As a result, if one considers a return-to-zero (RZ) transmission system, using a single optical pulse to represent a "1" bit and the absence of a pulse for a "0" bit, the bits can get corrupted in a number of ways. The shape of the "1" pulses can become distorted, generally becoming wider; their amplitudes can become different from pulse to pulse; their position in the reference-bit time slot can vary from pulse to pulse (timing jitter). Also, the "0"s may become corrupted by the wings of the distorted "1"s, spreading optical energy where none should be present. In addition, ASE and nonlinear effects may introduce random fluctuations in either level, leading to BER degradations.

It may thus be desirable to try to correct these defects at periodic intervals along the transmission line and to restore the initial pulse shapes as much as possible. This is referred to as regeneration. Many types of regenerator have been investigated, some correcting just one defect while others correct all defects. Depending on the number of defects being corrected, they are called 1R, 2R, or 3R regenerators.

1R is simply reamplification, and can be accomplished by any type of optical amplifier.

2R is reamplification plus reshaping. Ideally reshaping means a return to the original pulse shape. In practice, however, it is in general impossible to accomplish this and so reshaping means a partial reversal of some aspect of the pulse degradation: in one case it could be improvement of the overall pulse shape and in another case it could be reduction in noise in the "1"s. There are therefore many possible ways to accomplish reshaping, many of them involving nonlinear optical mechanisms.

3R is 2R plus retiming, i.e. reduction in the pulse timing jitter that may arise in a transmission line due to effects such as ASE, Raman gain, the Gordon–Haus effect, etc.

Since fiber OPAs can provide a large gain (for either the signal or the idler), they automatically provide 1R regeneration. Being fast nonlinear devices, they can also be used for providing reshaping and retiming, and hence 2R and 3R regeneration, in a number of ways.

For pulsed OPAs one idea is to use a pump whose intensity is periodically modulated at the same rate as the data and synchronized with it. (It is assumed that a clock is extracted from the signal by means such as a phase-locked loop.) Since the OPA gain is a strong function of the instantaneous pump power, if the pump pulses have short rise and fall times and if there is little or no intensity between the pulses, the amplified signal pulses will be strong only during the pump pulses. This will provide retiming and some reshaping of the "1" pulses. Regeneration should be performed before significant power leaks to the center of the "0" pulses, otherwise this light would also be amplified, which would be undesirable. This type of regeneration is attractive because the optical frequency of the pulses is not altered, which implies that such regenerators can be cascaded many times. However, one can also use the idler as the regenerated output; the performance is then similar except for the wavelength shift.

The above type of regeneration has been demonstrated experimentally [64]. A 10 Gb/s signal was amplified in a one-pump OPA. The pump was an actively mode-locked fiber laser driven by synchronous modulation, obtained from the signal via a phase-locked loop. This arrangement provided a reduction in timing jitter of 2.7 ps. A reduction in amplitude fluctuations was also obtained, by operating the OPA in saturation.

A rather different approach consists of using the amplified signal as the OPA pump itself. The pump modulation is then transferred to a CW input probe and emerges at the output on both the probe wavelength and the idler wavelength; either of these waves can then be used as the output for the regenerated signal. Because no clock is involved in this arrangement, no retiming is possible. The gain is not the usual OPA gain but takes into account the gain used for boosting the original signal as well as the conversion efficiency from the pump to the output signal or idler. This arrangement can provide reshaping, because of the nature of the transfer function from the pump to the probe or idler [65]. In experiments, data was regenerated with an extinction ratio as high as 14 dB, an extinction ratio enhancement of 5 dB, and a 5 dB negative power penalty [65].

It is interesting to note that a similar improvement in extinction ratio had previously been observed in a comparable arrangement, which was being viewed primarily as a wavelength converter rather than a regenerator [66].

The possibility of using fiber OPAs for the 2R regeneration of RZ-DPSK signals has been recently investigated. It was shown that saturated amplification in a non-degenerate fiber OPA is a potential candidate [67]. Note that this approach is not the same as that based on degenerate OPAs [20].

Optical sampling

Optical sources can generate short pulses at high repetition rates. These can be used for sampling in the optical domain. This requires a very fast optical device that can respond essentially instantaneously in comparison with the length of the sampling pulses. Fiber OPAs can respond on a femtosecond time scale, and thus can be used for optical sampling. The idea is similar to that of an optical DMUX except that now the pump consists of pulses much shorter than the period of the optical signal to be sampled. There is one pump pulse for every N_p periods, occurring at a slightly different time in each signal period. Then at the OPA output the idler consists of pulses coinciding with the pump pulses and with about the same duration. This train of pulses with a relatively low repetition rate can now be converted to electrical signals with relatively low frequencies, which can then be handled by electrical circuits and displayed on a conventional oscilloscope. The result is a display of the detailed structure of the periodic optical signal of interest, over a one-period interval. In particular, this method is useful for displaying periodic signals such as the bit patterns used for measuring bit-error rates. In addition, if sampling is used on an aperiodic signal, such as a typical communication signal, the eye pattern of the signal can be displayed; this provides very useful information about pulse distortion, timing jitter, Q-factor, etc.

This idea has been demonstrated experimentally [68]. By using OPA pump pulses of picosecond duration, the authors were able to implement optical sampling and to display a time-resolved periodic 300 GHz waveform. A commercial optical sampling oscilloscope based on this principle has recently been introduced [69]; it has a temporal resolution of 1 ps.

In related work [70] a Kerr switch was operated with pump pulses of 0.9 ps duration and 15 W peak power in a 20-m-long HNLF. The ZDW was at 1579 nm and the pump was placed very close to it. The signal wavelength could be anywhere in the 1535–1565 nm range. This arrangement provided optical switching with a parametric gain of 7.6 dB, which was used for demultiplexing signals at 160, 320, and 640 Gb/s. The power penalty was less than 0.2 dB at 160 Gb/s, and increased to 2.5 dB at 640 Gb/s. The same arrangement was also used for optical sampling of the optical signals at these rates, permitting the display of eye patterns for these very-high-speed signals.

(b) Continuous-wave OPAs

Some functions that can also be viewed as optical processing functions can be performed by OPAs operating on a CW basis.

Amplitude-noise reduction

This application exploits the facts that the gain of fiber OPAs saturates as the signal input power increases beyond a certain level P_{sat} and that this saturation is instantaneous. As a result, fluctuations around the high input level result in relatively small fluctuations around the high output level. Thus, if one considers a binary input signal, with a low level near 0 and a high level above P_{sat}, at the output the resulting binary waveform has noise at the high level that is compressed compared with that at the high input level. In other words, we have amplitude-noise reduction for the high level.

It is important to note that this is possible only because the OPA gain responds almost instantaneously to the input signal fluctuations. This is in contrast with what occurs in EDFAs, where the gain is modified only by reducing the population inversion, the latter being controlled by the average power not by short-term fluctuations. Hence in EDFAs the gain for the entire waveform is reduced by the same amount by saturation, and saturation does not lead to beneficial amplitude-noise reduction.

This mechanism has been studied theoretically and experimentally [71–73] and it has been verified that it does lead to a reduction in amplitude noise. The mechanism has also been used in combination with timing jitter reduction [74]. In a variation, the processed output is obtained from a higher-order FWM wave. It can be shown that these waves exhibit a stronger dependence on signal power, and therefore they have a transfer function that has a flatter top and is better suited to the reduction of amplitude fluctuations [75–77].

Phase- and amplitude-noise reduction

Differential phase-shift keying (DPSK) has emerged in recent years as an important modulation scheme for long-haul optical transmission [78–81]. If it is to be used over very long fiber links, it may need to be regenerated along the way. In that case, all-optical regeneration would be desirable. By using a degenerate PSA it is in principle possible to reduce phase fluctuations.

Recent experiments performed with a nonlinear Sagnac interferometer have shown that it is indeed possible to reduce phase error and that amplitude fluctuations can be reduced at the same time [81]. Maximum phase errors of 82° were reduced by a factor 2, and the SNR was simultaneously improved by nearly 6 dB. The performance of the phase-sensitive amplifier (PSA) was found to be robust with respect to input SNR and phase noise.

12.2.10 Monitoring of network parameters

A rather different area of application of OPAs in optical communication can be contemplated if one uses an amplified communication signal as a pump and a weak unmodulated probe as a "signal" for a fiber OPA. In such a situation one can use the idler as the output. Clearly the shape of the idler pulses will depend strongly on the shape of the pump pulses, and therefore any change in the latter will affect the former. Thus by monitoring the characteristics of the idler pulses one can in principle infer changes in the incoming pulses, which can be affected by a number of features of the transmission network.

This idea was first used to measure the accumulated residual dispersion of the input signal [82]. The concept was successfully demonstrated with 40 Gb/s RZ signals. This approach potentially works for data rates up to terabits per second and can be applied to provide a feedback for automatic dispersion compensation.

A major advantage of this type of technique is that it permits the indirect monitoring of some features of very short pulses (peak power, pulse length), without requiring the use of very fast detectors and electronics.

In an alternative approach a conventional OPA is employed, using a strong CW pump [83, 84]. An incoming communication signal now serves as the input "signal" of the OPA. By using high gain and fairly strong output signals, higher-order FWM terms are generated within the OPA. The amplitude of these terms depends very strongly on signal peak power and therefore on their duration, which is itself related to dispersion. Hence monitoring the high-order FWM terms can lead to a measurement of the dispersion accumulated in the transmission network by the signal pulses.

This method has attractive features. Since it uses an OPA employed in a conventional manner, it could be readily be implemented in systems using fiber OPAs for either signal amplification or wavelength conversion. And, of course, if the OPA is used for amplification then it generates an idler, which can potentially be used for other purposes.

In experiments, dispersion monitoring with 11 ps pulses that experience a net dispersion of ± 180 ps per nanometer has been demonstrated, with an OPA gain of more than 18 dB. These pulses are similar to those to be used in high-bandwidth (~ 40 Gb/s) communication systems; the device is compatible with 160 Gb/s systems.

A related method was introduced to monitor optical SNR changes in incoming transmission signals, again using the average power of a high-order FWM term in an OPA [85]. The basis for the idea is that, for a constant input power, the average FWM power is a function of the signal OSNR: a signal consisting of noiseless well-formed pulses will generate a higher average power via FWM than a signal consisting of noisy pulses. The method was validated, in controlled laboratory experiments, by adding a variable amount of noise to a signal in the form of ASE and measuring the corresponding changes in average power of the FWM term.

It also was proposed to measure chromatic dispersion and OSNR simultaneously with the same setup, since the average power of the FWM term depends on both these features [86].

While these network-monitoring techniques are in principle attractive and have been demonstrated in carefully controlled laboratory experiments, it is not clear whether they could provide accurate information in actual communication networks, where the signal quality might be affected by additional effects. In such situations it might be difficult to ascertain the origin of some variations in the FWM level and to determine the contribution of each possible mechanism accurately.

12.2.11 Quantum effects

In Chapter 3 we studied fiber OPAs using a purely classical formalism, based on Maxwell's equations and the expression for the nonlinear polarization density. A more advanced treatment, based on quantum electrodynamics, was presented in Chapter 9. We will now describe without proof a number of quantum-related features of fiber OPAs, which may find application in the future. Most of these features are related to the fact that the signal and idler are strongly connected at the quantum level.

A remarkable aspect of the signal and idler photon pairs generated by fiber OPAs is that they can be used to generate entangled quantum states. Since these photons have a common origin they behave like twins, in the sense that some of their properties remain identical even if they propagate in very different directions. In particular, if the signal and idler are in the same classical state of polarization (SOP) (say circular) and are analyzed by means of a vertical polarizer followed by a photon detector, one would find that all photon-detection events would be exactly the same in both arms. Hence, there would be a perfect correlation between the two measurements. This is different from what would happen with two independently generated laser beams, in which case there would be no correlation between the photon-detection events in the two arms. The fact that such photon pairs maintain such a strong correlation regardless of separation in space and time is regarded by some as being "spooky", because it appears to imply the instantaneous transfer of information between two locations. In fact, one can show that this does not occur, and that these correlations, while perhaps surprising, do not violate any fundamental laws of physics [87].

Such correlations can potentially be used in a number of applications, in systems based on the properties of quantum states of single particles. In optics, this means that the systems in principle should be able to emit, transmit and detect single photons, ideally with unit efficiency. While there are considerable practical difficulties in making such systems, a great deal of theoretical work has been done to understand their potential. Entangled states could be exploited in the following areas: cryptography, computing, teleportation, and dense coding [87].

Experiments have been performed to demonstrate the properties of the photon pairs generated by fiber OPAs. Pulsed twin beams of light have been generated in a Sagnac-loop non-degenerate fiber OPA [88]. For a gain of 2.2, the intensity noises of the amplified signal and the generated idler (conjugate) pulses were found to be correlated by 5.0 dB, and the subtracted noise dropped below the shot-noise limit by 1.1 dB (2.6 dB when corrected for losses). The generation of twin photon pairs in microstructure fiber has also been investigated [89].

In [90] a source of quantum-correlated photon pairs based on parametric fluorescence in a fiber Sagnac loop was reported. The photon pairs were generated in the 1550 nm fiber-optic communication band and detected with InGaAs–InP avalanche photodiodes operating in a gated Geiger mode. A generation rate greater than 103 pairs per second was observed, limited by the detection electronics. The non-classical nature of photon correlation in the pairs was also demonstrated.

Another interesting feature of fiber OPAs is their ability to reduce fluctuations in light intensity below the so-called shot-noise limit. The shot-noise limit is obtained with a laser output that is in a coherent state (CS). In this case, the light intensity fluctuates about its classical mean value with a standard deviation σ_c that can be calculated by quantum electrodynamics. This variance is the same at any time during an optical cycle. Optical parametric amplifiers, however, can generate squeezed states of light with noise variances σ that are different for the two quadratures: experiments show a minimum value σ_{min} and a maximum value σ_{max}, such that $\sigma_{min}\sigma_{max} = (\sigma_c)^2$. This implies that $\sigma_{min} < \sigma_c$, so that the fluctuations are reduced below the standard shot-noise limit for one quadrature; hence the name squeezed light. Squeezing approaching 10 dB has been demonstrated. Unfortunately, squeezing is very difficult to obtain, and to maintain in the presence of loss; hence, it is correspondingly difficult to use it in optical communication. At this

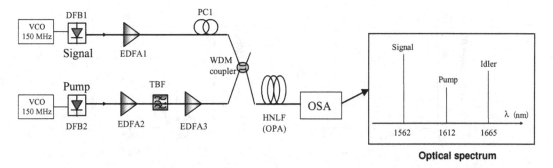

Fig. 12.9 On the left, the experimental setup: VCO, voltage-controlled oscillator; DFB, distributed-feedback laser; TBF, tunable bandpass filter; WDM, wavelength-division multiplexing; OSA, optical spectrum analyzer. On the input, a typical OPA output optical spectrum displayed on the OSA.

time it appears that squeezed light might be most useful for improving the sensitivity of interferometers such as those used for the detection of gravitational waves.

From a practical standpoint, squeezed light can be generated by degenerate fiber OPAs, in which all waves have the same frequency. Experiments have been performed with fiber Sagnac interferometers, and several decibels of squeezing has been demonstrated [91, 92].

12.3   OPAs in high-power wavelength conversion

Since single-mode fibers are capable of withstanding relatively high average powers (more than one kilowatt for large-core fibers), fiber OPAs may also be useful for applications requiring high average or CW powers. In particular, high-power wavelength conversion can potentially be an important area of application for fiber OPAs, as it could generate from existing sources high-power waves in spectral regions currently without such sources. For example, efficient high-power fiber lasers exist between 1 and 1.1 μm (ytterbium-doped fiber lasers, YDFLs) and between 1.53 and 1.62 μm erbium-droped fiber lasers (EDFLs) but not outside these bands. Fiber OPAs have the potential for making efficient wavelength converters, taking the power of one (or both) of these sources and transferring it to another region.

Experiments have been performed to verify some of the quantitative predictions made in Section 5.6 [93]. The schematic of the experimental setup is shown in Fig. 12.9.

The C-band signal was supplied by a DFB laser with a nominal wavelength 1562 nm, which could be tuned over a few nanometers by temperature control. It was amplified to a power of the order of 1 W by a C-band amplifier EDFA1. (The OPA could also be operated with a signal provided by an external cavity laser, tunable over the entire C-band; in this mode, the idler could be tuned from about 1660 to 1700 nm.)

The L-band pump was supplied by a distributed fiber Bragg (DFB) laser with a nominal wavelength 1611 nm. It was amplified to a power of the order of 1 W by the L-band amplifier EDFA2, and this was followed by a narrowband tunable bandpass filter TBF

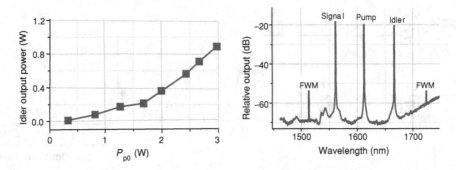

Fig. 12.10 On the left, the idler output power as a function of pump input power for 0.82 W signal input power. The idler wavelength was approximately 1665 nm. The pump wavelength was optimized near 1612 nm to maximize idler output power. On the right, typical optical output spectrum displayed on the OSA; the peaks labeled FWM result from four-wave mixing among the three main waves.

to remove ASE produced by EDFA2; EDFA3 then amplified the pump to a power of up to 3 W.

In order to obtain CW operation it was necessary to suppress stimulated Brillouin scattering (SBS) in the OPA HNLF. To do this, the linewidths of both lasers were increased by modulating their currents with the 150 MHz sinusoidal output of the VCOs. This produced some intensity modulation of the DFBs but, more importantly, substantial frequency modulation (FM). In this manner it was possible to increase the linewidths to several gigahertz, which was sufficient to suppress SBS efficiently. The SBS suppression was monitored as the powers were increased by means of a 1% tap inserted between the WDM coupler and the HNLF (not shown). A fused-fiber WDM coupler combined the signal and pump, with low loss, into the HNLF.

The HNLF was manufactured by Sumitomo Electric. It had the following nominal parameters: $\lambda_0 = 1612$ nm; $\gamma = 12$ W^{-1} km^{-1}; $D_\lambda = 0.024$ ps nm^{-2} km^{-1}; $\beta_4 = -1 \times 10^{-55}$ s^4 m^{-1}; $L = 40$ m.

Figure 12.10 (left) shows the measured idler output power P_i as a function of the pump power P_{p0}, for a signal input power $P_{s0} = 0.82$ W. The maximum value $P_i = 0.9$ W was obtained for $P_{p0} = 3$ W. This corresponds to an optical conversion efficiency of 23% (defined as the ratio of the idler power and the input pump power plus the signal power).

Figure 12.10 (right) shows a typical OPA output spectrum displayed by the OSA. The three main peaks correspond to the signal, pump, and idler. The two smaller peaks labeled FWM are due to spurious four-wave mixing between the three main waves. They are at least 30 dB lower than the main peaks. This shows that: (i) they have a negligible impact on the optical efficiency; (ii) the three-wave analysis used for the theoretical model is well justified. The ASE level between the main peaks is at least 40 dB less than the peak level, which shows that the three output waves have a high optical SNR, desirable in some applications.

Experiments were also performed with λ_i about 20 nm longer, and comparable results were obtained.

12.4 OPOs

Wavelength conversion can also be accomplished by using OPOs instead of OPAs. There are trade-offs involved in choosing one approach or the other. The most obvious is that an OPO does not require a signal source to seed it. Only the pump source is required, and this may lead to a simpler system as well as to a higher conversion efficiency (idler output power divided by total input power).

An important trade-off concerns the linewidth of the output. In an OPA, it is basically determined by the linewidths of the signal and the pump and with proper design can be as small as a few kilohertz. With an OPO, the idler linewidth depends again on the pump and signal linewidths but now the latter is determined by the OPO's optical cavity. A difficulty with fiber OPOs is that the cavity length is typically many meters long, so that the frequency spacing between the longitudinal modes is small, typically of the order of 1 MHz. As a result, it is difficult to make an intracavity optical filter that can select and track a single frequency. For example, the reflective FBGs that could be used in a cavity as end mirrors may have a bandwidth of a few GHz at best, which implies that many longitudinal modes will oscillate simultaneously. It is in principle possible to design more complex filtering systems that could achieve single-frequency operation, but they would be lossy. To date, this difficulty has prevented the single-frequency operation of fiber OPOs, and achieving it while maintaining desirable properties such as high efficiency and tunability poses a significant challenge.

The advantage of a pulsed OPO source is that one can have a high peak power with a moderate average power, in combination with a low duty cycle. The peak power is particularly useful for obtaining a large gain bandwidth, which can reach hundreds of nanometers (see subsection 11.2.1); this can in principle be used for making a light source with output tunable over this wide range.

The performances obtained with the first fiber OPOs are quite respectable and indicate that, with future improvements, such devices should eventually approach their theoretical performance and provide interesting new types of light source. Their performance was reviewed in Chapter 11.

References

1. "Interband wavelength conversion of 320 Gb/s (32×10 Gb/s) WDM signal using a polarization-insensitive fiber four-wave mixer," Watanabe, S., Takeda, S., Chikama, T. In *Proc. European Conf. on Optical Communication*, 1998; vol. 3, pp. 83–7.
2. "Polarization independent frequency conversion by fiber four-wave mixing with a polarization diversity technique," Hasegawa, T, Inoue, K., Oda, K. *IEEE Photon. Technol. Lett.*; 1993; vol. 5, pp. 947–9.
3. "Polarization-independent fiber optical parametric amplifier," Wong, K. Y., Marhic, M. E., Uesaka, K., Kazovsky, L. G. In *Proc. OECC/IOOC* 2001, Sydney, July 2001.
4. "Polarization-independent one-pump fiber-optical parametric amplifier," Wong, K. K. Y., Marhic, M. E., Uesaka, K., Kazovsky, L. G. *IEEE Photon. Technol. Lett.*, 2002; vol. 14, pp. 1506–8.

5. "Fiber parametric amplifiers for wavelength band conversion," Islam, M. N., Boyraz, O. *IEEE J. Selected Topics in Quantum Electron.*; 2002; vol. 8, pp. 527–37.
6. "Polarization-independent two-pump fiber optical parametric amplifier with polarization diversity technique," Kalogerakis, G., Marhic, M. E., Kazovsky, L. G. In *Proc. Optical Fiber Communication Conf.*, March 2006, Anaheim CA; paper OWT4.
7. "Polarisation-independent phase conjugation of light-wave signals," Jopson, R. M., Tench, R. E. *Electron. Lett.*; 1993; vol. 29, pp. 2216–7.
8. "Polarization independent wavelength conversion using fiber four-wave mixing with two orthogonal pump lights of different frequencies," Inoue, K. *J. Lightwave Technol.*; 1994; vol. 12, pp. 1916–20.
9. "Polarization-independent two-pump fiber optical parametric amplifier," Wong, K. K. Y., Marhic, M. E., Uesaka, K., Kazovsky, L. G. *IEEE Photon. Technol. Lett.*; 2002; vol. 14, pp. 911–3.
10. "Polarization-independent, highly-efficient optical fiber wavelength converter without spectral spread using synchronous phase/frequency modulations," Yamashita, S., Torii, K. *IEICE Transactions on Electron.*; 2003; vol. E86–C, pp. 1370–3.
11. "Two-pump fiber parametric amplifiers," Radic, S., McKinstrie, C. J. *Optical Fiber Technology: Materials, Devices and Systems*; 2003; vol. 9, pp. 7–23.
12. "Polarization independent, all-fiber phase conjugation incorporating inline fiber DFB lasers," Yamashita, S., Set, S. Y., Laming, R. I. *IEEE Photon. Technol. Lett.*; 1998; vol. 10, pp. 1407–9.
13. "Polarization dependent parametric gain in amplifiers with orthogonally multiplexed optical pumps," Radic, S., Mckinstrie, C. J., Jopson, R. In *Proc. Optical Fiber Communication Conf.*, Atlanta GA, March 2003; *Technical Digest*: vol. 2, paper ThK3, pp. 508–9.
14. "Fiber-optic parametric amplifiers in the presence of polarization-mode dispersion and polarization-dependent loss," Yaman, F., Lin, Q., Radic, S., Agrawal, G. P. *J. Lightwave Technol.*; 2006; vol. 24, pp. 3088–96.
15. "Raman-induced polarization-dependent gain in parametric amplifiers pumped with orthogonally polarized lasers," Lin, Q., Yaman, F., Agrawal, G. P. *IEEE Photon. Technol. Lett.*; 2006; vol. 18, pp. 397–9.
16. "Combating dispersion with parametric amplifiers," Li, R. D., Kumar, P., Kath, W. L., Kutz, J. N. *IEEE Photon. Technol. Lett.*; 1993; vol. 5, pp. 669–72.
17. "Long-distance pulse propagation in nonlinear optical fibers by using periodically spaced parametric amplifiers," Kutz, J. N., Kath, W. L., Li, R. D., Kumar, P. *Opt. Lett.*; 1993; vol. 18, pp. 802–4.
18. "Pulse propagation in nonlinear optical fiber lines that employ phase-sensitive parametric amplifiers," Kutz, J. N., Hile, C. V., Kath, W. L., Li, R. D., Kumar, P. *J. Opt. Soc. Amer. B*; 1994; vol. 11, pp. 2112–23.
19. "Reduction of quantum noise in soliton propagation by phase-sensitive amplification," Deutsch, I. H., Abram, I. *J. Opt. Soc. Amer. B*; 1994; vol. 11, pp. 2303–13.
20. "All-optical regeneration of differential phase-shift keying signals based on phase-sensitive amplification," Croussore, K., Kim, C., Li, G. *Opt. Lett.*; 2004; vol. 29, pp. 2357–9.
21. "Compensation of the soliton self-frequency shift with phase-sensitive amplifiers," Goedde, C. G., Kath, W. L., Kumar, P. *Opt. Lett.*; 1994; vol. 19, pp. 2077–9.
22. "Parametric interaction of a modulated wave in a single-mode fiber," Bar Joseph, I., Friesem, A. A., Waarts, R. G., Yaffee, H. H. *Opt. Lett.*; 1986; vol. 11, pp. 534–6.
23. "Near-noiseless amplification of light by a phase-sensitive fibre amplifier," Levandovsky, D., Vasilyev, M., Kumar, P. *Pramana*; 2001; vol. 56, pp. 281–5.

24. "Phase-sensitive amplifier based on two-pump four-wave mixing in an optical fiber," Takano, K., Tanemura, T., Kikuchi, K. In *Proc. European Conf. on Optical Communication*, Cannes, September 2006; vol. 4, paper Tu1.3.6., pp. 23–4.
25. "Phase-sensitive amplification in a fiber," McKinstrie, C. J., Radic, S. *Optics Express*; 2004; vol. 12, pp. 4973–9.
26. "In-line frequency-nondegenerate phase-sensitive fiber-optical parametric amplifier," Tang, R., Devgan, P., Voss, P. L., Grigoryan, V. S., Kumar, P. *IEEE Photon. Technol. Lett.*; 2005; vol. 17, pp. 1845–7.
27. "High-repetition-rate pulsed-pump fiber OPA for amplification of communication signals," Kalogerakis, G., Shimizu, K., Marhic, M. E., Wong, K. K. Y., Uesaka, K., Kazovsky, L. G. *J. Lightwave Technol.*; 2006; vol. 24, pp. 3021–7.
28. "Influence of the pump spectrum on three-wave mixing parametric amplification," Helmfrid, S., Arvidsson, G. *J. Opt. Soc. Amer. B*; 1991; vol. 8, pp. 2477–80.
29. "Continuous-wave fiber optical parametric wavelength converter with +40-dB conversion efficiency and 3.8-dB noise figure," Wong, K. K. Y., Shimizu, K., Marhic, M. E., Uesaka, K., Kalogerakis, G., Kazovsky, L. G. *Opt. Lett.*; 2003; vol. 28, pp. 692–4.
30. "Highly efficient four-wave mixing in an optical fiber with intensity dependent phase matching," Yamamoto, T., Nakazawa, M. *IEEE Photon. Technol. Lett.*; 1997; vol. 9, pp. 327–9.
31. "Wavelength conversion bandwidth in fiber based optical parametric amplifiers," McKerracher, R. W., Blows, J. L., de Sterke, C. M. *Optics Express*; 2003; vol. 11, pp. 1002–7.
32. "Transparent wavelength conversion in fibre with 24 nm pump tuning range," Westlund, M., Hansryd, J., Andrekson, P. A., Knudsen, S. N. *Electronics Lett.*; 2002; vol. 17; vol. 38, no. 2, pp. 85–6.
33. "Narrow linewidth wavelength converter with 70 nm of signal tuning band using strain distribution to suppress SBS," Marconi, J. D, Chavez Boggio, J. M., Fragnito, H. L. In *Proc. 31st European Conf. on Optical Communications*, September 2005, Glasgow; paper Mo4.5.6.
34. "Wide-band tuning of the gain spectra of one-pump fiber optical parametric amplifiers," Marhic, M. E., Wong, K. K. Y., Kazovsky, L. G. *IEEE J. Select. Topics in Quantum Electron.*; 2004; vol. 10; pp. 1133–41.
35. "Multiple wavelength conversion with gain by a high-repetition-rate pulsed-pump fiber optical parametric amplifier," Kalogerakis, G., Marhic, M. E., Kazovsky, L. G. *J. Lightwave Technol.*; 2005; vol. 23, pp. 2954–60.
36. "Multiple-band bit-level switching in two-pump parametric devices," Radic, S., McKinstrie, C. J., Jopson, R. M., Centani, J. C., Chraplyvy, A. R. *IEEE Photon. Technol. Lett.*; 2004; vol. 16, pp. 852–4.
37. "Selective suppression of idler spectral broadening in two-pump parametric architectures," Radic, S., McKinstrie, C. J., Jopson, R. M., Centanni, J. C., Chraplyvy, A. R., Jorgensen, C. G., Brar, K., Headley, C. *IEEE Photon. Technol. Lett.*; 2003; vol. 15, pp. 673–5.
38. "Compensation for channel dispersion by nonlinear optical phase conjugation," Yariv, A., Fekete, D., Pepper, D. M. *Opt. Lett.*; 1979; vol. 4, pp. 52–4.
39. "Cancellation of third-order nonlinear effects in amplified fiber links by dispersion compensation, phase conjugation, and alternating dispersion," Marhic, M. E., Kagi, N., Chiang, T. K., Kazovsky, L. G. *Opt. Lett.*; 1995; vol. 20, pp. 863–5.
40. "Dispersion compensation with SBS-suppressed fibre phase conjugator using synchronized phase modulation," Tani, M., Yamashita, S. *Electron. Lett.*; 2003; vol. 39, pp. 1375–7.
41. "Wavelength division multiplexed transmission over standard single mode fiber using polarization insensitive signal conjugation in highly nonlinear optical fiber," Radic, S., Jopson,

R. M., McKinstrie, C. J., Gnauck, A. H., Chandrasekar, S., Centanni, J. C. In *Proc. Optical Fiber Communication Conf.*, Atlanta GA, March 2003; paper PD12.

42. "Parametric amplifier for mid-span phase conjugation with simultaneous compensation of fiber loss and chromatic dispersion at 10 Gb/s," Boggio, J., Chavez M., Guimaraes, A., Callegari, F. A., Marconi, J. D., Rocha, M. L., DeBarros, M. R. X., Fragnito, H. L. *Microwave and Optical Technology Lett.*; 2004; vol. 42, pp. 503–5.

43. "Cancellation of spectral spread in SBS-suppressed fiber wavelength converters using a single phase modulator," Yamashita, S., Tani, M. *IEEE Photon. Technol. Lett.*; 2004; vol. 16, pp. 2096–8.

44. "Reduced timing jitter in dispersion-managed light-wave systems through parametric amplification," Santhanam, J., Agrawal, G. P. *J. Opt. Soc. Amer. B*; 2003; vol. 20, pp. 284–91.

45. "Performance analysis of variable optical delay circuit using highly nonlinear fiber parametric wavelength converters," Sakamoto, T., Okada, A., Moriwaki, O., Matsuoka, M., Kikuchi, K. *J. Lightwave Technol.*; 2004; vol. 22, pp. 874–81.

46. "Optical delay line using four-wave mixing for repeated wavelength shifting," Marhic, M. E. proposal to US Army Research Office, 1999.

47. "Tunable all optical delay via slow and fast light propagation in a Raman assisted fiber optical parametric amplifier: a route to all optical buffering," Dahan, D., Eisenstein, G. *Optics Express*; 2005; vol. 13, pp. 6234–49.

48. "All-optical, wavelength and bandwidth preserving, pulse delay based on parametric wavelength conversion and dispersion," Sharping, J., Okawachi, Y. van Howe, J., Xu, C., Willner, A., Wang, Y., Gaeta, A. *Optics Express*; 2005; vol. 13, pp. 7872–7.

49. "12.47 ns continuously-tunable two-pump parametric delay," Ren, J., Alic, N., Myslivets, E., Saperstein, R. E., McKinstrie, C. J., Jopson, R. M., Gnauck, A. H., Andrekson, P. A., Radic, S. In *Proc. ECOC'06*, Cannes, September 2006; postdeadline paper Th.4.4.3, pp. 45–6.

50. "All-optical storage of a picosecond-pulse packet using parametric amplification," Bartolini, G. D., Serkland, D. K., Kumar, P., Kath, W. L. *IEEE Photon. Technol. Lett.*; 1997; vol. 9, pp. 1020–2.

51. "All-optical picosecond-pulse packet buffer based on four-wave mixing loading and intracavity soliton control," Wang, L., Agarwal, A., Su, Y., Kumar, P. *IEEE J. Quantum Electron.*; 2002; vol. 38, pp. 614–19.

52. "All-optical loadable and erasable storage buffer based on parametric nonlinearity in fiber," Agarwal, A., Wang, L. J., Su, Y. K., Kumar, P. *J. Lightwave Technol.*; 2005; vol. 23, pp. 2229–38.

53. "Active optical pulse compression with a gain of 29.0 dB by using four-wave mixing in an optical fiber," Yamamoto, T., Nakazawa, M. *IEEE Photon. Technol. Lett.*; 1997; vol. 9, pp. 1595–7.

54. "10-GHz return-to-zero pulse source tunable in wavelength with a single- or multiwavelength output based on four-wave mixing in a newly developed highly nonlinear fiber," Clausen, A. T., Oxenlowe, L., Peucheret, C., Poulsen, H. N., Jeppesen, P., Knudsen, S. N., Gruner-Nielsen, L. *IEEE Photon. Technol. Lett.*; 2001; vol. 13, pp. 70–2.

55. "Wavelength tunable 40GHz pulse source based on fibre optical parametric amplifier," Hansryd, J., Andrekson, P. A. *Electron. Lett.*; 2001; vol. 37, pp. 584–5.

56. "40-Gb/s transmission using RZ-pulse source based on fiber optical parametric amplification," Torounidis, T., Sunnerud, H., Hedekvist, P. O., Andrekson, P. A. *IEEE Photon. Technol. Lett.*; 2003; vol. 15, pp. 1159–61.

57. "Signal generation and transmission at 40, 80, and 160 Gb/s using a fiber-optical parametric pulse source," Torounidis, T., Westlund, M., Sunnerud, H., Olsson, B. E., Andrekson, P. A. *IEEE Photon. Technol. Lett.*; 2005; vol. 17, pp. 312–14.
58. "Fibre optical parametric amplifier pulse source: theory and experiments," Torounidis, T., Karlsson, M., Andrekson, P. A. *J. Lightwave Technol.*; 2005; vol. 23, pp. 4067–73.
59. "Regeneratively modelocked dual-wavelength soliton-pulse fibre-optical parametric oscillator in C- and L-bands," Lasri, J., Devgan, P., Tang, R., Grigoryan, V., Kath, W. L., Kumar, P. *Electron. Lett.*; 2004; vol. 40, pp. 622–3.
60. "Wavelength-tunable all-optical clock recovery using a fiber-optic parametric oscillator," Wang, L. Y., Agarwal, A., Kumar, P. *Optics Comm.*; 2000; vol. 184, pp. 151–6.
61. "Clock multiplication in a singly resonant fiber parametric oscillator," Franco, P., Fontana, F., Cristiani, I., Zenobi, M., Midrio, M., Romagnoli, M. *Opt. Lett.*; 1996; vol. 21, pp. 788–90.
62. "Fiber four-wave mixing demultiplexing with inherent parametric amplification," Hedekvist, P. O., Karlsson, M., Andrekson, P. A. *J. Lightwave Technol.*; 1997; vol. 15, pp. 2051–8.
63. "O-TDM demultiplexer with 40-dB gain based on a fiber optical parametric amplifier," Hansryd, J., Andrekson, P. A. *IEEE Photon. Technol. Lett.*; 2001; vol. 13, pp. 732–4.
64. "Simultaneous 3R regeneration and wavelength conversion using a fiber-parametric limiting amplifier," Su, Y. Wang, L. Agrawal, A. Kumar, P. In *Proc. Optical Fiber Communication Conf.*, Anaheim CA, March 2001; paper MG4.
65. "All-optical 2R regeneration using data-pumped fibre parametric amplification," Li, Y., Croussore, K., Kim, C., Li, G. *Electron. Lett.*; 2003; vol. 39, pp. 1263–4.
66. "Tunable wavelength converter using cross-gain modulation in a fiber optical parametric amplifier," Sakamoto, T., Wong, K. K. Y., Uesaka, K., Marhic, M. E., Kazovsky, L. G. In *Proc. Optical Fiber Communication Conf.*, Anaheim CA, March 2002; paper TuS4.
67. "Regeneration of RZ-DPSK signals by fiber-based all-optical regenerators," Matsumoto, M. *IEEE Photon. Technol. Lett.*; 2005; vol. 17, pp. 1055–7.
68. "300-Gb/s eye-diagram measurement by optical sampling using fiber-based parametric amplification," Li, J., Hansryd, J., Hedekvist, P. O., Andrekson, P. A., Knudsen, S. N. *IEEE Photon. Technol. Lett.*; 2001; vol. 13, pp. 987–9.
69. www.picosolve.com.
70. "Novel fiber Kerr-switch with parametric gain demonstration of optical demultiplexing and sampling up to 640 Gb/s," Watanabe, S., Okabe, R., Hainberger, R., Schmidt-Langhorst, C., Schubert, C., Weber, H. G. In *Proc. European Conf. on Optical Communication*, 2004; paper Th4.1.6.
71. "Optical level equalization based on gain saturation in fibre optical parametric amplifier," Inoue, K. *Electron. Lett.*; 2000; vol. 36, pp. 1016–7.
72. "All-optical limiter using gain flattened fibre parametric amplifier," Su, Y., Wang, L., Agarwal, A., Kumar, P. *Electron. Lett.*; 2000; vol. 36, pp. 1103–5.
73. "Experimental study on noise characteristics of a gain-saturated fiber optical parametric amplifier," Inoue, K., Mukai, T. *J. Lightwave Technol.*; 2002; vol. 20, pp. 969–74.
74. "Timing-jitter and amplitude-noise reduction by a chirped pulsed-pump fiber OPA", Shimizu, K., Kalogerakis, G., Marhic, M. E., Kazovsky, L. G. In *Proc. Optical Fiber Communication Conf.*, Atlanta GA, March 2003; paper TuH5, pp. 197–8.
75. "All-optical signal reshaping via four-wave mixing in optical fibers," Ciaramella, E., Trillo, S. *IEEE Photon. Technol. Lett.*; 2000; vol. 12, pp. 849–51.

76. "Suppression of level fluctuation without extinction ratio degradation based on output saturation in higher order optical parametric interaction in fiber," Inoue, K. *IEEE Photon. Technol. Lett.*; 2001; vol. 13, pp. 338–40.
77. "All-optical regeneration in one- and two-pump parametric amplifiers using highly nonlinear optical fiber," Radic, S., McKinstrie, C. J., Jopson, R. M., Centanni, J. C., Chraplyvy, A. R. *IEEE Photon. Technol. Lett.*; 2003; vol. 15, pp. 957–9.
78. "All-optical regeneration of differential phase-shift keying signals based on phase-sensitive amplification," Croussore, K., Kim, C., Li, G. *Opt. Lett.*; 2004; vol. 29, pp. 2357–9.
79. "Demonstration of phase-regeneration of DPSK signals based on phase-sensitive amplification," Croussore, K., Kim, I., Han, Y., Kim, C., Li, G. *Optics Express*; 2005; vol. 13, pp. 3945–50.
80. "All-optical regeneration of differential phase-shift keyed signals based on phase-sensitive amplification," Croussore, K., Kim, C., Li, G. in *Proc. SPIE Defense and Security Symposium*, 2005, vol. 5814, pp. 166–75.
81. "Phase-and-amplitude regeneration of differential phase-shift keyed signals using a phase-sensitive amplifier," Croussore, K., Kim, I., Kim, C., Han, Y., Li, G. *Optics Express*; 2006; vol. 14, pp. 2085–94.
82. "A novel dispersion monitoring technique based on four-wave mixing in optical fiber," Li, S., Kuksenkov, D. V. *IEEE Photon. Technol. Lett.*; 2004; vol. 16, pp. 942–4.
83. "Simultaneous residual chromatic dispersion monitoring and frequency conversion with gain using a parametric amplifier," Ng, T. T., Blows, J. L., Mok, J. T., Hu, P., Bolger, J. A., Hambley, P., Eggleton, B. J. *Optics Express*; 2003; vol. 11, pp. 3122–7.
84. "Cascaded four-wave mixing in fiber optical parametric amplifiers: application to residual dispersion monitoring," Ng, T. T., Blows, J. L., Mok, J. T., McKerracher, R. W., Eggleton, B. J. *J. Lightwave Technol.*; 2005; vol. 23, pp. 818–26.
85. "In-band OSNR monitoring using fibre optical parametric amplifier," Ng, T. T., Blows, J. L., Mok, J. T., McKerracher, R. W., Eggleton, B. J. *Electron. Lett.*; 2005; vol. 41; pp. 352–3.
86. "Simultaneous in-band OSNR and chromatic dispersion monitoring using a fibre optical parametric amplifier," Ng, T. T., Blows, J. L., Rochette, M., Bolger, J. A., Littler, I., Eggleton, B. J. *Optics Express*; 2005; vol. 13; pp. 5542–52.
87. "Quantum information processing: cryptography, computation, and teleportation," Spiller, T. P. *Proc. IEEE*; 1996; vol. 48, pp. 1719–46.
88. "Observation of twin-beam-type quantum correlation in optical fiber," Sharping, J. E., Fiorentino, M., Kumar, P. *Opt. Lett.*; 2001; vol. 26, pp. 367–9.
89. "Quantum-correlated twin photons from microstructure fiber," Sharping, J. E., Chen, J., Li, X., Kumar, P., Windeler, R. S. *Optics Express*; 2004; vol. 12, pp. 3086–94.
90. "All-fiber photon-pair source for quantum communications," Fiorentino, M., Voss, P. L., Sharping, J. E., Kumar, P. *IEEE Photon. Technol. Lett.*; 2002; vol. 14, pp. 983–5.
91. "Squeezing in fibers with optical pulses," Bergman, K., Haus, H. A. *Opt. Lett.*; 1991; vol. 16, pp. 663–5.
92. "Amplitude squeezing of light by means of a phase-sensitive fiber parametric amplifier," Levandovsky, D., Vasilyev, M., Kumar, P. *Opt. Lett.*; 1999; vol. 24, pp. 984–6.
93. "Tunable fiber parametric wavelength converter with 900 mW of CW output power at 1665 nm," Marhic, M. E., Williams, G. M., Goldberg, L., Delavaux, J. M. P. In *Conf. Photonics West*, San Jose C A, January 2006; *Proc. SPIE*, vol. 6103, pp. 165–76.

13 Nonlinear crosstalk in fiber OPAs

13.1 Introduction

Fiber OPAs have a number of features that are potentially attractive for optical communication systems. In particular there is the prospect of using amplifiers with several hundred nanometers of gain bandwidth for the construction of wavelength-division multiplexing (WDM) systems with tens of terabits per second of capacity. However, before this potential can be realized a number of effects that can degrade the signal-to-noise ratio (SNR) must be addressed. Specifically these effects are: cross-gain modulation, mediated via pump depletion; four-wave mixing (FWM) between signals or between signals and a pump; cross-phase modulation (XPM) between signals. The physical origin of these effects is simple: efficient OPA operation generally requires that all the waves of interest be close to the zero-dispersion wavelength (ZDW) of the fiber, and this in turn implies that a number of other unwanted nonlinear interactions will also be well phase matched and so may generate large undesired effects. This origin indicates that these effects are fundamental in nature and may therefore be difficult to suppress.

Because of the importance of this issue, these effects have been investigated in depth by several research groups [1–12]. It has been found, through simulations and experimentally, that some of these effects can indeed be quite large under certain circumstances, to the point where they bring into question the viability of using fiber OPAs in WDM communication systems. It is therefore important to analyze in depth the origin of these effects and to quantify their dependence on the various parameters, in order to design strategies to minimize their impact on system performance.

In this chapter we provide simplified analyses of FWM, cross-gain modulation (XGM), and cross-phase modulation (XPM). This approach has the advantage of showing how the various terms scale with the system parameters, so that strategies can be designed to minimize them. We begin with a study of FWM crosstalk, which has received the most attention in the literature. We then consider XGM crosstalk, which has been the object of fewer studies. Finally, we consider the possibility of XPM crosstalk.

13.2 Four-wave mixing

It is well known that multiple optical carriers in long transmission fibers can interact with each other by means of signal–signal FWM, generating new waves that can fall

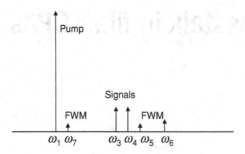

Fig. 13.1 Spurious FWM in a one-pump OPA: ω_5 is due to signal–signal interaction and ω_6 and ω_7 are due to pump–signal interaction.

at the same frequencies as other carriers and so degrading their SNR. It is also known that the same phenomenon can occur in amplifiers such as erbium-doped fiber amplifiers (EDFAs), semiconductor optical amplifiers (SOAs), and Raman amplifiers, because of the relatively high nonlinearity of the gain media and the good phase matching that prevails over their short lengths.

Fiber OPAs are subject to the same phenomenon, for the same reasons. The problem is exacerbated by the fact that, in order to have a large bandwidth, fiber OPAs need to be operated in the vicinity of the zero-dispersion wavelength (ZDW). This in turn implies that chromatic dispersion is small, so that signal–signal FWM between closely spaced signals can be very well phase matched. When combined with the high nonlinearity of the fiber medium, this results in the generation of FWM terms that can exceed the levels encountered in EDFAs.

In addition, in a fiber OPA with moderate to large gain, idlers are generated that have almost the same amplitude as the corresponding signals over most of the amplifier length. These idlers will in turn give rise to FWM terms, by interactions among themselves, with the pumps, and with the signals.

13.2.1 One-pump OPA

To clarify the nature of the possible FWM terms, let us first consider a one-pump fiber OPA with a pump at ω_1, which generates gain for signals at ω_3 and ω_4; see Fig. 13.1. For simplicity, the idlers are not shown; they would be to the left of the pump. Figure 13.1 shows some of the FWM terms that are generated by the signals and the pump.

In Table 13.1 are listed all the possible types of undesired FWM term that arise from two signals and their associated idlers, in either a one-pump or a two-pump OPA. The initial signals are denoted by S_1 and S_2, the idlers by I_1 and I_2, and the pumps by P_1 and P_2. An "x" in a box signifies that the corresponding wave is involved in the generation of the FWM term corresponding to that line. On each line there are either two x's (degenerate FWM) or three x's (non-degenerate FWM). For each line with two (three) x's, there are actually two (three) FWM frequencies. Thus the total number of FWM terms is simply equal to the number of x's in the table. (However, for a one-pump OPA one of the two FWM terms associated with lines 12 and 13 is actually the idler corresponding to that

Table 13.1. FWM terms due to two signals in one-pump and two-pump OPAs

	S_1	S_2	I_1	I_2	P_1	P_2
1	x	x				
2			x	x		
3	x	x	x			
4	x	x		x		
5	x		x	x		
6		x	x	x		
7	x		x			
8		x	x			
9	x			x		
10		x		x		
11	x	x			x	
12	x				x	
13		x			x	
14			x	x	x	
15			x		x	
16				x	x	
17	x		x		x	
18	x			x	x	
19		x	x		x	
20		x		x	x	
21	x	x				x
22	x					x
23		x				x
24			x	x		x
25			x			x
26				x		x
27	x		x			x
28	x			x		x
29		x	x			x
30		x		x		x
31	x				x	x
32		x			x	x
33			x		x	x
34				x	x	x

signal and should not be counted as a spurious FWM term, but for a two-pump OPA it should be kept in that category.)

For any OPA, lines 1–10 are applicable as they contain only signals and idlers; this corresponds to a total of 24 terms. For a one-pump OPA, with pump P_1, lines 11–20 are added, corresponding to an additional 26 terms that involve P_1. Hence a one-pump OPA has a total of $50 - 4 = 46$ spurious FWM terms.

For a two-pump OPA, new FWM terms arise from the interaction of P_2 with signals and/or idlers (but not with P_1). These corresponds to lines 21–30, which are the same as lines 11–20 but with P_1 replaced by P_2. In addition, there are four terms involving both

P_1 and P_2, and one signal or one idler; these correspond to lines 31–34. Subtracting the four terms that simply correspond to the signals and idlers, we find that that there are a total of $88 - 4 = 84$ spurious FWM terms.

Simulations do not show this many spurious frequencies, for either the one-pump or the two-pump case. The reason is that some FWM terms are degenerate, in the sense that they have exactly the same frequency.

In typical communication applications the OPA pump powers used are much larger than the signal powers (about 100–1000 times larger). Hence it is clear that FWM interactions that involve a pump have the potential for yielding undesired FWM terms that are much stronger than interactions involving only signals and/or idlers. Therefore we can classify all possible FWM interactions into two groups, depending on whether they involve a pump: interactions involving a pump will be referred to as being of first order while those not involving a pump will be said to be of second order. The reason for this terminology will be made more explicit below when we calculate the magnitude of the FWM terms in certain situations.

In all OPAs with moderate to large gain the idlers have almost the same magnitude as the signals, and so the signals, idlers, and pump(s) always have a spectrum consisting of symmetric frequencies. As a result, the spectrum of the resulting FWM terms will have the same property (note that the amplitudes of symmetric components may be slightly different). Symmetric pairs of FWM terms will behave as a signal and idler pair and will be amplified by the OPA gain as such. It is important to note that this amplification will be phase sensitive, because both FWM terms initially arise independently; this phase dependence may provide opportunities for selective FWM reduction.

Finally, it is worth noting that when calculating the output spectrum of an OPA by numerical means such as the split-step Fourier method (SSFM), one may see a number of terms that are not accounted for in the list of Table 13.1. These terms are the result of further FWM of terms listed in the table, which may mix with each other and/or possibly with a pump. Generally these terms are smaller than those discussed so far, and for that reason we will not consider them further.

(a) Second-order FWM

Let us consider for example the case where the two initial signals of Fig. 13.1 mix, to yield a FWM term at $\omega_5 = 2\omega_3 - \omega_4$. We assume that the signals exhibit maximum parametric gain, so that $g_3 = g_4 = g_5 = g = \gamma P_{10}$, and also that the gains are nearly exponential [i.e. the power gain $G \approx e^{2gz}/4$], which is a good approximation for moderate to large gains. We also assume that the FWM interaction is perfectly phase matched, which provides a worst-case situation.

We use here an approximate analysis in order to estimate the magnitude of the FWM terms. This approach has been used previously for studying FWM in other types of amplifier, such as EDFAs [13], erbium-doped waveguide amplifiers (EDWAs) [14, 15], and Raman amplifiers [16]. A key assumption is that the FWM terms are amplified by the parametric gain, just as signal and idler, but we do not consider in detail the equation governing the idler associated with the FWM term. We let P_{k0}, $k = 1, 3, 4$, denote the powers of the three initial waves. With these assumptions the field at ω_s, A_5, is governed by

$$\frac{dA_5}{dz} = gA_5 + i\gamma A_3^* A_4^2 = gA_5 + \frac{i\gamma A_{30}^* A_{40}^2}{8} e^{3gz}. \tag{13.1}$$

The solution is

$$A_5 = \frac{i\gamma A_{30}^* A_{40}^2}{16g} e^{3gL} = \frac{iA_{30}^* A_{40}^2}{16P_{10}} e^{3gL}. \tag{13.2}$$

In terms of powers, we find that the ratio of the FWM crosstalk power to the signal output power is

$$XT_{ss} = \frac{P_5}{GP_{30}} = \frac{1}{4}\left(\frac{P_{40}}{P_{10}}\right)^2 G^2 = \frac{1}{4}\left(\frac{GP_{40}}{P_{10}}\right)^2 = \frac{1}{4}\left(\frac{P_{4,\text{out}}}{P_{10}}\right)^2, \tag{13.3}$$

where $P_{4,\text{out}} = GP_{40}$ is the signal output power.

Equation (13.3) predicts that *this type of crosstalk depends only on the pump power and the signal output power*. It does not explicitly depend on the gain or the signal input power. It is also completely independent of the type of fiber used and is the same whether dispersion-shifted fiber (DSF) or highly nonlinear fiber (HNLF) is employed. This is a very important design tool, because it tells us how we should choose the pump power and output signal level to achieve a particular level of crosstalk. This should be done without considering the OPA gain required or the signal input level; these should be dealt with separately, choosing the OPA gain necessary to overcome the span loss and/or give the necessary bandwidth.

We also note that the FWM power depends on the *square* of the ratio of the signal output power and pump power (which should be a small number in a typical OPA for communication applications). This is why we classify this type of FWM as being of second order.

It is interesting to note that the quadratic dependence of XT_{ss} on $P_{4,\text{out}}$ predicted by Eq. (13.3) is in good agreement with experiment. In [12] it was found that when plotted on a logarithmic scale the graph of crosstalk power versus signal output power was nearly linear, with a slope of 2.2; this is close to the value 2 expected for a quadratic relationship.

It is also often convenient to use decibels for the various quantities in Eq. (13.3); this leads to

$$(XT_{ss})_{\text{dB}} = 2(P_{4,\text{out}} - P_{10})_{\text{dB m}} - 6. \tag{13.4}$$

This form is particularly useful when dealing with experiments, for which the output pump and signal powers can simply be read off an optical spectrum analyzer (OSA), directly in dB m. It is also interesting to note that because only the difference between these two powers enters Eq. (13.4), it is actually not necessary for the OSA readings to be calibrated to provide absolute powers; relative powers are sufficient.

Let us consider an example. In a long-haul communication system we may want to have $P_{4,\text{out}} = 0$ dB m. If we want a crosstalk level $(XT_{ss})_{\text{dB}} = -46$ dB, which may be adequate for a communication system, we find that we need $P_{10} = 20$ dB m. This is a reasonable pump power for a fiber OPA.

It should be noted that Eqs. (13.3) and (13.4) correspond to a single FWM term, generated by two signals. In a wideband WDM system the number of signals N_s may be as large as 100. These signals generate a large number, of the order of N_s^3, of FWM terms. Hence some FWM terms may fall on a particular signal; this number is of the order of N_s^2. Assuming that these FWM terms add up on an incoherent (power) basis then the crosstalk level given by Eq. (13.3) should be multiplied by a term of the order of N_s^2. For example, if N_s is an even number and the signals are symmetrically distributed about an empty channel then it can be shown that the number of FWM components falling on the empty channel is well approximated by

$$N_{\text{FWM}} \approx \frac{3}{2}\frac{N_s}{2}\left(\frac{N_s}{2}-1\right)+1. \tag{13.5}$$

Then Eq. (13.4) should be modified as follows:

$$(XT_{ss})_{\text{dB}} = 2(P_{4,\text{out}} - P_{10})_{\text{dB m}} - 6 + 10\log_{10}(N_{\text{FWM}}). \tag{13.6}$$

The predictions of Eq. (13.6) are in reasonable agreement with the experimental results. In [12] results of crosstalk measurements as a function of N_s are presented. As N_s is increased from 4 to 8, $(XT_{ss})_{\text{dB}}$ increases by 6 dB; this is to be compared with the increase of 7 dB predicted by Eq. (13.6). This good agreement indicates that Eq. (13.6) can be used for obtaining an approximate idea of how the signal–signal crosstalk scales with the number of signals.

Equation (13.6) indicates that for $N_s = 100$ the signal–signal FWM crosstalk would be larger by about 20 dB than for two signals. This is a considerable increase, which should be taken into account when choosing the desired values of the pump power and the signal output power.

However, one should remember that Eq. (13.6) is pessimistic, in that it is based on the assumption that all the signal–signal FWM interactions are perfectly phase matched. While this may well be correct for closely spaced signals, in a 100-channel system some FWM interactions will be between signals that are at least 10 nm apart, and these could well be poorly phase matched. Hence Eq. (13.6) should be viewed as a conservative upper limit.

As in transmission systems, a number of other approaches can be used to reduce the impact of signal–signal FWM in OPAs. For example, one could use uneven spacing of the carriers, to make some unwanted FWM terms fall between carriers.

Also, one could use alternating polarization states for the carriers; this has the effect of significantly reducing the FWM terms due to channels with orthogonal SOPs. Overall, this leads to a reduction in total signal–signal crosstalk by about a factor 2. (Note that for this approach to work, the OPA has to be polarization-independent; the benefit associated with this approach has been experimentally demonstrated in such an OPA [11].)

We also see from Eq. (13.3) that using a higher pump power can significantly reduce signal–signal FWM. Using high power in turn reduces the fiber length required to obtain a given gain, and this reduces problems associated with ZDW fluctuations and random birefringence. Therefore increasing the pump power to reduce FWM brings about additional benefits, related to the reduced fiber length.

(b) First-order FWM

Let us now consider an example where the pump mixes with one or two signals. The pump at ω_1 and the signals at ω_3 and ω_4 mix to yield new waves at $\omega_6 = \omega_3 + \omega_4 - \omega_1$ and $\omega_7 = \omega_1 + \omega_4 - \omega_3$, corresponding to two types of FWM; the locations of these terms are shown in Fig. 13.1. The first type of first-order FWM also yields degenerate terms (e.g. with $\omega_3 = \omega_4$), but the other type does not. The terms of the first type falls away from the pump, beyond the signals, whereas the other term falls between the pump and the signals (see Fig. 13.1).

With assumptions similar to the equation for the signal–signal case, the growth rate of either type of pump–signal FWM term is governed by an equation of the form

$$\frac{dA_6}{dz} = gA_6 + 2i\gamma A_1^* A_3 A_4$$

$$= gA_6 + \frac{i\gamma A_{10}^* A_{30} A_{40}}{2} e^{2gz}. \qquad (13.7)$$

This leads to an expression for the relative pump–signal crosstalk:

$$XT_{\text{ps}} = \frac{4P_6}{GP_{30}} = \frac{4P_{40}}{P_{10}} G = \frac{4P_{4,\text{out}}}{P_{10}}; \qquad (13.8)$$

here $P_{4,\text{out}} = GP_{40}$ is the signal output power. Thus the relative pump–signal FWM crosstalk is proportional to the ratio of the powers of the signal and pump at the output. This is why we refer to this type of FWM as being of first order.

We note that $XT_{\text{ps}} \approx 8\sqrt{XT_{\text{ss}}}$, and so the pump–signal crosstalk is generally larger than the signal–signal crosstalk. In fact it appears that pump–signal crosstalk could be very large and hence possibly harmful. Consider for example a situation like before, where $(XT_{\text{ss}})_{\text{dB}} = -46$ dB. Since $(XT_{\text{ps}})_{\text{dB}} = (XT_{\text{ss}})_{\text{dB}}/2 - 9$, the previous relation predicts that $(XT_{\text{ps}})_{\text{dB}} = -14$ dB, which is quite large. Such a value however, is not necessarily observed in practice. One reason is that pump–signal interactions can be poorly phase matched and therefore will fail to reach the magnitude predicted by the above theory, which assumes perfect phase matching. Also, the gain for the FWM term itself may be lower than that for the signals, which violates the conditions behind Eq. (13.8) and leads to a lower result.

An aspect of pump–signal FWM that tends to make it less dangerous than its signal–signal counterpart is that the number of such terms grows more slowly with the number of channels. Specifically, it can be shown that the total number of such terms is of the order of N_s^2, whereas it is N_s^3 for signal–signal terms.

These FWM terms can also be relatively easily separated from the carriers in a one-pump OPA. One way is to place the pump halfway between two grid frequencies if the signals are on a regular grid: then all the FWM terms will fall between two carriers [1].

Another way is to use a sub-octave signal spectrum, i.e. a band of carriers that occupy the region $[\omega_c + \Delta\omega, \omega_c + 2\Delta\omega]$. Then pump–signal FWM terms will fall in the region $[\omega_c + 2\Delta\omega, \omega_c + 4\Delta\omega]$ or the region $[\omega_c, \omega_c + \Delta\omega]$, i.e. they will all fall outside the signal band.

Equation (13.8) shows that pump–signal FWM will also be reduced if the pump power can be increased.

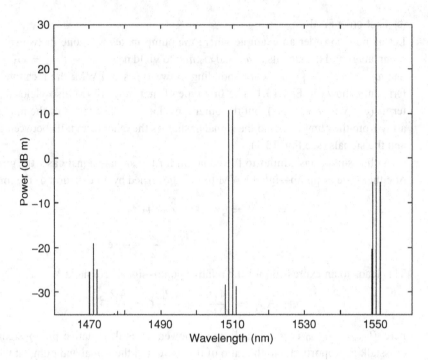

Fig. 13.2 Output spectrum for a one-pump fiber OPA, with parameters $\gamma = 11$ W^{-1} km^{-1}, $\lambda_0 = 1550$ nm, $\beta^{(4)} = -5.8 \times 10^{-56}$ s^4 m^{-1}, $D_\lambda = 0.03$ ps nm^{-2} km^{-1}, $L = 500$ m, $P_{10} = 0.5$ W, $\lambda_1 = 1550$ nm, $\lambda_3 = 1508$ nm, and $\lambda_4 = 1509$ nm; $P_{30} = P_{40} = 0.1$ mW. The signal–signal FWM is near 1509 nm. The pump-signal FWM terms are near 1468 nm as well as adjacent to the pump.

Let us now compare the results predicted by the approximate relations derived above with the accurate results of a split-step Fourier method (SSFM) simulation. Figure 13.2 shows the output spectrum of a one-pump fiber OPA, calculated by the SSFM. The parameters were adjusted to yield a maximum gain of about 20 dB near 1505 nm. Two signals were introduced at 1508 and 1509 nm with powers $P_{30} = P_{40} = 0.1$ mW, while the pump power was $P_{10} = 0.5$ W. The figure shows that $P_{4,\text{out}} \approx 10$ mW. According to the preceding calculations, we would expect that $XT_{\text{ps}} \approx -11$ dB and $XT_{\text{ss}} \approx -40$ dB.

From Fig. 13.2 we obtain $XT'_{\text{ss}} \approx -43$ dB. So the agreement for signal–signal crosstalk is quite good.

For the pump–signal crosstalk away from the pump, we obtain from Fig. 13.2 the value $XT'_{\text{ps}} \approx -67$ dB, which is far lower than that predicted by the approximate result given in Eq. (13.8). There are two reasons for this: (i) the interaction is poorly phase matched, as the wavelengths are far apart and not centered on the ZDW; (ii) the OPA gain for the FWM term is much lower than for the signals, which violates the assumptions behind Eq. (13.7).

For the pump–signal crosstalk close to the pump, Fig. 13.2 yields the value $XT'_{\text{ps}} \approx -23$ dB, which is closer to the -11 dB predicted by Eq. (13.8). This is as expected

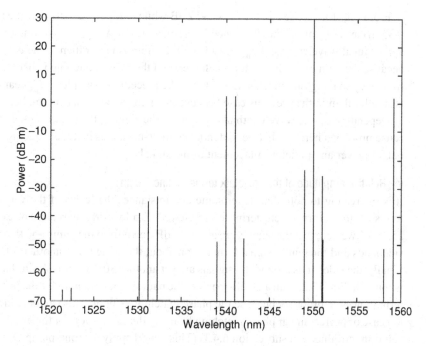

Fig. 13.3 Output spectrum for a one-pump fiber OPA, with the same parameters as in Fig. 13.2. The signal wavelengths are now $\lambda_3 = 1540$ nm and $\lambda_4 = 1541$ nm. The signal–signal FWM is near 1540 nm. The pump-signal FWM is near 1531 nm as well as adjacent to the pump.

because: (i) the interaction is well phase matched, as the extreme wavelengths are close together; (ii) the OPA gain for the FWM term is not much lower than for the signals, in reasonable agreement with the assumptions behind Eq. (13.7).

Overall then, in this situation two types of FWM crosstalk (signal–signal and pump–signal far from the pump) are below −30 dB even though the signal output power is 10 mW, which is quite high by communication standards. The pump–signal FWM term close to the pump is significant and potentially dangerous; if in reality it is a threat then it can be reduced by lowering the signal output power or simply by not placing signals where such FWM terms can be generated. This example shows that, with realistic parameter values and suitable wavelength assignments, FWM crosstalk can be kept within tolerable bounds for a one-pump OPA.

Figure 13.3 shows results obtained with the same OPA, but with the signal wavelengths changed to 1540 and 1541 nm. Because in this region the OPA phase matching is not perfect, the gain is not maximal and the signal growth is linear rather than exponential. The gain is about 10 dB (compared with 20 dB at the maximum). Under these circumstances the assumptions behind the preceding crosstalk derivations are not valid, and we do not expect the conclusions to apply closely. Nevertheless, we expect qualitative agreement with the general trends. Specifically, we expect that, since the signal output power is lower than in the previous example, the crosstalk levels should be lower.

From Fig. 13.3 we see that $XT'_{ss} \approx -52$ dB while $XT'_{ps} \approx -28$ dB near the pump and -35 dB away from it. The difference in these values of XT'_{ss} is due to finite dispersion at the signal wavelengths: XT'_{ps} away from the pump is larger than in the previous case because the gain of the FWM term is larger, and the phase matching is better, since the frequency spacings are much smaller that in the preceding example; XT'_{ps} near the pump is smaller than in the previous case because the gain of the signals is smaller.

Depending on the wavelength allocations of the signals, the pump–signal terms may cause problems here. As before, potential counter-measures include reducing the signal output power and judicious placement of the signals.

(c) Relative amplitude of the crosstalk terms for linear gain

It is interesting to note that under some circumstances the levels of the various FWM terms exhibit a form of clustering when viewed on a logarithmic scale. For example, in Fig. 13.3 we note that the signals are about 30 dB down from the pump and that the pump sidebands and the pump–signal term are on about the same level, another 30 dB down. Finally, the sidebands around the signals are yet another 30 dB down. A similar structure is found in Fig. 13.2, with a 20 dB spacing; the pattern does not hold for the pump–signal term, however, probably because the phase matching is poor. This can be understood in the case of perfect linear phase matching by using the expression for the exact total-field solution introduced in subsection 6.4.1. (This would apply to a one-pump OPA with the signals close to the pump, where the signal amplitude gain is linear with distance.)

Let us consider that the total input field consists of one pump and two signals. The slowly varying envelope (SVE) of the total field has the form

$$A(0, t) = A_{10}[e^{-i\omega_1 t} + u(e^{-i\omega_3 t} + e^{-i\omega_4 t})], \tag{13.9}$$

where $u = A_{40}/A_{10}$ is the signal input amplitude relative to that of the pump; we will assume for simplicity that u is real. Then the field at a distance z and time t is given by

$$A(z, t) = A_{10}[e^{-i\omega_1 \tau} + u(e^{-i\omega_3 \tau} + e^{-i\omega_4 \tau})]$$
$$\times \exp\{i\Phi|e^{-i\omega_1 \tau} + u(e^{-i\omega_3 \tau} + e^{-i\omega_4 \tau})|^2\}, \tag{13.10}$$

where $\Phi = \gamma P_{10} z$ and $\tau = t - z/v$. We can rewrite Eq. (13.10) as

$$A(z, t) = A_{10} \exp(i\Phi)[e^{-i\omega_1 \tau} + u(e^{-i\omega_3 \tau} + e^{-i\omega_4 \tau})] \exp\{i\Phi[u(e^{-i(\omega_1 - \omega_3)\tau} + e^{-i(\omega_1 - \omega_4)\tau})$$
$$+ u(e^{-i(-\omega_1 + \omega_3)\tau} + e^{-i(-\omega_1 + \omega_4)\tau})] + u^2(e^{-i\omega_3 \tau} + e^{-i\omega_4 \tau})(e^{i\omega_3 \tau} + e^{i\omega_4 \tau})\}.$$
$$\tag{13.11}$$

We now expand the exponential in powers of its argument and keep only the dominant terms, i.e. those that are significant for our discussion of FWM in OPAs. Specifically, we find that

$$\frac{A(z, t)}{A_{10} \exp(i\Phi)} \approx e^{-i\omega_1 \tau} + u(e^{-i\omega_3 \tau} + e^{-i\omega_4 \tau}) + i\Phi u(e^{-i\omega_3 \tau} + e^{-i\omega_4 \tau})$$
$$+ i\Phi u(e^{-i(2\omega_1 - \omega_3)\tau} + e^{-i(2\omega_1 - \omega_4)\tau})$$
$$- \frac{\Phi^2}{2} u^2 (e^{-i(2\omega_3 - \omega_1)\tau} + e^{-i(2\omega_4 - \omega_1)\tau} + 2e^{-i(\omega_3 + \omega_4 - \omega_1)\tau})$$
$$- \frac{i\Phi^3}{2} u^3 (e^{-i(2\omega_3 - \omega_4)\tau} + e^{-i(2\omega_4 - \omega_3)\tau}). \tag{13.12}$$

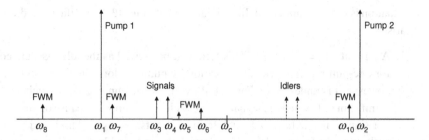

Fig. 13.4 Some spurious FWM terms in a two-pump OPA: the idlers associated with the two signals are to the right of ω_c (dotted lines); ω_8 is due to degenerate FWM interaction between the pump at ω_1 and the signal at ω_3; ω_{10} is due to the idler associated with ω_7 and to FWM between the idlers corresponding to ω_3 and ω_4 and the pump at ω_2.

The first three exponentials correspond to the original fields. Next two exponentials constitute an amplified version of the input signals. Combining these contributions, we see that the signal gain for either signal is $1 + i\Phi$, i.e. it is linear with distance; hence the power gain is $G = 1 + \Phi^2$, as expected near the pump for a one-pump OPA. The terms in the second line of the equation correspond to the two idlers. As expected, they have a power conversion gain $G_i = \Phi^2$. These results are the same as those obtained in subsection 6.4.1.

The terms in the third line correspond to pump–signal FWM while those in the fourth line correspond to signal–signal FWM. Converting the amplitudes of these FWM terms into crosstalk, we find that

$$XT_{\text{ps}} \approx \frac{1}{4}\frac{P_{\text{s,out}}}{P_{10}} \quad \text{and} \quad XT_{\text{ss}} \approx \frac{1}{4}\left(\frac{P_{\text{s,out}}}{P_{10}}\right)^2 \qquad (13.13)$$

It is interesting to note that the forms of these crosstalk terms are quite similar to those arrived at by integration in the case of exponential gain, even though we here we have linear gain. In particular, we find as before that the pump–signal and signal–signal terms are respectively of first and second order in the ratio $P_{\text{s,out}}/P_{10}$.

We see from Fig. 13.3 that these relations are in fact in good agreement with the numerical simulations. In that figure we have $P_{\text{s,out}}/P_{10} \approx 2 \times 10^{-3}$ and we see that $XT_{\text{ps}} \approx -28$ dB (near the pump) and $XT_{\text{ss}} \approx -52$ dB, close to the values expected from Eq. (13.13).

13.2.2 Two-pump OPA

In two-pump fiber OPAs the generation of spurious FWM terms is somewhat more complex than in a one-pump OPA. Although the nature of the FWM terms is quite similar to those for a one-pump OPA (because terms simultaneously involving the two pumps correspond to signals and idlers, not to new FWM terms), the complication arises from the location of the pumps with respect to the ZDW and hence the differences in phase matching and location of the FWM terms with respect to the signals and idlers. The situation is illustrated in Fig. 13.4. For the signal–signal FWM the situation and the countermeasures are similar to the one-pump case. For the pump–signal FWM the situation is more complicated. In comparison with Fig. 13.1, significant additional terms arise.

1. A term at $\omega_8 = 2\omega_1 - \omega_3$. This term can be viewed as the idler associated with ω_3 for one-pump parametric amplification by pump 1 alone, just as in Fig. 13.1. Here, however, it is undesirable. For an OPA with a wide pump spacing, such terms will be significant only when the signals are very close to pump 1, because of phase matching. For signals inside $[\omega_1, \omega_2]$, such terms fall outside $[\omega_1, \omega_2]$ and so do not interfere with the signals. In addition, it is generally undesirable to have a signal close to a pump because in such regions the flatness of the OPA gain cannot be maintained in any case [8].
2. Terms associated with interactions with pump 2. Some of these terms are of the form $\omega_9 = 2\omega_3 - \omega_2$; when the signals are between ω_1 and ω_c they fall outside the region $[\omega_1, \omega_2]$ and are not shown in Fig. 13.4, as they lie too far to the left. Other terms are of the form $\omega_{10} = \omega_2 + \omega_3 - \omega_4$; ω_{10} is due to the idler associated with ω_7 and to FWM between the idlers corresponding to ω_3 and ω_4 and the pump at ω_2.

For some FWM terms the counter-measures are similar to those in the one-pump case. Placing the pumps between grid points, when the signals are on a grid, makes all such terms fall halfway between grid points. However, if a sub-octave spectrum is used, with say signals restricted to $[\omega_1 + (\omega_2 - \omega_1)/4, \omega_c]$, then some FWM terms due to pump 1 (i.e. ω_6) fall in $[\omega_c, \omega_2]$, which overlaps with the idler spectrum. To prevent this, one would need to restrict signals to the range $[\omega_1 + (\omega_2 - \omega_1)/3, \omega_c]$. It is important to note that for ω_6-type terms the phase matching can be very good, because the center frequency is close to the ZDW; hence it is difficult to use phase mismatch to weaken these terms.

As in the one-pump case, increasing the pump power will reduce the relative influence of all the spurious FWM terms.

Figure 13.5 shows the output spectrum of a two-pump OPA calculated using the SSFM. The pump–signal FWM is near 1570 nm and the pump–idler FWM near 1530 nm. Since these terms fall between the pumps, they could interfere with a very broad wavelength-division multiplexing (WDM) signal spectrum; however, they are over 50 dB lower than the signals and could thus be manageable. The signal–idler FWM terms are near 1510 and 1590 nm; they are more than 70 dB lower than the signals and thus insignificant for most purposes.

13.3 Cross-gain modulation

Let us now consider another nonlinear phenomenon that has been identified as a possible source of impairment in fiber OPAs, namely cross-gain modulation (XGM). This mechanism exists in all types of amplifier in which the gain responds very quickly to signal variations, such as Raman amplifiers and semiconductor optical amplifiers (SOAs); in contrast, EDFAs are essentially immune to XGM. In fiber OPAs, XGM corresponds to the mechanism shown schematically in Fig. 13.6. An IM signal at λ_3 is amplified by the

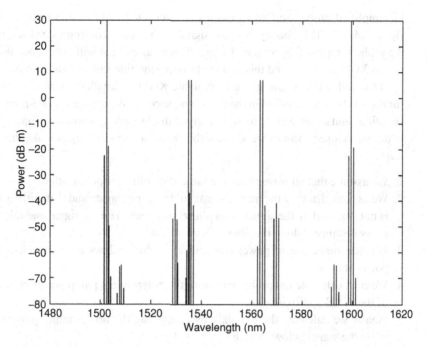

Fig. 13.5 Output spectrum for a two-pump fiber OPA, with the parameters $\gamma = 10 \text{ W}^{-1} \text{ km}^{-1}$, $\lambda_0 = 1550.0$ nm, $\beta_4 = -1 \times 10^{-55}$ s^4 m^{-1}, $D_\lambda = 0.0615$ ps nm^{-2} km^{-1}, $L = 250$ m, $P_{10} = P_{20} = 1$ W, $\lambda_1 = 1502.6$ nm, $\lambda_2 = 1599.4$ nm, $\lambda_3 = 1535.0$ nm, $\lambda_4 = 1536.0$ nm, and $P_{30} = P_{40} = 1$ µW.

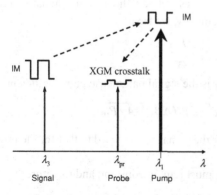

Fig. 13.6 Schematic illustration of the signal quality degradation due to cross-gain modulation (XGM) in a one-pump OPA configuration.

parametric gain due to a pump at λ_1. As it grows, it draws power away from the pump, since the total optical power must remain constant. Hence the pump power itself has IM, inverted with respect to that of the signal. If we now consider a probe at λ_{pr} without initial IM, traveling with the pump and signal, it will experience different gains at different times, depending on whether it travels with a depleted part of the pump. Therefore the amplified probe itself will exhibit IM, attenuated and inverted with respect to the signal IM. This IM transfer from the signal to the probe constitutes XGM crosstalk. If the probe is replaced by a second signal then its amplitude will experience fluctuations due to XGM crosstalk and this will lead to deterioration in its signal quality.

In the following we are going to evaluate XGM crosstalk on a basis similar to that for the FWM terms studied in the preceding section. We consider a one-pump OPA for simplicity and model XGM crosstalk analytically, by making a number of approximations to derive a closed-form expression for the crosstalk. Specifically, we take the following steps.

1. We assume that all waves have the same state of linear polarization.
2. We assume that only the pump, signal, and idler propagate and that initially the pump is not depleted by the signal. This allows us to calculate the signal and idler power at every distance z along the fiber.
3. We then make use of power conservation, which allows us to calculate the pump power at every z.
4. We next calculate the probe gain from the z-dependent pump power by making use of the WKB approximation.
5. Finally we calculate the crosstalk by comparing the probe output powers obtained when the signal is low or high.

Let P_{30} and $P_3(z)$ respectively denote the signal power at the input and at the distance z. We have

$$P_3(z) = G_3(z)P_{30} = [G_4(z) + 1]P_{30}, \qquad (13.14)$$

where $G_4(z)$ and $G_3(z) = G_4(z) + 1$ are respectively the idler and signal power gains for an undepleted OPA of length z. Specifically, we have

$$G_4(z) = \left[\frac{\gamma P_{10}}{g_3}\sinh(g_3 z)\right]^2, \qquad (13.15)$$

where P_{10} is the pump power and g_3 is the signal parametric gain coefficient, given by

$$g_3 = \frac{1}{2}\sqrt{-\Delta\beta_3(\Delta\beta_3 + 4\gamma P_{10})}; \qquad (13.16)$$

$\Delta\beta_3$ is the wavevector mismatch for the signal (as opposed to that for the probe, which will be introduced later).

In a lossless fiber, the total power must remain constant, and therefore

$$P_1(z) + P_3(z) + P_4(z) = P_1(z) + P_{30} + 2P_4(z) = P_{10} + P_{30}, \qquad (13.17)$$

where $P_1(z)$ and $P_4(z)$ are the total pump power and the idler power at z, respectively. Hence

$$P_1(z) = P_{10} - 2P_4(z) = P_{10} - 2G_4(z)P_{30} = P_{10} + \Delta P(z), \qquad (13.18)$$

where

$$\Delta P(z) = -2P_{30}\left[\frac{\gamma P_{10}}{g_3}\sinh(g_3 z)\right]^2 \qquad (13.19)$$

is the pump–power change due to depletion. Equation (13.18) shows how the total pump power is depleted as a result of the signal amplification.

Let us now introduce an unmodulated probe. We assume that its amplitude is small, so that it does not modify the preceding calculations significantly. Using Eq. (13.19), we see that the probe experiences a parametric gain coefficient that is a function of z, namely

$$g_{pr}(z) = \frac{1}{2}\sqrt{-\Delta\beta_{pr}[\Delta\beta_{pr} + 4\gamma P_1(z)]}$$

$$= \frac{1}{2}\sqrt{-\Delta\beta_{pr}[\Delta\beta_{pr} + 4\gamma(P_{10} + \Delta P)]}$$

$$= g_{pr}(0)\sqrt{1 + \frac{4\gamma \Delta P \Delta\beta_{pr}}{[g_{pr}(0)]^2}}$$

$$\approx g_{pr}(0) - \frac{\Delta\beta_{pr}\gamma \Delta P}{2g_{pr}(0)}$$

$$= g_{pr}(0) + \frac{\Delta\beta_{pr}\gamma P_{30}}{g_{pr}(0)}\left[\frac{\gamma P_{10}}{g_3}\sinh(g_3 z)\right]^2$$

$$= g_{pr}(0) + K \sinh^2(g_3 z), \tag{13.20}$$

where $\Delta\beta_{pr}$ is the wavevector mismatch for the probe and

$$K = \frac{\Delta\beta_{pr}\gamma P_{30}}{g_{pr}(0)}\left(\frac{\gamma P_{10}}{g_3}\right)^2. \tag{13.21}$$

Assuming that the signal and the probe both experience maximum parametric gain, we have $\Delta\beta_{pr} = -4\gamma P_{10}$, $g_{pr}(0) = g_3 = \gamma P_{10}$, and so $K = -2\gamma P_{10}$.

Equation (13.20) shows that $g_{pr}(z)$ is a relatively simple function of z, which can easily be integrated with respect to z. Hence this is a case where we can make use of the WKB approximation for the overall power gain, by integrating $g_{pr}(z)$ (See Appendix 2). The WKB approximation in this case states that, at the fiber end, the idler associated with the probe has a total power-conversion gain

$$G_{pr,i}(L) = \frac{r_{pr}(0)r_{pr}(L)}{g_{pr}(0)g_{pr}(L)}\sinh^2\left\{\int_{z=0}^{L} g_{pr}(z)dz\right\}$$

$$\approx \sinh^2\left\{g_{pr}(0)L + \frac{K}{2}\left[\frac{\sinh(2g_3 L)}{2g_3} - L\right]\right\}$$

$$= \sinh^2\left\{\gamma P_{10} L - \gamma P_{30}\left[\frac{\sinh(2\gamma P_{10} L)}{2\gamma P_{10}} - L\right]\right\}, \tag{13.22}$$

where we have assumed that the probe experiences maximum parametric gain at every point along the fiber, so that $g_{pr}(0) = r_{pr}(0)$ and $g_{pr}(L) = r_{pr}(L)$.

Equation (13.22) can now be used to calculate the crosstalk. We define the initial probe power as $P_{pr}(0) = P_{pr,0}$. Then the probe idler output power when the signal is low

$(P_{30} = 0)$ is

$$P_{pr}^{lo}(L) = P_{pr,0}[G_{pr,i}(L)]_{lo} \approx P_{pr,0} \sinh^2(\gamma P_{10} L). \tag{13.23}$$

When the signal is high ($P_{30} > 0$), the probe idler output power becomes

$$P_{pr}^{hi}(L) = P_{pr0}[G_{pr,i}(L)]_{hi}$$
$$\approx P_{pr0} \sinh^2 \left\{ \gamma P_{10} L - \gamma P_{30} L \left[\frac{\sinh(2\gamma P_{10} L)}{2\gamma P_{10} L} - 1 \right] \right\}. \tag{13.24}$$

We define the difference in probe output power between these two levels as the crosstalk experienced by the probe,

$$\Delta P_{XGM} = P_{pr,0} \lfloor G_{pr,i}(L) \rfloor_{lo} - P_{pr,0} \lfloor G_{pr,i}(L) \rfloor_{hi}$$
$$\approx -2\gamma P_{30} P_{pr,0} \sinh(2\gamma P_{10} L) \left[\frac{\sinh(2\gamma P_{10} L)}{2\gamma P_{10}} - L \right]. \tag{13.25}$$

To obtain the final expression in Eq. (13.25) we have assumed that the magnitude of the second term within the braces in Eq. (13.24) is small compared with 1. It can be shown that this is valid for typical OPA parameters, as follows. Let A denote the second term; it is given by

$$A = -2\gamma P_{30} L \left[\frac{\sinh(2\gamma P_{10} L)}{2\gamma P_{10} L} - 1 \right]. \tag{13.26}$$

When the gain G is large,

$$\sinh(2\gamma P_{10} L) \approx \exp(2\gamma P_{10} L)/2$$
$$= 2[\exp(2\gamma P_{10} L)/2]^2 \approx 2G$$

(where G denotes either the signal or the idler gain, since they are almost the same for large gain). Then

$$A \approx -2\gamma P_{30} L \left(\frac{G}{\gamma P_{10} L} - 1 \right)$$
$$= 2\frac{P_{30}}{P_{10}} (\gamma P_{10} L - G) \approx -2G \frac{P_{30}}{P_{10}} = -2\frac{P_{3,out}}{P_{10}}. \tag{13.27}$$

If strong pump depletion is avoided, $P_{3,out} \ll P_{10}$ and therefore $|A| \ll 1$; this justifies the assumption made to obtain the final form of Eq. (13.25).

Then ΔP_{XGM} can be simplified as follows:

$$\Delta P_{XGM} \approx -\frac{P_{pr,0}}{4} \sinh(2\gamma P_{10} L) 2\gamma P_{30} \left[\frac{\sinh(2\gamma P_{10} L)}{2\gamma P_{10}} \right]$$
$$= -2\frac{(GP_{30})^2}{P_{10}} = -2\frac{(P_{3,out})^2}{P_{10}}, \tag{13.28}$$

where we have assumed that $P_{pr,0} = P_{30}$.

To place this crosstalk in the same context as that of the FWM terms studied in Section 13.2, we now assume that the signal modulation is sinusoidal. Then XGM of the probe generates symmetric sidebands about the probe. Let P_a denote the power of each sideband. The probe's temporal power variation is then of the form

$$P_{\text{pr}}(t) = \left[\sqrt{GP_{30}} + \sqrt{P_a}\cos(\omega_a t)\right]^2 \approx GP_{\text{pr},0} + 2\sqrt{GP_a P_{30}}\cos(\omega_a t). \quad (13.29)$$

To relate this picture of XGM to the preceding FWM picture, we set

$$\Delta P_{\text{XGM}} = \sqrt{GP_a P_{30}} = \frac{(GP_{30})^2}{P_{10}}. \quad (13.30)$$

From this we find that the relative crosstalk due to XGM is

$$XT_{\text{XGM}} = \frac{P_a}{GP_{30}} = \left(\frac{GP_{30}}{P_{10}}\right)^2 = \left(\frac{P_{3,\text{out}}}{P_{10}}\right)^2 \quad (13.31)$$

or, in decibels,

$$(XT_{\text{XGM}})_{\text{dB}} = 2(P_{3,\text{out}} - P_{10})_{\text{dBm}}. \quad (13.32)$$

According to our preceding definition for FWM, this says that the crosstalk due to XGM is of second order and hence rather small. For example, if the signal output power is 25 dB lower than the pump then the XGM sidebands flanking the probe are about 50 dB lower than the probe.

This conclusion is confirmed by numerical simulations. Figure 13.7 corresponds to the same input conditions as Fig. 13.5 except that a probe has been added at 1545 nm. The sum of the two signals is equivalent to a carrier located halfway between the signals, with 100% IM. This IM in turn causes XGM, i.e. IM of the pump and the probe. Indeed, we can see that the pump will exhibit sidebands, already seen in Fig. 13.5, which can be interpreted as corresponding to pump IM. Furthermore, we see that the probe is also flanked by two sidebands, which are about 50 dB down as expected from Eq. (13.32).

(Actually this needs to be qualified. The reason is that the sidebands around the probe also have another origin, namely direct FWM between the signals and the probe. From what we saw in Section 13.2, this FWM is also of second order and therefore the sidebands associated with it are also about 50 dB down compared with the probe. Thus, without a more detailed investigation we can only conclude that the sidebands flanking the probe represent an upper limit for the XGM crosstalk. At any rate, it is clear that XGM generates rather small crosstalk contributions in this case.)

In the preceding discussion we have defined XGM crosstalk as the ratio of the power of the XGM sidebands and the probe power. The advantage of this definition is that it is the same as that adopted for FWM crosstalk in Section 13.2. As a result, we have a uniform basis for the comparison of crosstalk due to either FWM or XGM.

Since XGM crosstalk may be detrimental, it is important to find ways to minimize it. If a two-pump OPA is used, XGM crosstalk is three times smaller when the pumps are orthogonal than when they are parallel [9]. The reason for this can be understood as follows.

Consider two two-pump OPAs, one with parallel pumps (XXXX), and the other with two orthogonal pumps (XYXY) (this notation is explained in Chapter 4). We assume that they have the same maximum gain, which requires the total pump power for XYXY

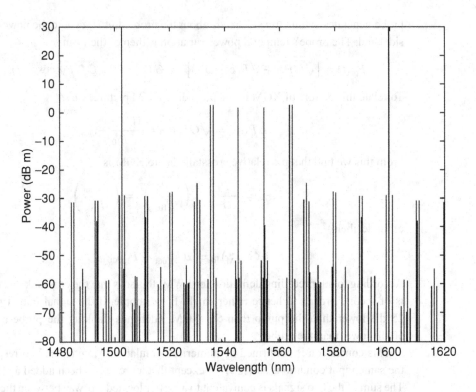

Fig. 13.7 The conditions are the same as for Fig. 13.5, except that a probe has been added at 1545.0 nm. The sidebands around the probe are due in part to XGM. Their low power indicates that XGM crosstalk is low in this case.

to be three times as large as for XXXX. A signal injected into either amplifier grows at the same rate in both. Because in an OPA the total power is conserved, the power corresponding to the growth in the signal must come from the pumps. Hence the total pump power is reduced by the same amount in both cases. However, because the XYXY pump power is larger by a factor of 3 (in a fiber without polarization-mode dispersion (PMD) and by a factor 2 in a fiber with PMD), it experiences an amount of depletion that is *relatively* smaller than for XXXX. Therefore, the gain experienced by a probe, that depends on the pump power, is less affected in XYXY than in XXXX, i.e. we expect XGM crosstalk to be smaller in XYXY than in XXXX, by a factor of 3 (5 dB) in a fiber without PMD. The benefit of using orthogonal pumps for reducing XGM has been verified experimentally [10, 11].

13.4 Coherent crosstalk

Some FWM terms studied in Section 13.2 can have the same frequency as a carrier. For XGM, we saw that IM sidebands are created around a carrier. In both cases the total field in the vicinity of the carrier frequency thus needs to be calculated as the *coherent* sum of the carrier and these crosstalk terms.

In the preceding sections we investigated FWM and XGM crosstalk by calculating the power of the new frequency components created by these nonlinear mechanisms. We saw that the powers of the second-order terms can be quite low under typical OPA conditions, i.e. they can be about 60 dB lower than the signal output power.

On the basis of such numbers we might think that the impact on system performance will therefore be low. This, however, must be re-examined in view of what is important from a system standpoint, which is eventually the SNR. To do this we need to look at the signal variations in the time domain and see how they fluctuate owing to detrimental nonlinear effects.

We have seen that XGM corresponds to signal IM caused by pump IM. Since IM sidebands are always in phase with the associated carrier, the total field is obtained by the coherent addition of in-phase components, generating substantial fluctuations. For example, IM sidebands that are 60 dB down from the carrier will lead to temporal power variations which are only about 30 dB down from the average power and may therefore be non-negligible. Thus the XGM contributions will need to be carefully assessed to determine their impact on system performance.

For FWM terms the situation is more complex, as the FWM terms that fall on a particular carrier generally have random phases, i.e. phases that have no particular relation with that of the carrier or with each other. Thus while the addition of these terms and the carrier is coherent, since they have the same frequency, the fact that these FWM terms have random phases implies that the amplitude of their sum will have zero mean with a standard variation proportional to the square root of the number of terms. At any rate, a single FWM term that is down by 60 dB from a carrier may cause power fluctuations of the signal that are only 30 dB down from the signal-power variations.

These numbers will be even larger for first-order FWM, and it may thus be necessary to prevent these terms from overlapping with signals, by a proper choice of signal frequencies.

13.5 Cross-phase modulation

Cross-phase modulation (XPM) between signals in optical amplifiers is a concern in the design of WDM systems. Erbium-doped fiber amplifiers can generate XPM levels comparable with those due to long transmission fibers, because XPM can be very well phase matched over short lengths [15]. This concern is further heightened in amplifiers using highly nonlinear media, such as tellurite EDFAs [13] or discrete Raman amplifiers [16]. Here we investigate signal–signal XPM in fiber OPAs. We find that signals are not affected by XPM as severely as one might anticipate if it were calculated in the same manner as for EDFAs; the reason is that a probe experiences additional XPM due to the idler and the pump, instantaneously depleted by the signal gain. The net XPM is then about the same as if the OPA had no gain.

Consider an unmodulated probe and a signal with IM fed into an optical amplifier. We assume that there is no walkoff due to dispersion. For an EDFA, the XPM of the probe at the output is then

$$\Phi_1(L, t + L/v) = 2\gamma \int_0^L P_3(z, t')dz = 2\gamma P_3(0, t) \int_0^L G(z)dz, \quad (13.33)$$

where v is the speed of light in the fiber, $t' = t + z/v$; $P_3(0, t)$ is the signal input power, and $G(z)$ is the signal power gain between 0 and z. If the pump power is increased then $G(z)$ increases everywhere and Φ_1 increases.

In addition, in fiber OPAs: (i) an idler is generated, with the same IM amplitude as the signal; (ii) the photons gained by signal and idler come from the pump and thus the pump acquires IM, twice as large as the signal, but with opposite sign.

Hence, the probe XPM at the output is

$$\Phi_2 = 2\gamma \int_0^L [P_3(z, t') + P_4(z, t') + P_1(z, t')]dz. \quad (13.34)$$

Since total power is conserved along the fiber, we have

$$P_3(z, t) + P_4(z, t) + P_1(z, t) = P_0, \quad (13.35)$$

where P_0 is the total input power. As a result $\Phi_2 = 2\gamma P_0 L$, i.e. the XPM of the probe is independent of the amplifier gain.

This remarkable result sets OPAs apart from other types of fiber amplifier, as far as XPM generation between co-propagating signals is concerned. Of course, the fact that the signal XPM does not increase with gain does not mean that there is no XPM, just that it has the same value as if the pump were turned off. Depending on the length and type of the fiber, a significant amount of XPM may still be generated in a fiber OPA. If XPM is a concern in a particular system, its magnitude should be calculated.

Since the signal–signal XPM is the same as in a passive piece of fiber, the calculation is simple. In fact, it is quite clear that in some OPAs XPM is expected to be negligible compared with what it is in a transmission line. For example, if a high-power pump is used with a short length of dispersion-shifted fiber (DSF), which has about the same γ value as the transmission line then the OPA XPM will be only a fraction of that in the transmission line.

Experiments have been performed to verify this theory [17]. The measurements were complicated by the presence of cross-gain modulation (XGM), which interfered with the coherent detection system used for measuring the phase. This was circumvented by using two signals with opposite IM to eliminate the pump IM. The net results showed good agreement with theory: while XPM in an EDFA showed a strong increase with gain, the signal–signal XPM in an OPA was found to be essentially independent of gain.

13.6 Conclusion

In this chapter we have considered the main third-order nonlinear effects in fiber OPAs that can potentially have a detrimental effect on the performance of communication systems using such amplifiers. Specifically, we considered two types of FWM, XGM, and XPM. We found that XPM is the least troublesome effect, but that XGM and FWM can both lead to significant signal degradation under certain circumstances. We showed that the power of the crosstalk terms scales either linearly or quadratically with the ratio of the signal output power and the pump power. Hence one way to reduce all these terms is to use as low a value as possible for this ratio, which means increasing the pump power and/or reducing the signal output power.

Both XGM and signal–signal FWM generate second-order crosstalk, which is difficult to separate spectrally from the signals but can be of moderate amplitude. However, pump–signal FWM generates first-order crosstalk which can have a large amplitude but which can be separated from the signals by a proper choice of pump and carrier frequencies. These considerations show that the transmission of WDM signals with low impairment through fiber OPAs is not a simple matter. While one should be able to transmit a few signals through a few OPAs while avoiding significant penalties, transmitting many signals through several OPA stages will be challenging. In order to exploit the large bandwidth of fiber OPAs by transmitting dense WDM spectra over long distances, careful OPA and system design will have to be undertaken.

Another avenue for reducing some of these nonlinear effects is to use a fiber with a large value of $\beta^{(3)}$, because this will have the effect of increasing D for wavelengths that are not very close to the zero-dispersion wavelength (ZDW). This in turn will reduce FWM between waves away from the ZDW. An interesting aspect of this approach is that in principle it can be used without affecting the overall OPA gain spectrum, since the latter depends only on the even orders of dispersion, $\beta^{(2)}$, $\beta^{(4)}$, etc. (see Chapter 5). Ideally, then, one would like to have a fiber with a high value of $\beta^{(3)}$ and simultaneously low values of $\beta^{(2)}$ and $\beta^{(4)}$. This is feasible in principle, but it may be difficult to find suitable designs for such fibers and actually to manufacture them. This poses an interesting challenge for fiber designers and manufacturers.

References

1. "Fiber parametric amplifiers for wavelength band conversion," Islam, M. N., Boyraz, O. *IEEE J. Selected Topics in Quantum Electron.*; 2002; vol. 8, pp. 527–37.
2. "Amplification of WDM signals in fiber-based optical parametric amplifiers," Torounidis, T., Sunnerud, H., Hedekvist, P. O., Andrekson, P. A. *IEEE Photon. Technol. Lett.*; 2003; vol. 15, no. 8, pp. 1061–3.
3. "Spurious four-wave mixing in two-pump fiber-optic parametric amplifiers," Callegari, F. A., Boggio, J. M. C., Fragnito, H. L. *IEEE Photon. Technol. Lett.*; 2004; vol. 16, no. 2, pp. 434–6.
4. "Cross-talk-induced limitations of two-pump optical fiber parametric amplifiers," Blows, J. L., Hu, P. F. *J. Opt. Soc. Amer. B*; 2004; vol. 21, no. 5, pp. 989–95.
5. "Design strategy for controlling four-wave mixing-induced crosstalk between channels in a fibre optical parametric amplifier," Blows, J. L. *Optics Comm.*; 2004; vol. 236, nos. 1–3, pp. 115–22.
6. "Influence of zero dispersion wavelength variations on cross-talk in single-pumped fiber optic parametric amplifiers," Boggio, J. M. C., Callegari, F. A., Marconi, J. D., Guimaraes, A., Fragnito, H. L. *Optics Comm.*; 2004; vol. 242, nos. 4–6, pp. 471–8.
7. "Four-wave mixing crosstalk in optical fibre parametric amplifiers with orthogonal pumps," Hu, P., Lows, J. L. *Optics Comm.*; 2005; vol. 250, nos. 4–6, pp. 421–7.

8. "Double-pumped fiber optical parametric amplifier with flat gain over 47-nm bandwidth using a conventional dispersion-shifted fiber," Boggio, J. M. C., Marconi, J. D., Fragnito, H. L. *IEEE Photon. Technol. Lett*; 2005; vol. 17, pp. 1842–4.
9. "Reduction of WDM signal crosstalk in two-pump fiber optical parametric amplifiers," Wong, K. K. Y., Marhic, M. E. In *Proc. Conf. on Lasers and Electro-Optics*, Tokyo, July 2005; paper CLI2-2, pp. 1562–3.
10. "Experimental studies of the WDM signal crosstalk in two-pump fiber optical parametric amplifiers," Wong, K. K. Y., Lu, G. W., Chen, L. K. *Optics Comm.*; 2007; vol. 270, pp. 429–32.
11. "Polarization-interleaved WDM signals in a fiber optical parametric amplifier with orthogonal pumps," Wong, K. K. Y., Lui, G. W., Chen, L. K. In *Proc. Conf. on Lasers and Electro-Optics*, Long Beach CA, May 2006; paper CMW2.
12. "Crosstalk in double-pumped fiber optic parametric amplifiers for wavelength division multiplexing systems," Chavez Boggio, J. M., Marconi, J. D., Fragnito, H. L. *Optics Comm.*; 2006; vol. 259, pp. 94–103.
13. "Large cross phase modulation and four wave mixing in tellurite EDFA's," Marhic, M. E., Morita, I., Ho, M.-C., Akasaka, Y., Kazovsky, L. G. *IEEE Electron. Lett.*; 1999; vol. 35, pp. 2045–7.
14. "Low third-order glass-host nonlinearities in erbium-doped waveguide amplifiers," Marhic, M. E., Nikonov, D. E. *Proc. SPIE*; 2002; vol. 4645, pp. 193–202.
15. "Crosstalk in WDM systems caused by cross-phase modulation in erbium-doped fiber amplifiers," Shtaif, M., Eiselt, M., Tkach, R. W., Stolen, R. H., Gnauck, A. H. *IEEE Photon. Technol. Lett.*; 1998; vol. 10, pp. 1796–8.
16. "Cross phase modulation in discrete Raman amplifiers and its reduction," Akasaka, Y., Morita, I., Marhic, M. E., Ho, M. C., Kazovsky, L. G. In *Proc. Optical Communication Conf.*, Baltimore MD, May 2000; vol. 3, pp. 197–9.
17. "Signal–signal cross-phase modulation in fiber optical parametric amplifiers," Kalogerakis, G., Marhic, M. E., Kazovsky, L. G. In *Proc. OECC/COIN2004*, Yokohama, Japan, July 2004; paper 15D1-3.

14 Distributed parametric amplification

14.1 Introduction

During recent years much research has been performed with the aim of developing fiber Raman amplifiers. Some work was done on discrete Raman amplifiers, but now it appears that the best way to utilize Raman gain is in distributed amplification, i.e. the amplification of communication signals along transmission fibers rather than in discrete amplifiers located between such fibers. Some system manufacturers have developed long-haul systems based on distributed Raman amplification (DRA). Such systems, however, generally require high pump powers, sometimes in excess of 1 W, and this raises concerns about safety and reliability.

Concerning fiber OPAs, to date most efforts have concentrated on discrete devices, i.e. subsystems that could eventually be used as a substitute for other discrete optical amplifiers, such as erbium-doped fiber amplifiers (EDFAs) [1, 2]. Raman and parametric fiber amplification are related third-order nonlinear phenomena, which occur in common silica fibers in the presence of strong optical pumps. Whether a particular amplification mechanism is best suited for discrete or distributed amplification depends on a number of parameters, such as the gain bandwidth, the required pump power, the noise characteristics, etc.

Distributed parametric amplification (DPA) has previously been investigated as a detrimental effect, because strong carriers can further amplify in transmission fibers the amplified spontaneous emission (ASE) generated by the discrete EDFAs used to amplify signals between fiber spans. This has been done for one-pump [3–19] and two-pump [20] parametric amplification. Here we take the opposite view and investigate a possibly beneficial application of DPA, namely the parametric amplification of signals along a transmission fiber by using a co-propagating pump.

In comparison with DRA, DPA has the following potential advantages:

1. **Pump power** DPA requires less pump power than DRA to obtain the same gain.
2. **Double Rayleigh scattering (DRS)** Parametric gain in fibers occurs only for co-propagating pump(s) and signals, and so DPA is free from DRS, a significant problem for DRA.
3. **Pump attenuation** DPA pumps can have about the same wavelength as the signals, and so the pumps and signals have about the same loss. In contrast, DRA pumps

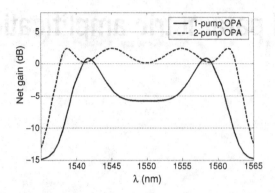

Fig. 14.1 Theoretical DPA gain spectra for one pump, at $\lambda_1 = \lambda_0 = 1550$ nm, and for two pumps, at $\lambda_1 = 1540.04$ nm and $\lambda_2 = 1560.00$ nm. The total pump power is 66 mW in both cases. (After [21], © 2005 IEEE.)

generally have wavelengths about 100 nm shorter than the signals; hence the pumps typically have higher losses than the signals.

4. **Idler generation** In DPA an idler is generated that is not present in DRA. The idler has several important properties, which make it potentially interesting for improving the signal-to-noise ratio, etc.

Even with long transmission spans (which are generally >40 km) the pump power required to compensate for fiber loss is relatively small, of the order of 100 mW. This implies that the gain bandwidth will be relatively small, at most about 10 nm. Therefore DPA cannot compete with approaches providing larger bandwidths, such as DRA, tellurite EDFAs, etc. However, DPA may be a useful supplement for these techniques: for example, DPA could be used for transmission near 1300 nm in a system using standard single-mode fiber (SMF) for transmission in the C-band. Also, DPA could be used for coarse wavelength-division multiplexing (CWDM), by employing several pumps separated by tens of nanometers. It is also possible that fibers with two or three zero-dispersion wavelengths (ZDWs) could be used for increasing the total DPA bandwidth.

In Section 14.2 we discuss a theoretical and experimental investigation of DPA performed with a 75-km-long DSF. We then discuss possible extensions of this DPA approach in several directions, particularly the possibility of using phase-sensitive amplification to obtain low noise figures (NFs).

14.2 DPA experiment in 75 km of DSF

Fiber loss in DPA is relatively large (e.g. about 20 dB for 100 km of dispersion-shifted fiber (DSF)). Hence we must use the solutions for lossy fibers presented in subsections 3.5.4 and 3.6.3 to calculate the gain spectra.

Figure 14.1 shows theoretical gain spectra for one- and two-pump DPAs using DSF, with the following parameters [21]: $L = 75$ km; $\alpha = 0.2$ dB km^{-1}, $\beta^{(4)} = -5.0 \times$

10^{-55} s^4 m^{-1}; $\lambda_0 = 1550$ nm; $D_\lambda = 0.07$ ps nm^{-2} km^{-1}; $\gamma = 2$ W^{-1} km^{-1}; total input pump power $P_0 = 66$ mW.

The gain spectrum for the two-pump case is interesting, being similar to the Chebyshev spectra that can be obtained with lossless fibers [22], even though here the total fiber loss is considerable (about 15 dB). Both gain spectra are relatively wide; in particular, the two-pump gain spectrum covers most of the C-band and is almost as wide as that of an EDFA.

However, the gain shapes of Fig. 14.1 are very sensitive to the pump wavelength(s). This is particularly true for the two-pump OPA case: decreasing λ_1 by 0.01 nm leads to a drop of about 5 dB of the center of the spectrum. This is due to the fact that we assumed that λ_0 is constant along the fiber. In real fibers, however, λ_0 varies along the fiber with a standard deviation σ_0 of the order of 1 nm. Then the gain spectrum cannot be affected by changes in λ_1 much smaller than σ_0. This has been verified experimentally [20]: a 2.5 nm change in λ_1 was required for the gain to drop by 5 dB. Therefore the gain shape in real fibers, while not exactly predictable, should be relatively insensitive to pump wavelength. Variations in λ_0 can also be helpful for obtaining a flat-topped spectrum, as found in [20].

Distributed parametric amplification experiments were performed with a single pump to check the theoretical predictions. The transmission fiber consisted of three 25 km DSF spools, with average ZDW about 1550 nm, $D_\lambda = 0.07$ ps nm^{-2} km^{-1}, $\gamma = 2$ W^{-1}km^{-1}, and $\alpha = 0.2$ dB km^{-1}. The fiber had a total loss of 14.5 dB. A tunable laser set at $\lambda_1 = 1551.13$ nm served as the pump source. The CW pump was phase modulated by two modulators in cascade to suppress stimulated Brillouin scattering (SBS). They were both driven by a 2.5 Gb/s^1 pseudo-random bit sequence (PRBS) [24].

Polarization controllers aligned the SOP of the pump with the phase modulators. The pump was then amplified by a C-band EDFA and filtered by a 1.3 nm band-pass filter to reduce the EDFA ASE. Three other tunable laser sources, set at 1549.14, 1549.46, and 1549.76 nm, served as signal sources. They were modulated by a 10 Gb/s^{-1} non-return-to-zero (NRZ) $2^{23} - 1$ PRBS; a single Mach–Zehnder modulator was used for the first two signals and the third was modulated separately. Signals 2 and 3 served to generate crosstalk for signal 1. Maximum gain was achieved by aligning the signal and pump SOPs. The signals and pump were then combined by a 3 dB coupler and entered the DSF. The signal input powers were below −18 dB m, while the input pump power was 18.23 dB m (66.5 mW). This ensured that the pump was not significantly depleted and that nonlinear effects related to the signals, such as cross-gain modulation (XGM) [25] and four-wave mixing (FWM), were kept to a minimum. The optical spectrum at the output of the DSF was observed with an optical spectrum analyzer (OSA). A bandpass filter selected the desired signal wavelength before detection by the O/E converter.

Figure 14.2(a) shows the amplified EDFA ASE spectrum only. It has a 10 dB bandwidth of about 2 nm on each side of the pump, which can be used for amplifying modulated signals. As expected, the gain bandwidth is smaller than that predicted by theory (Fig. 14.1); this is probably due to substantial longitudinal ZDW variations in this three-segment link. A non-destructive measurement of the fiber dispersion map [26, 27] would be essential for designing a DPA scheme.

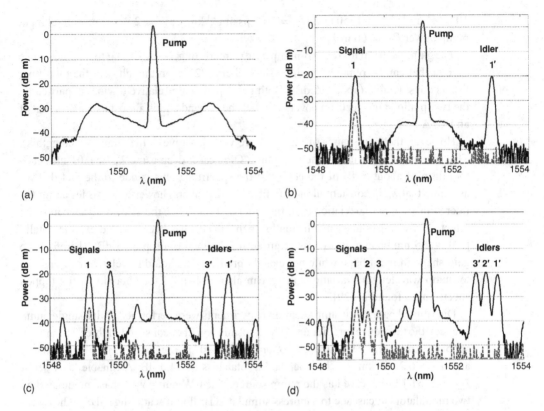

Fig. 14.2 (a) Amplified spontaneous emission (ASE) output spectrum at the output of a distributed parametric amplifier (DPA) with 18.23 dB m of input pump power. The spectrum at the DPA output for (b) a single channel on (channel 1), (c) two channels on (channels 1 and 3), and (d) three channels on (channels 1, 2, and 3). (After [21], © 2005 IEEE.)

A bandpass filter was then inserted to suppress EDFA ASE, and the signals were turned on one by one. Three cases were examined: (i) only channel 1 on; (ii) channels 1 and 3 on, in which case XGM was generated; (iii) all three channels on. The choice of the channel wavelengths was such that channels 2 and 3 generated an FWM component at the wavelength of channel 1. Therefore, both XGM and FWM effects were generated on channel 1 in case (iii).

The powers of channels 1, 2, and 3 at the DSF input were −18.74, −20.71, and −20 dB m, respectively. The corresponding on–off signal gains were 14.4, 16.9, and 16.5 dB. These gains were sufficient to compensate the transmission loss of the DSF (15.52 dB), in good agreement with the theoretical predictions. The optical spectra observed at the output of the DSF for these three cases are shown in Figs. 14.2(b), 14.2(c), and 14.2(d), respectively, when the pump was on (solid lines) and when it was off (broken lines).

The amplified signal at 1549.14 nm (channel 1) was then selected by a filter and detected by an optical receiver. The eye patterns were displayed on an oscilloscope for the above three cases. Examples are shown in Fig. 14.3(a); the clear opening indicates the high quality of the received signal. By varying the received power using an attenuator, the BER was also measured; BER plots are shown in Fig. 14.3(b), together with a

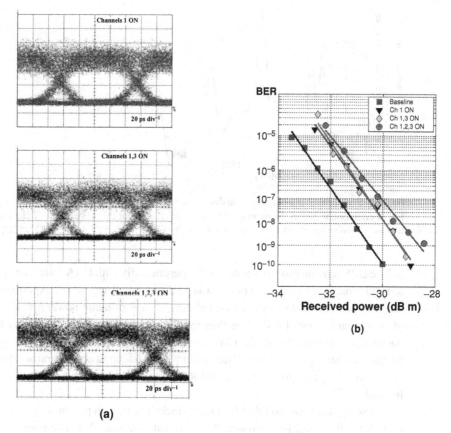

Fig. 14.3 (a) Eye patterns and (b) BER plots for the received signal at 1549.14 nm for the three cases examined and for the back-to-back configuration (the baseline). (After [21], © 2005 IEEE.)

reference baseline for the transmitter and receiver back-to-back. It can be seen that the DPA leads to a power penalty of the order of 1.2 dB in cases (i) and (ii) and 2.1 dB in case (iii). The signal degradation in case (i) was mainly due to intensity noise transfer from pump to signal [28] and the conversion of pump phase modulation to signal intensity modulation along the DSF [29, 30]. Both these effects degraded the mark level of the signal eye. Another possible source of degradation was stimulated Brillouin scattering (SBS), which is generally more severe for a distributed OPA because of the long fiber. Figure 14.3(a) confirms that the eye opening was limited by excess noise in the mark level. The power penalty for (ii) was the same as that for (i); the lower input signal power levels in comparison with the pump power guarantee substantially low cross-gain modulation (XGM) effects. For (iii) the penalty increased to 2.1 dB. Figure 14.4 shows the spectrum at the output of the DSF when only channels 2 and 3 were present. These channels created a four-wave mixing (FWM) component at the wavelength of channel 1, which was found to be 27 dB below the power level of channel 1. Therefore, the increased penalty when all three channels were on was mainly due to imperfect filtering of the interfering channel 2 at the receiver.

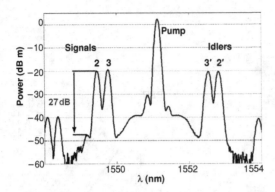

Fig. 14.4 Spectrum at the output of the distributed OPA when channels 2 and 3 were on. The FWM component due to the mixing of those channels falls at the wavelength of channel 1. It was measured to be 27 dB lower than the power level of the signals. (After [21], © 2005 IEEE.)

The DPA scheme was then compared experimentally with DRA, using single-channel transmission. The signal input power was −5 dB m in both cases. The experimental setup for DPA was similar to that discussed above; 79 mW of pump power provided 14.1 dB of signal gain, about 1.4 dB less than the total loss of the DSF. Increasing the pump power to compensate for the fiber loss substantially degraded the signal quality; this was attributed to strong SBS effects. The output was then filtered and the transmitted channel was detected by an optical receiver; BER plots indicated a penalty of 1.2 dB at a BER level of 10^{-9}.

The setup was then modified for DRA. Backward Raman pumping was used, since it provides better noise performance than forward pumping [31]. The pump was a high-power fiber Raman laser. A pump power equal to 282 mW provided 14.1 dB of signal gain. A BER plot indicated a 1.09 dB penalty at BER level of 10^{-9}, about the same as for DPA.

The experimental results verified that it is possible to transmit communication signals over a typical fiber link by DPA with relatively low pump power. Additional channels did not cause large degradation due to nonlinear crosstalk. Direct comparison with a backward-pumped DRA system showed that DPA required only about one-third the pump power of DRA, which could be a significant advantage.

Because the pump power P_0 required to compensate fiber loss is relatively low (less than 100 mW), the theoretical DPA gain bandwidth is fairly small, of the order of 10 nm or less. Furthermore, in practical fibers the bandwidth depends on the λ_0 distribution along the fiber length. In [21] it was found to be reduced from a theoretical value of about 20 nm (assuming constant λ_0 along the fiber span) to about 2 nm (for an actual three-spool span). This is much smaller than for DRA, and so DPA will not compete with DRA but could be operated to provide supplementary capacity.

An important aspect that should be taken into account is polarization dependence. The gain for DRA showed a dependence on the input signal polarization of less than 1 dB, much less than that in the DPA case (of the order of 10 dB). In the relatively polarization-independent DRA, only half the pump power contributes to signal amplification. Therefore, for a fair comparison a polarization-independent DPA scheme [32] should be employed. While such a scheme has not yet been demonstrated in long fibers, the fact that parametric amplification with two parallel pumps works well in a 25-km-long DSF [20] indicates that there is a good possibility that the same would be true with two orthogonal pumps. As discussed in subsection 4.4.3, in such a case twice as much pump power would be needed (i.e. 158 mW). Nevertheless, this comparison confirmed that DPA is capable of performing similarly to DRA while using less pump power. These considerations would be important in a multiple-segment link, where it would be necessary to use polarization-independent amplification to handle the random polarization of signals.

14.3 Possible extensions of DPA

The experiments described above were preliminary, and there are a number of avenues that can be explored for improving these results further.

(a) Cascaded stages with idler removal

The simplest possibility is to cascade additional links using DPA as in the first link. In order to have the same initial conditions as for the first link, the idler must be removed after each stage by means of a band filter. Because the idlers have been removed, the gains of the links do not depend on the relative phases of the pumps and the signals and so there is no need for phase tracking, locking, etc. Furthermore, the pump frequencies in the various stages do not have to be exactly the same because the signal frequency is not affected by the pump frequency. In this situation, all links will perform in the same manner. Therefore the gain of the chain will simply be the product of the gains of the individual links. Also, noise accumulation in all the links will be the same as in the first link; hence the noise power will grow linearly with the number of links. This arrangement has the advantage that it is easy to understand and to implement once the first link has been implemented.

(b) Cascaded stages without idler removal

The preceding arrangement, however, has the disadvantage that when the idlers are discarded power is lost that could otherwise potentially be used for improving the SNR at the receiver. The reason for this is that an idler carries essentially the same amount of information as its associated signal (albeit in a slightly different form if phase modulation is used). We could thus consider letting the idlers propagate all the way along with the signals. Now all stages after the first would have both signal and idler present at the input, and their amplification would therefore be phase sensitive. In effect stages 2 and beyond would become phase-senoitive amplifiers (PSAs). If the pump power were kept the same as in the first stage, the signal power gain would be four times larger in each following stage; hence to compensate for the same fiber loss, it would be necessary to reduce the pump power in stage 2 and beyond.

While this is in principle feasible, it poses some technical problems. In particular, in order to obtain maximum gain there would be a need to track and control the relative phases of the waves. In addition, the pump frequency would have to be exactly the same in all stages because it has to remain precisely halfway between the signal and idler frequencies. It would then be necessary to use a means for precisely adjusting the frequency and phase, such as laser seeding, a phase-locked loop, etc.

The main benefit to be derived from this approach is that, because there are fewer open ports, less vacuum fluctuations enter and so the NF of the signal at the end of the system would be better than in (a). In fact, in a long system most of the noise will be due to the losses in the fibers, and the contribution of the open idler port of the first fiber will be relatively negligible.

An experimental demonstration has been performed with two stages cascaded in this manner [34], albeit with short fibers exhibiting little loss. The second stage exhibited the anticipated PSA behavior, as its gain was found to be affected by the phases of the input waves; these could be altered by changing the length of a different type of fiber connecting the two stages.

(c) Signal and idler fed into cascaded PSA stages

The system of (b) will in effect behave in almost the same way as if the signal and idler had been present at the input of the first fiber, each with half the amplitude of the input signal in (b). This suggests another possibility: make all the stages PSAs, by having the signal and idler simultaneously present at the input of the first fiber and at the inputs of all the following stages. Then there will be no open port at all and the only noise will be due to fiber loss. Hence this arrangement will have a better NF than (b), but of course it will be somewhat more difficult to implement than the latter because of the need for additional phase adjustment at the input of the first stage.

A related experiment was recently performed on a single-stage PSA, with a 25-km-long DSF [34]. Input signal and idler were obtained by using the two sidebands of a modulated carrier, a method previously used for studying phase-sensitive FWM. The gain obtained was in good agreement with the theory for phase-sensitive amplification. The emphasis here was not on distributed amplification, since the fiber loss was relatively low.

This work was then extended to a 60-km-long fiber [35]. In this case the fiber length was sufficiently large that it could be considered to correspond to a typical communication link. Therefore this work represents the first demonstration of a one-stage distributed PSA link, suitable for communication applications. Cascading several such stages as described above could lead to a long practical phase-sensitive distributed parametric amplification (PS-DPA) system.

(d) PS-DPA system with constant pump power

So far we have considered DPA in which the pump is attenuated along with the signal as they co-propagate through the fiber. Vasilyev recently proposed a form of phase-sensitive DPA in which the pump power is kept essentially constant along each fiber by means of distributed Raman amplification [36]; the signals are still subject to fiber loss, which is compensated by DPA.

This has several advantages. By reducing the required signal and pump launch power, the signal–signal interactions (i.e. FWM and XPM) and pump–signal interactions (FWM) are reduced. Also, the scheme yields a further improvement in SNR at the receiver. In fact this scheme in principle can provide the best SNR that can be obtained at the end of a series of lossy fibers, using any type of amplifier.

In addition to low-noise amplification of the amplitude, PSA by definition performs a complete regeneration of binary phase information, such as that in differential or binary phase-shift keying. Since interchannel and intrachannel nonlinearities tend to distort the signal phase and frequency, leading to timing jitter and ghost-pulse formation, continuous regeneration of the phase by PSA counteracts these effects.

An important practical consideration is that the Raman gain brings about a reduction in the required pump launch power, which seems to imply that SBS suppression should become easier. However, since the pump power is maintained at the same level over the entire fiber, there may actually be no benefit. In fact, if the same pump were allowed to propagate through several consecutive fibers then the SBS threshold would be further reduced. Hence a careful investigation of SBS in this type of PS-DPA will be necessary to determine its impact.

Another practical difficulty is that it is assumed that the Raman pump SOP is arranged to be parallel to the parametric pump and orthogonal to the signal and the idler, in order to avoid Raman amplification of the signal and deterioration of the NF. Such SOP control is difficult to maintain in typical communication fibers, that are not polarization-maintaining (See Chapter 6).

It is of interest to calculate the parametric pump power required to compensate signal fiber loss in this case. This can be done simply by equating the signal field gain and loss coefficients, which means that $g = \alpha/2$, where g is the usual parametric gain coefficient. Under optimum phase-matching conditions we have $g = \gamma P_0$, and so we find that the required pump power is $P_0 = \alpha/(2\gamma)$. For a typical communication fiber, $\alpha \approx 0.2$ dB km^1 or 0.045 nepers km^1, and $\gamma \approx 2$ W^{-1} km^{-1}. This yields $P_0 \approx 11$ mW, a remarkably low power, which has important practical implications. On the positive side this low pump power can readily be supplied by inexpensive semiconductor lasers, at any wavelength in the 1200–1700 nm window. On the negative side, however, a low power implies that the gain bandwidth is going to be small, i.e. even smaller than the 1 nm obtained in recent DPA experiments with a pump power of the order of 60 mW [21].

14.4 Conclusion

The experimental investigations described in Chapters 11 and 12 dealt primarily with the operation of fiber OPAs that are relatively compact in size and can be considered to be discrete or lumped devices compared with the vast length of fiber transmission lines.

By contrast, in this chapter we have considered the occurrence of parametric amplification in long transmission fibers, which we then considered to be distributed. It has been known for a long time that DPA can occur in transmission fibers, simply as a result of the presence of one or more strong signals that then lead to the amplification of ASE produced by the EDFAs used periodically for boosting the signals. This amplified noise can cause SNR degradation, and it has therefore been the object of intense research by several groups. We have chosen not to dwell on this topic in this book, because it is rather specialized and constitutes a negative effect of parametric amplification that must be dealt with in those systems affected by it.

We have dealt at length, however, with the more positive aspects of DPA, which can potentially be exploited by designing communication systems rather differently. We have seen that by using pumps of moderate power (100 mW, or less) it is possible to generate enough gain in typical communication fibers to overcome their loss, in the same manner as with distributed Raman amplification. We have described the results of experiments that have verified the possibility of transmitting several 10 Gb/s^{-1} signals in this manner. We have also discussed the possibility of extending this work in several directions, concluding with what would be the ultimate communication system from the point of view of SNR, namely PS-DPA with a Raman-amplified pump. While there are considerable practical difficulties associated with making further progress in this area, it may be that improvements in fiber-optic technology will render the implementation of these advanced DPA schemes possible. If so, this will bring about what could be a new generation of optical communication systems.

References

1. "Fiber-based optical parametric amplifiers and their applications," Hansryd, J., Andrekson, P. A., Westlund, M., Li, J., Hedekvist, P. O. *IEEE J. Select. Topics Quantum Electron.*; 2002; vol. 8, pp. 506–20.
2. "Toward practical fiber optical parametric amplifiers and oscillators," Marhic, M. E., Wong, K. K. Y., Kalogerakis, G., Kazovsky, L. G. *Optics and Photonics News*; 2004; vol. 15, pp. 20–5.
3. "A new design arrangement of transmission fiber dispersion for suppressing nonlinear degradation in long-distance optical transmission systems with optical repeater amplifiers," Henmi, N., Aoki, Y., Ogata, T., Saito, T., Nakaya, S. *J. Lightwave Technol.*; 1993; vol. 11, pp. 1615–21.
4. "Parametric instability of optical amplifier noise in long-distance optical transmission systems," Lorattanasane, C., Kikuchi, K. *IEEE J. Quantum Electron.*; 1997; vol. 33, pp. 1068–74.
5. "New analytical results on fiber parametric gain and its effects on ASE noise," Nuyts, R. J., Tzeng, L. D., Mizuhara, O., Gallion, P. *IEEE Photon. Technol. Lett.*; 1997; vol. 9, pp. 535–7.
6. "On the joint effects of fiber parametric gain and birefringence and their influence on ASE noise," Carena, A., Curri, V., Gaudino, R., Poggiolini, P., Benedetto, S. *J. Lightwave Technol.*; 1998; vol. 16, pp. 1149–57.
7. "Parametric gain in multiwavelength systems: a new approach to noise enhancement analysis," Bosco, G., Carena, A., Curri, V., Gaudino, R., Poggiolini, P., Benedetto, S. *IEEE Photon. Technol. Lett.*; 1999; vol. 11, pp. 1135–7.
8. "Novel analytical method for the BER evaluation in optical systems affected by parametric gain," Bosco, G., Carena, A., Curri, V., Gaudino, R., Poggiolini, P., Benedetto, S. *IEEE Photon. Technol. Lett.*; 2000; vol. 12, pp. 152–4.

9. "Impact of parametric mixing of ASE and signal on high-power festoon systems with random dispersion variation," Mauro, J. C., Chowdhury, D. Q. *IEEE Photon. Technol. Lett.*; 2001; vol. 13, pp. 212–4.
10. "A novel analytical approach to the evaluation of the impact of fiber parametric gain on the bit error rate," Bosco, G., Carena, A., Curri, V., Gaudino, R., Poggiolini, P., Benedetto, S. *IEEE Trans. on Comms.*; 2001; vol. 49, pp. 2154–63.
11. "Power threshold due to parametric gain in dispersion-mapped communication systems," Serena, P., Bononi, A. *IEEE Photon. Technol. Lett.*; 2002; vol. 14, pp. 1521–3.
12. "Parametric gain in fiber systems with periodic dispersion management," Consolandi, F., De Angelis, C., Capobianco, A. D., Nalesso, G., Tonello, A. *Optics Comm.*; 2002; vol. 208, pp. 309–20.
13. "A parametric gain approach to DPSK performance evaluation in presence of nonlinear phase noise," Serena, P., Orlandini, A., Bononi, A. In *Proc. European Conf. on Optical Communication*, Stockholm, Sweden, September 2004; paper We4, p. 124.
14. "Observation of parametric noise amplification owing to modulation instability in anomalous dispersion regime," Saunders, R. A., Garthe, D., Patel, B. L., Lee, W. S., Epworth, R. E. *Electron. Lett.*; 1995; vol. 31, pp. 1088–90.
15. "Parametric gain in the strongly nonlinear regime and its impact on 10-Gb/s NRZ systems with forward-error correction," Serena, P., Bononi, A., Antona, J. C., Bigo, S. *J. Lightwave Technol.*; 2005; vol. 23, pp. 2352–63.
16. "Effective simulation method for parametric signal–noise interaction in transmission fibers," Vanin, E., Jacobsen, G., Berntson, A. *Opt. Lett.*; 2006; vol. 31, pp. 2272–4.
17. "Optical fiber parametric-gain-induced noise coloring and amplification by modulated signals," Xu, B., Brandt-Pearce, M. *J. Opt. Soc. Amer. B*; 2004; vol. 21, pp. 499–513.
18. "Analysis of noise amplification by a CW pump signal due to fiber nonlinearity," Xu, B., Brandt-Pearce, M. *IEEE Photon. Technol. Lett.*; 2004; vol. 16, pp. 1062–4.
19. "Parametric noise amplification inherent in the coherence of fundamental optical soliton sequence propagating in fiber," Inoue, T., Namiki, S. *IEEE J. Select. Topics in Quantum Electron.*; 2004; vol. 10, pp. 900–5.
20. "Amplification of broadband noise pumped by two lasers in optical fibers," Chavez Boggio, J. M., Tenenbaum, S., Fragnito, H. L. *J. Opt. Soc. Am. B*; 2001; vol. 18, pp. 1428–35.
21. "Transmission of optical communication signals by distributed parametric amplification," Kalogerakis, G., Marhic, M. E., Wong, K. K. Y., Kazovsky, L. G. *J. Lightwave Technol.*; 2005; vol. 23, pp. 2945–53.
22. "Broadband fiber optical parametric amplifiers and wavelength converters with low-ripple Chebyshev gain spectra," Marhic, M. E., Park, Y., Yang, F. S., Kazovsky, L. G. *Opt. Lett.*; 1996; vol. 21, pp. 1354–6.
23. "Polarization-independent two-pump fiber optical parametric amplifier," Wong, K. K. Y., Marhic, M. E., Uesaka, K., Kazovsky, L. G. *IEEE Photon. Technol. Lett.*; 2002; vol. 14, pp. 911–3.
24. "Amplification of WDM signals in fiber-based optical parametric amplifiers," Torounidis, T., Sunnerud, H., Hedekvist, P. O., Andrekson, P. A. *IEEE Photon. Technol. Lett.*; 2003; vol. 15, pp. 1061–3.
25. "Nondestructive position-resolved measurement of the zero-dispersion wavelength in an optical fiber," Eiselt, M., Jopson, R. M., Stolen, R. H. *J. Lightwave Technol.*; 1997; vol. 15, pp. 135–43.

26. "Measurement of the dispersion map of installed G.653 fiber links using four-wave mixing," Artiglia, M., Caponi, R., Grazioli, E., Pagano, A., Panella, A., Potenza, M., Riccardi, E., Sordo, B. In *Proc. Optical Fiber Communication Conf.* San Jose CA, 1998; paper WM2.
27. "Pump-to-signal transfer of low-frequency intensity modulation in fiber optical parametric amplifiers," Marhic, M. E., Kalogerakis, G., Wong, K. K. Y., Kazovsky, L. G. *J. Lightwave Technol.*; 2005; vol. 23, pp. 1049–55.
28. "Pump FM to signal IM conversion in fiber OPAs," Marhic, M. E., Kalogerakis, G., Kazovsky, L. G. In *Proc. of OECC/COIN2004*, Yokohama, Japan, July 2004; paper 13P-87.
29. "Impact of pump phase modulation on the gain of fiber optical parametric amplifier," Mussot, A., Durecu-Legrand, A., Lantz, E., Simonneau, C., Bayart, D., Maillotte, H., Sylvestre, T. *IEEE Photon. Technol. Lett.*; 2004; vol. 16, pp. 1289–91.
30. "Pump to signal RIN transfer in Raman fiber amplifiers," Fludger, C. R. S., Handerek, V., Mears, R. J. *J. Lightwave Technol.*; 2001; vol. 19, pp. 1140–8.
31. "Polarization-independent one-pump fiber-optical parametric amplifier," Wong, K. K. Y., Marhic, M. E., Uesaka, K., Kazovsky, L. G. *IEEE Photon. Technol. Lett.*; 2004; vol. 14, pp. 1506–8.
32. "Gain characteristics of a frequency nondegenerate phase-sensitive fiber-optic parametric amplifier with phase self-stabilized input," Tang, R., Lasri, J. S. Devgan, P. S., Grigoryan, V., Kumar, P., Vasilyev, M. *Optics Express*; 2005; vol. 13, pp. 10483–93.
33. "In-line frequency-nondegenerate phase-sensitive fiber-optical parametric amplifier," Tang, R., Devgan, P., Voss, P. L., Grigoryan, V. S., Kumar, P. *IEEE Photon. Technol. Lett.*; 2005; vol. 17, pp. 1845–7.
34. "Inline frequency-non-degenerate phase-sensitive fibre parametric amplifier for fibre-optic communication," Tang, R., Devgan, P., Grigoryan, V. S., Kumar, P. *Electron. Lett.*; 2005; vol. 41, pp. 1072–4.
35. "Distributed phase-sensitive amplification," Vasilyev, M. *Optics Express*; 2005; vol. 13, pp. 7563–71.

15 Prospects for future developments

15.1 Introduction

In the preceding chapters we have reviewed the state of the art in fiber OPAs, in terms of the current theoretical understanding and of their experimental performance and progress toward applications. Along the way we have identified a number of obstacles that stand in the way of further progress. In this chapter we review some areas of technology where advances would be desirable for the development of fiber OPAs or related devices. We also look at areas that have not been investigated much to date and may yet present interesting research and development opportunities.

15.2 Fibers

Conventional fiber structures, consisting of a cylindrical core surrounded by a cladding made from a lower-index material, offer limited opportunities for optimizing the fiber properties of fiber OPAs. In contrast, as mentioned in Section 2.1, holey fibers (HFs) have the potential for providing new generations of fibers suitable for fiber OPAs. In principle all HF properties, including loss, nonlinear index, mode effective area, and dispersion, can be optimized. To improve OPA performance significantly beyond the level reached with currently available fibers, all properties will eventually need to be simultaneously optimized in the same HF. This will pose an interesting challenge for the designers and manufacturers of HFs.

15.2.1 High-γ fibers

A large value of the fiber nonlinearity coefficient γ (see Eq. (2.24)) is beneficial in a number of ways. It makes it easier to obtain a large gain bandwidth at moderate pump power. It also reduces the product $P_0 L$ required to achieve a given gain. Since γ is proportional to n_2/A_{eff}, in order to achieve a large γ it is desirable to have a large n_2 as well as a small A_{eff}. The normalized frequency of a cylindrical fiber is $V = 2\pi a \sqrt{n_{\text{co}}^2 - n_{\text{cl}}^2}/\lambda$, where a is the core radius, λ is the vacuum wavelength, n_{co} is the core refractive index, and n_{cl} is the cladding refractive index. For single-mode fibers, V is typically 2. Thus $A_{\text{eff}} \propto a^2 \propto 1/(n_{\text{co}}^2 - n_{\text{cl}}^2)$ and so there is a limit to how small A_{eff} can be, because n_{co} is never very large for low-loss material: it is at most about 5 in the near infrared, a typical value for semiconductors.

However, n_2 can be increased by two or three orders of magnitude in comparision with its value in silica by using highly nonlinear materials. Many such materials have been investigated for nonlinear optics applications and could potentially be considered for fiber OPAs.

Among these materials is tellurite, with an n_2 value larger than that of silica by a factor of about 25. It has been theoretically investigated as a possible OPA medium [1]. A step-index fiber has been used for obtaining Raman amplification over a 160 nm bandwidth [2]; it had a maximum Raman gain coefficient 16 times larger than that of a comparable silica-based fiber. Highly nonlinear HFs have also been fabricated [3, 4] with γ values of the order of 600 W^{-1} km^{-1}. One of these had a small core diameter such that the zero-dispersion wavelength (ZDW) was shifted to 1573 nm [3] and a relatively low loss, 0.18 dB m^{-1}. It would be interesting to test its performance as an OPA [1] with an L-band pump. By using a 5 m length, the fiber loss would be below 1 dB and therefore negligible as a first approximation. A pump power of 1 W would then yield $\Phi \approx 3$, sufficient for obtaining 20 dB gain. The pump power would be low enough that CW experiments could be performed.

Other materials of possible interest are As_2S_3, Si, GaAs, metal-doped glasses, lead glass, Bi_2O_3 glass, etc. It has been projected that fibers made from gallium lanthanum sulphide (GLS) glass could provide values of γ 10 000 times larger than those of conventional communication fibers [5]. A lead silicate HF with $A_{eff} = 2.6$ and $\gamma = 640$ W^{-1} km^{-1} was recently reported [6]. By making fibers holey the ZDW of these fibers can be shifted into the C-band [7].

15.2.2 Fibers with tailored dispersion

A challenging aspect of fiber design for OPAs is obtaining the desired dispersion properties. The first requirement is to have the fiber's λ_0 value in the desired wavelength range. This can be a challenge with some highly nonlinear materials that have material λ_0 values very far from the C-band; for example, As_2S_3 has $\lambda_0 \approx 4.5$ μm. Then the fiber must be designed to exhibit a great deal of waveguide dispersion, in order to shift λ_0 by the desired amount. Generally this can be achieved by having a very small A_{eff}; this requires a cladding with a very low effective index, i.e. a very large air fraction.

To obtain large gain bandwidths, it is then necessary to have low values of $\beta^{(4)}$. Recently progress has been made with HNL-DSFs made by the conventional step-index design. Low values of $\beta^{(4)}$, with either sign, have been obtained. In particular, with a fiber with $\beta^{(4)} = 2 \times 10^{-56}$ s^4 m^{-1}, wavelength conversion by FWM was obtained over a bandwidth of nearly 200 nm, which represents an improvement of nearly a factor 2 over previous results [8].

With HFs, to date no work has been done in this area, but it can be anticipated that, by adjusting the shape and size of the air holes, in principle one should be able to control this parameter as well. Work done on controlling $\beta^{(3)}$ for HFs in such a manner could be used as a model [9–11].

The gain bandwidth of OPAs does not directly depend on $\beta^{(3)}$, and so it seems that there may be no need to be concerned with $\beta^{(3)}$ when designing fibers for OPAs. However, we have seen in Chapter 13 that detrimental four-wave mixing (FWM) can be reduced if their wavevector mismatch $\Delta\beta$ is large. Since the $\Delta\beta$ value such terms increases with $\beta^{(3)}$, it might be advantageous to design fibers with as large a $\beta^{(3)}$ value as possible while maintaining the desired value of $\beta^{(4)}$.

For OPAs designed to act as wavelength converters between very distant regions, the dispersion properties obtained with fibers having a single zero-dispersion wavelength (ZDW) may not be adequate, and one may require a fiber with two or even three ZDWs. Holey fibers designed to have such dispersion properties have been used for this type of wavelength conversion [12].

15.2.3 Fibers for novel wavelength ranges

To date, most work with fiber OPAs has been done in the C-band, primarily because of the availability of fibers and pump sources in that range. One advantage of fiber OPAs is that in principle they can operate about arbitrary center wavelengths, a feature that could be used for making amplifiers or wavelength converters operating in ranges where suitable devices are currently not available. To do this it is be necessary to extend the design of highly nonlinear fibers (HNLFs) or HFs to the desired ranges.

As an example, wavelength conversion was recently achieved by four-wave mixing from the C-band to a wavelength as low as 540 nm, by using a silica photonic crystal fiber (PCF) having two ZDWs, at 750 nm and 1260 nm [12]. While the conversion efficiency achieved in this experiment was low, improvements in fiber fabrication may eventually make it possible to operate fiber OPAs efficiently in such distant wavelength ranges.

15.2.4 Polarization-maintaining fibers (PMFs)

As we have seen in previous chapters, in fibers that are not designed to maintain polarization states there exists weak randomly varying birefringence, which causes states of polarization (SOPs) to vary in a random manner along the fibers. This random evolution, related to the phenomenon of polarization-made dispersion, is the source of significant difficulties for the design and operation of fiber OPAs. In particular, it can cause distortions of the gain spectra as one tries to achieve very broadband operation. In addition, it makes it difficult to obtain the polarization-independent operation of two-orthogonal-pump OPAs, via the introduction of wavelength-dependent polarization-dependent gain (PDG).

These difficulties exist for either long or short fiber OPAs as we now discuss.

1. For long OPAs, significant averaging over SOPs can take place, which makes the shape of the gain spectra fairly predictable, provided that the wavelength spacings are not very large. For large wavelength spacings, though, this predictability disappears and there can be very large differences in gain spectra from one fiber to another.

2. For short fiber OPAs the SOPs may well retain their relative orientations (i.e. identical or orthogonal); however, there is little opportunity for any averaging over SOPs. Hence one finds again that the gain spectra can vary greatly with the birefringence maps of particular fibers, regardless of the wavelength spacings considered.

Under these circumstances, it becomes apparent that unless random birefringence can be completely eliminated from fibers, the design and fabrication of fiber OPAs with specific gain spectra is going to be hindered by the effects of polarization-mode dispersion.

One way to avoid the detrimental effects of random birefringence is to use polarization-maintaining fibers (PMFs). With such fibers, if the input waves are linearly polarized along the axes of birefringence, they remain so along the entire fiber.

If all the waves have the same SOP then the analysis of the OPA is greatly simplified, as it is essentially a scalar problem, as studied in Chapter 3. Variations in gain spectra can still occur due to ZDW variations, but the unpredictable variations due to polarization-mode dispersion are eliminated. In addition, polarization-independent operation can be obtained by using a polarization-diversity scheme, for either one- or two-pump OPAs.

Polarization-maintaining (PM) HNL-DSF can be fabricated [13]. The internal stress required to obtain PM operation makes it difficult to use very small fiber cores, and so PM fibers tend to have smaller γ values than their non-PM counterparts [14]. Nevertheless, the benefits that can be derived from PM operation may compensate for the drawbacks of a reduced γ value.

It should also be noted that the fabrication of small-core large-γ HFs often leads naturally to fibers with very large birefringence, which can therefore be used as PMFs. Such fibers could in principle be used to combine PM operation with a large γ value, which would be very desirable for making short high-performance fiber OPAs.

15.2.5 Control of longitudinal ZDW variations

Even if SOP randomness can be eliminated, it will remain true that all fibers will still exhibit random longitudinal variations in ZDW. Since OPA operation depends critically on the precise location of the ZDW, such variations will cause an unpredictable distortion of the gain spectrum from one fiber to another.

While manufacturing techniques keep improving and have significantly reduced ZDW variations in HNLFs, it appears that further progress in this direction may be very difficult. One may then face the task of designing fiber OPAs based on imperfect fibers.

Possibilities do exist for manipulating the ZDW map of a particular fiber *after* it has been manufactured. Basically, any physical mechanism that affects the refractive index is also likely to shift the ZDW. Well-known examples are temperature, strain, and ultraviolet exposure. The ZDW of HNLF shifts by about 0.06 nm per degree centigrade [15]. It shifts by as much as 40 nm for 1% of strain in HNLFs with a thin cladding [16]. Ultraviolet exposure can be used for permanently shifting the ZDW by as much as 100 nm [17]. So, these techniques could in principle be used for creating arbitrary ZDW maps in fibers after fabrication. While this is theoretically possible, the practical implementation of any of these techniques will entail significant challenges. Certainly the demonstration of uniform shifts in a few segments should be feasible, but achieving arbitrary shifts in many segments will be very challenging.

15.2.6 Dispersion-map measurements

Actually, being able to modify dispersion maps accurately, as described in the preceding section, would be effective only if one also has the ability to measure the dispersion map accurately in the first place. If that is not possible then one will not know what the initial map is, or, therefore, the modification needed to obtain a particular gain spectrum.

It is therefore highly desirable to have a non-destructive technique which can provide highly accurate measurements of the ZDW that have a high spatial resolution. A number of methods have been introduced for measuring the ZDW map, or related properties, for the fiber in a fiber OPA but their longitudinal resolution is of the order of 10–20 m [18–20]; ideally it would be desirable to improve on this, perhaps by as much as one order of magnitude. If such a technique became available, it could in principle be coupled with a technique for modifying the ZDW map, in order to obtain essentially arbitrary ZDW maps and therefore to tailor gain spectra at will. Such a combination of these two methods would make it possible to fabricate OPAs with gain profiles suitable for various applications, in particular for the broadband amplification of WDM signals for optical communication.

15.3 Pumps

The development of cost-effective fiber OPAs will require the availability of moderate-cost pump lasers with suitable characteristics, namely high power, narrow linewidth, and low relative intensity noise (RIN). Ideally, they would be available throughout the 1200–1700 nm range for the development of subsystems for optical communications. On the one hand, probably the best prospects rest with semiconductor lasers, since they can in principle be made at any wavelength in that range, by a suitable choice of material. Single-frequency semiconductor lasers with a single-transverse-mode output power in excess of 2.2 W at 864 nm have been made using a master-oscillator power-amplifier structure (MOPA) [21]. Similar devices are now commercially available at other wavelengths [22, 23].

On the other hand, if the design of fiber OPAs with very large bandwidths becomes possible then the availability of suitable pumps at just a few wavelengths would be sufficient. In that case it might be possible to use fiber lasers, which can generate high powers with high efficiency and low amplified spontaneous emission (ASE). As an extreme example of this, one might envision a two-pump fiber OPA, with pumps at 1300 nm and 1620 nm, that has bandwidth of about 300 nm covering most of the current telecommunication bands. The 1620 nm pump could be provided by an L-band erbium-doped fiber laser (EDFL), while the 1300 nm pump could be provided by a praesodymium-doped fiber laser (PDFL).

An interesting possibility is to use a fiber laser made from a short highly doped fiber. Such lasers are just a few centimeters long but can generate up to 5 W power, with a linewidth of just a few kilohertz [24].

Also, an ASE-free laser has been reported [25]. It is a semiconductor laser made with a Sagnac ring interferometer within an external cavity. It has an output power of several milliwatts and an ASE level away from the pump below the internal noise level of an optical spectrum analyzer (OSA). Its low-frequency RIN is not described, but if it is also greatly reduced then this could provide a means for making a low-RIN pump.

It is also good to keep in mind that if high-quality pumps are available in some regions of the spectrum, such as the C- and L-bands and the 1000–1100 nm range, then in principle other pumps can be derived from these with good efficiency by nonlinear means such as periodically poled lithium niobate (PPLN) devices or fiber OPAs. With fiber OPAs, conversion efficiencies approaching 50% can be achieved. This method has already been used for particular experiments [26, 27] and could perhaps be extended to other situations.

In principle it is possible to consider the use of Raman amplification, which can also be used to convert the power of high-power pumps available in certain wavelength regions. One approach would be to make a standard Raman fiber laser by attaching a fiber Bragg grating (FBG) at each end of a fiber to make a Fabry–Pérot cavity. Unfortunately, such sources have a linewidth typically of the order of 1 nm, determined by the bandwidth of the FBGs used. This linewidth corresponds to IM and FM of the output and would not be suitable for an OPA pump, particularly for optical-communication applications. An alternative approach would be to use a narrow-linewidth seed, such as can be provided by a laser diode, and to amplify it to a high level in a Raman amplifier (RA). However, there are difficulties associated with this. In particular a high gain is needed to amplify the seed to a high level, sufficient for depleting the pump, in order to obtain a high conversion efficiency. But high gain is difficult to obtain with RAs, because (i) the Raman gain coefficient is relatively weak (about 1/3 of the parametric gain coefficient) (ii) double Rayleigh scattering sets an upper limit to the gain.

For these reasons, to date Raman amplification has not been used successfully for generating the kind of high-power high-quality pumps required for parametric amplification. However, perhaps with novel approaches the above difficulties could be circumvented, so that RAs could be used for providing versatile pumps for OPA pumping. It should be noted that a great deal of work has been done on so-called Raman-assisted parametric processes; some ideas introduced in that area might be applicable to the generation of high-power pumps.

15.4 SBS suppression

Table 15.1 shows a variety of techniques that in principle can be used for stimulated Brillouin scattering (SBS) suppression. In the following we discuss their potential for providing efficient SBS suppression for future fiber OPAs and OPOs.

Table 15.1. Techniques for SBS suppression

Technique	Threshold increase	Loss	Idler broadening	Dispersion change	Permanent fiber change
PM/FM	>20 dB	5 dB (PM)	Yes	No	No
Two pumps	3 dB		No	No	No
Isolators (N)	10 log(N) dB	about N dB	No	No	Yes
Gaps (N)	10 log(N) dB	< N dB	No	No	Yes
Stress	>10 dB	0 dB	No	Yes	Yes
Tapered core	7 dB		No	No	—
Core design	6 dB		No	No	—
Temperature	3 dB 100 °C^{-1}	0 dB	No	Yes	No
Bi$_2$O$_3$ glass		3 dB m^{-1}	No	—	—
Al$_2$O$_3$ core	6 dB	37 dB km^{-1}	—	—	—

To date most work with fiber OPAs has used pump-spectrum broadening by either phase or frequency modulation for SBS suppression. While this method is effective, it is not simple or inexpensive, as it requires substantial RF power and electronics. The use of external phase modulators also introduces significant losses, typically 5 dB per modulator. Being able to replace this active technique by one or more passive methods could help the development of practical fiber OPAs. Some known passive techniques, such as air gaps, stress distributions, tapered cores, or isolators, could perhaps be used for this purpose. However, the use of HFs could lead to solutions that are not possible with current fibers. For example, it is possible that suitably designed microstructured fibers could exhibit bandgaps for phonons, i.e. frequency ranges in which sound waves cannot propagate; this would lead to SBS-free operation in that range [28]. Alternatively, one might envision HFs with holes filled with liquid having a high attenuation for acoustic waves but not for optical waves.

A promising approach is to design fibers for which the overlap between the optical and acoustic modes of interest is smaller than for conventional fibers. In this manner an increase in SBS threshold as large as 6 dB has recently been achieved [29].

Another interesting approach is the use of stress variations along fibers. A specially fabricated cable, creating a sinusoidal strain variation in a fiber, achieved a 7 dB increase in SBS threshold and could achieve more with a higher strain [30]. Recently a threshold increase of more than 10 dB was obtained by creating a staircase strain distribution in an HNLF [31]. One difficulty with this approach is that strain also modifies the refractive index and dispersion: 1% of strain causes a ZDW shift of 4–5 nm [31] (in another type of HNLF the shift was about 10 times larger [16]). Since the gain spectrum of fiber OPAs is very sensitive to longitudinal ZDW variations, it appears that this method would be incompatible with generating the gain spectra associated with uniform fibers; nevertheless, one possibility exists for using strain for SBS suppression while maintaining such a gain spectrum. The idea is to use periodic strain variations, with a short spatial period. Since the gain spectrum is not affected by short-scale ZDW fluctuations, it would be essentially the same as if the fiber were uniform, with a ZDW equal to the average

ZDW for one period of the actual fiber. It has been shown theoretically and experimentally that periodic ZDW variations do not affect the spectrum of a typical OPA if the period is of the order of one-hundredth of the fiber length or less [15]. For an OPA that is a few hundred meters long, one would need to have a period shorter than a few meters. The cable of [30] had a period of 0.6 m and would therefore satisfy this condition. This type of approach thus offers the possibility of strong SBS suppression while maintaining the gain spectrum and is therefore very attractive.

It should be noted that cabling may not be the only way to produce such periodic strain distributions. For example, one could consider winding a fiber on a circular drum, attaching it uniformly to the drum by means of epoxy or similar material; then, by deforming the drum one could create regions of positive and negative strain, located periodically along the fiber.

Also, one should not overlook the fact that a two-pump OPA has an SBS threshold which is 3 dB higher than that for a one-pump OPA with the same total gain.

It is also interesting to note that bismuth oxide fibers have an advantage compared to silica-based fibers when it comes to SBS threshold [32]. Step-index fibers have a very large group velocity dispersion (GVD) in the C-band but, by using a small-core HF design, it is possible to bring the ZDW near 1550 nm for communication applications [33, 34].

Similarly, heavy core doping with aluminum oxide can provide a 6 dB increase in SBS threshold; unfortunately, this is accompanied by a large increase in loss [35].

15.5 Integrated optics

Silica-based single-mode integrated-optic waveguides can exhibit losses below 0.01 dB cm^{-1}, and they can thus be used for making waveguides that are a few meters long and have a loss of just a few decibels. If similar losses could be achieved with high-γ waveguides, such lengths would be sufficient for making OPAs with moderate pump powers.

In comparison with fiber technology, integrated optics presents opportunities for other features, such as: wafer-scale fabrication; integration with other components such as couplers, Bragg gratings, etc.; achieving small A_{eff} values with ridge waveguides; the creation of air gaps for SBS suppression; the design of slow-wave structures for enhancing nonlinear interactions; producing layered structures from different materials; the deposition of materials not suitable for making fibers; ion implantation with highly nonlinear dopants; producing tapered transitions for low-loss connection to fibers. Thus integrated optics could provide an interesting alternative to optical fibers for making OPAs based on the third-order optical susceptibility $\chi^{(3)}$.

One regime for which integrated-optic OPAs might be of interest is that of short pulses with high peak powers. In this case the lengths required for obtaining substantial gains could be modest, and so waveguide structures might be attractive.

In recent years the possibility of obtaining optical amplification in silicon, by means of stimulated Raman scattering (SRS), has attracted considerable attention [36, 37]. The appeal is obvious, as the development of silicon-based optical devices could eventually lead to the integration of optical and electronic components in the silicon platform; this possibility helps to drive the enormous semiconductor industry. Significant SRS gain has been obtained in short waveguides, because of the large value of the Raman gain in silicon.

The Kerr nonlinear index n_2 of silicon is more than 100 times larger than that of silica. Also, silicon-on-insulator (SOI) technology permits the fabrication of ridge waveguides with effective areas smaller than 1 µm^2. This combination yields values of γ as large as 3×10^4 W^{-1} km^{-1}. It has also been shown that the ZDW of such ridge waveguides can be located near 1550 nm by proper choice of the waveguide dimensions [38, 39]. As a result, 5-cm-long waveguides can support soliton propagation near 1550 nm, with a peak power of the order of 2 W [39].

Since parametric amplification requires power levels similar to those for soliton formation ($\gamma P_0 L \geq 1$), it follows that parametric amplification should also occur in such waveguides, with similar parameters.

In fact parametric amplification has recently been observed in Si waveguides [40]. Near-zero dispersion was obtained at 1550 nm by using Si waveguides buried in silica, with a cross-section of the order of 300 nm $\times$ 600 nm. In a 17-mm-long guide, pumped with 4 W peak power, an on/off gain of 2.9 dB was observed. The gain was limited by a number of effects, including a waveguide loss of the order of 1.1–1.4 dB /cm^{-1} at 1550 nm and a nonlinear loss of 9.7 dB outside the phase-matching region. In addition an input coupling loss in excess of 10 dB limited the net input–output gain.

The results of these first experiments are promising and indicate that parametric amplification in integrated optics based on Si, or other nonlinear materials, may indeed provide another direction for developing practical parametric amplifiers based on the third-order optical susceptibility.

Recently there has also been considerable interest in developing so-called slow-light structures. As the name indicates, in such structures the group velocity of light is substantially reduced compared with its value *in vacuo*, c. A major area of application is the creation of large time delays that can be used for processing information.

Another aspect of slow light, however, is that it leads to an enhancement of nonlinear interactions. One way to understand this is to consider the propagation of a light pulse carrying a certain amount of energy. If the pulse travels more slowly, its length in space is reduced, and therefore the energy per unit volume that it carries is increased and so is the electric field. As a result, nonlinear effects are enhanced.

Slow-light structures with group velocities as low as $10^{-2}c$ to $10^{-3}c$ have been demonstrated [41, 42]. The possibility of enhancing SRS in an SOI slow-light photonic crystal waveguide has been investigated theoretically [43]. It was found that the Raman gain can be enhanced by a factor of almost 10^5 compared with its value in bulk Si. If similar enhancements could be obtained for the Kerr effect, parametric amplification in short structures might become possible. Of course, the need for proper phase matching between the waves introduces another constraint, which will make the design process more challenging.

15.6 Pump resonators

A fiber OPA with 400 nm bandwidth was recently demonstrated [44]. This bandwidth is now about ten times larger than that of an erbium-doped fiber amplifier (EDFA), which is the main type of amplifier used in today's optical communication systems.

A practical difficulty with the experiments described above, however, is that the required pump powers were fairly high, being in excess of 10 W. Since CW pumps with such high power were not available, pulsed pumps with the required peak powers, but with a duty cycle of only about 1%, were used. This allowed operation with pump sources of moderate average powers yet with the desired peak power. Such pulsed amplifiers, however, are of limited practical interest in optical communication, where one really needs CW amplifiers, providing a gain that is constant in time.

In principle, passive optical resonators can provide a means for combining the virtues of a high pump power inside the OPA and a moderate pump power provided externally by a pump laser. In Kerr-related nonlinear work with fiber optics, to our knowledge a resonating pump has been used only once, for making a CW modulation-instability laser [45].

To enhance optical pump power by means of an optical resonator, a wave generated by an external pump laser is directed onto a partially reflecting mirror that forms part of the cavity. A proportion of the light enters the cavity and bounces between the mirrors, becoming trapped. If the light frequency matches one of the resonance frequencies of the cavity, the optical power in the cavity can build up to a level exceeding that of the externally incident light by a factor proportional to the "finesse" of the cavity. The finesse is a function of the losses in the cavity, and in fiber cavities it can reach tens or even hundreds, if the cavity losses are low.

This idea can in principle be applied to fiber OPAs, and significant benefits can be derived from it. These are:

1. **Large CW gain bandwidth with moderate pump power** For example, one should be able to obtain a bandwidth in excess of 300 nm with a CW laser power of less than 1 W. To our knowledge, at this time fiber OPAs are the only type of optical amplifier that can provide such large CW gain bandwidths.
2. **Reduced fiber length** Because of the high intracavity power, high gains (> 20 dB) can be achieved in relatively small lengths of HNLF (of the order of 20 m). Such fiber lengths are similar to those used for EDFAs. A small length would result in reduced fiber cost and would also have several major additional consequences, as follows.
3. **Improved fiber uniformity** Short fibers would have more uniform longitudinal properties, namely lower variations of the zero-dispersion wavelength λ_0 as well as reduced random birefringence and polarization-mode dispersion. The latter aspects are very important for obtaining gain spectra with well-defined shapes, particularly for very wideband amplification.
4. **Reduced spurious nonlinear interactions** A relatively small fiber length would also reduce nonlinear effects between signals in the OPA, which arise when the OPA

is used with multiple optical carriers. These nonlinear effects are signal–signal and pump–signal four-wave mixing (FWM) as well as cross-gain modulation (XGM). As discussed in Chapter 14, they can be alarmingly large in OPAs using longer fibers, to the point where fiber OPAs can be at a considerable disadvantage compared with other types of amplifier (EDFAs or Raman amplifiers).

5. **Orthogonal signal and pump** If the intracavity power is very high, it may actually be possible to have the signals orthogonal to the pump and still have sufficient gain. There are benefits attached to this. First, since the transverse Raman gain is very small, the OPA noise factor (NF) will be almost unaffected by the Raman gain and so in principle could approach 3 dB (which is not the case when the SOPs are the same). In addition, since the required pump power will be larger by a factor 3 for a given gain, pump depletion and hence XGM will be reduced; the same is true of pump–signal FWM (see Chapter 13).

6. **Elimination of external ASE** Most current OPA experiments make use of high-power EDFAs for boosting the pump to a high level, and these inevitable add much ASE to the amplified pump. This ASE can be reduced by the use of bandpass filters but it cannot be totally eliminated. With a resonator pumped directly by a narrow-linewidth laser, this type of ASE does not enter into the system and this should lead to a higher-quality pump.

7. **Operation with arbitrary center wavelength** If the use of resonators can reduce the required pump laser power to tens of milliwatts then semiconductor lasers could be used directly. Such lasers can in principle be fabricated anywhere in the 1200–1700 nm region, i.e. through the entire range of interest for optical communication.

8. **Reduced hazards** By reducing the power required from the pump laser, the probability of optical damage to components such as connectors, filters, and couplers, and of injuries to operators, should be significantly reduced. This would be an important point for the utilization of OPAs in practical systems.

All these important advantages show that the use of pump resonators to boost the effective pump power could improve the performance of fiber OPAs in several very important respects. This approach to the design of fiber OPAs may thus provide a leap in performance that could greatly accelerate the development of practical fiber OPAs. While resonators hold considerable promise for improving fiber OPAs, there are, however, a number of challenges that would need to be overcome in order to fulfill this promise. The most obvious is the need to reduce cavity losses in order to achieve a high quality factor Q. Another linear-optics issue is that of the cavity linewidth: if it is too narrow, achieving and maintaining resonance will place very stringent requirements on the laser source. Other aspects associated with nonlinear optical phenomena are that SBS will need to be suppressed and that the high circulating power may introduce resonance shifts, modification of the gain spectrum, pump depletion, or modulation instability. In particular, for SBS suppression it would be good to avoid the need for pump dithering by using static means such as stress and novel core design, as discussed in Section 15.4; however, if required pump dithering can be rendered compatible with cavity resonance [45].

While improving fiber OPAs by making use of fiber resonators appears challenging, one may draw some inspiration from the fact that parametric oscillation has recently been observed in microcavities just tens of microns in size, with a pump power below 1 mW and a conversion efficiency of 36% [46]. This compact source illustrates some benefits that can be derived from operating an OPA with a very-high-Q pump resonator.

15.7 Discrete or distributed parametric amplification?

In the preceding sections we have discussed a number of areas of research that could help the development of practical *discrete* fiber OPAs, for applications in optical communication.

While the use of discrete fiber OPAs in essentially conventional optical communication systems is a possible avenue for the utilization of parametric amplification, we saw in Chapter 14 that it is also possible to envision a radically different type of fiber communication system, based on *distributed* parametric amplification (DPA). In this approach the signal amplification necessary for compensating fiber loss takes place in the transmission fiber itself, and there is no need for amplification between segments of communication fibers. The use of distributed parametric gain instead of distributed Raman gain has two advantages: (i) the required pump power is lower; (ii) the NF can also be lower.

Furthermore, one can in principle use the phase-sensitive version of DPA, which yields additional improvements in pump power and NF. Finally, the ultimate version of phase-sensitive DPA involves the use of a parametric pump whose intensity is kept nearly constant by means of Raman amplification. In that case the fiber loss is balanced locally by the parametric gain, and in principle this can yield the best performance in terms of SNR that can be achieved at the receiver. If such a system could be built that could also handle a large number of high-frequency signals, this could potentially offer a very different way of designing high-capacity optical communication systems, with a performance significantly better than that obtainable by current designs. Naturally, there are considerable obstacles that would need to be overcome if this promise is ever to be realized. These are as follows.

Bandwidth

This is a major concern, because the usable bandwidth demonstrated in DPA experiments in a typical communication fiber, with a pump power of the order of 60 mW [47], is only about 1 nm. This is very small compared with the 35 nm bandwidth of EDFAs used in current systems or the potential bandwidth of 200 nm or so that could be obtained with discrete OPAs.

One reason for this small bandwidth is the low value of the product γP_0 required for compensating loss. For a given type of fiber, i.e. for given α and γ, P_0 is fixed and cannot be increased to obtain a larger bandwidth.

The other reason for the small bandwidth is longitudinal variations in λ_0. These reduce the bandwidth from an attractive theoretical value of several tens of nanometers to about

1 nm. If one could manufacture fibers with greatly improved λ_0 variation then it would be possible to obtain a much larger bandwith.

Another possibility might be to use fibers that have multiple ZDWs. For example, fibers with ZDWs at both 1310 nm and 1550 nm have been fabricated. Also, holey fibers with three ZDWs have been made. It appears that there is no fundamental reason preventing the construction of fibers with a large number of ZDWs. If such fibers were available, one could place a pump near each ZDW and thereby increase the aggregate bandwidth available.

A simpler alternative would be to place the pumps away from the ZDW. In this case the theoretical bandwidth would be smaller but the practically available bandwidth might still be of the order of 1 nm, as it would be less affected by ZDW fluctuations. This approach would have the merit of working with potentially any type of fiber. An interesting aspect of this approach is that bandwidth can be added as needed, by turning on additional pumps; this could present a significant advantage for the deployment and activation of such systems.

Detrimental nonlinear effects

Another area of concern is undesired nonlinear effects such as those investigated in Chapter 14: signal–signal and pump–signal FWM, as well as XGM. These effects were investigated experimentally in [47] and were found to be tolerable for the number of channels used (three) and the distance covered (75 km); however, since they effects increase with the number of channels and the number of consecutive stages, their impact will have to be carefully investigated to determine whether they will impose a limit on system performance.

Also, if a multi-pump arrangement is used (as contemplated above), the presence of several relatively strong pumps will also lead to the generation of relatively strong FWM products, by pump–pump FWM. This will set a limit on how closely such pumps could be located and therefore on what the total capacity could be.

If Raman gain is used to maintain a constant pump power along the fiber, and the pump phase does not jump between fiber segments, the effective fiber length becomes the entire fiber (from the pump's point of view). Thus the SBS threshold can become much lower than in a lossy fiber, for which the effective length is of the order of 20 km. So, even though the pump power may be low, it may still be significantly above threshold. Therefore the usual SBS-suppression techniques may still have to be employed.

Polarization independence

In conventional optical communication systems it is desirable to have devices such as amplifiers that are polarization-independent, i.e. whose performance (e.g. amplifier gain) is independent of the SOP of the incident signal. This is to accommodate the fact that the SOPs of signals coming from different parts of a network may be different and, furthermore, may vary in time owing to environmental fluctuations. For this reason a great deal of work has been done to develop polarization-independent discrete fiber OPAs. Up to the present time, all the thetheoretical and experimental work with DPAs

Table 15.2. Communication systems based on parametric amplification

	OPA	Idler kept?	Raman	NF	Bandwidth
1	Discrete	No	—	Highest	Large
2	Discrete	Yes	—	—	Large
3	Distributed	No	No	—	Small
4	Distributed	Yes	No	—	Small
5	Distributed	No	Yes	—	Smallest
6	Distributed	Yes	Yes	Lowest	Smallest

has been done for the case where all waves have the same SOP. While this is a useful simplification, it may not be sufficient to deal with realistic input signals.

One possibility is to use two orthogonal pumps. Since pumps and signals may need to be close together, the impact of polarization-mode dispersion may be relatively small and so this approach may work well over fairly long distances.

It should be noted that the quantum-noise properties of DPAs with orthogonal pumps have not been investigated, and this would need to be done to see what the ultimate performance of such systems can be.

Polarization tracking is another way to achieve polarization-independent operation. In this case the signal and pump could be forced to have the same SOP along the entire fiber. This approach would also have the advantage of being potentially compatible with Raman amplification of the parametric pump(s) without affecting the signal (which does not appear to be possible if orthogonal parametric pumps are used). The approach would also be compatible with polarization-maintaining fibres (PMFs); in fact PMFs may be required to avoid polarization-mode dispersion, to prevent the Raman pump from directly affecting the signal.

Table 15.2 lists the main categories of fiber communication systems based on parametric amplification that we have discussed so far, as well as some additional possibilities. They are classified according to the following characteristics: whether discrete or distributed OPAs are used; whether idlers are kept after each stage (i.e. whether operation of the system is phase-sensitive); for the distributed cases, whether Raman gain is used to maintain the parametric pump power constant.

For all the phase-sensitive systems there are two possibilities for the first stage: one may have only signals present at the input or one may inject idlers along with the corresponding signals. For a long chain of amplifying stages, the performance of these two types of system will be essentially the same.

We have previously discussed schemes 1, 3, 4, and 6 (Table 15.2). Schemes 2 and 5 have been added, as they are logical entries in this table and constitute additional possible candidates for the design of advanced communication systems based on parametric amplification.

There are many trade-offs involved in the different designs, and a thorough comparison is beyond the scope of this book. However, it will have to be undertaken if parametric amplification is going to be considered seriously for the implementation of optical communication systems. Among the important features that will need to be compared are the following: NF; bandwidth; spurious nonlinear effects; aggregate traffic capacity; systems reach; SBS suppression; polarization independence; stability; frequency, phase, and polarization control systems.

15.8 Conclusion

The basic theory of fiber OPAs is fairly well understood and indicates that under ideal circumstances these devices can exhibit interesting characteristics in terms of bandwidth, gain, conversion efficiency, noise figure, etc. They can thus in principle be candidates for application in many areas, including optical communication, high-speed signal processing, wavelength conversion, spectroscopy, high-power generation, etc.

Progress in these areas, however, is hampered by the fact that existing fibers are non-ideal in several respects, in particular because they exhibit longitudinal dispersion and birefringence variations and a low SBS threshold. In addition, the pump sources that are currently available are also far from being ideal. Again, the impact of these limitations is fairly well understood. Nevertheless, improvements can be made in all these aspects and it is expected that, with progress in these areas, fiber OPAs will come within reach of more applications.

While the path to some applications is fairly clear, there are still many questions that remain to be answered about others. For example, much work remains to be done on systems using distributed and/or phase-sensitive amplification.

Recent developments in $\chi^{(3)}$ parametric amplification in silicon waveguides and silica microspheres also indicate that interesting solutions may be found by using $\chi^{(3)}$ in media that are not strictly speaking fiber-based. This broadening of the scope of the field could provide additional exciting research opportunities in the future.

References

1. "Design of highly-nonlinear tellurite fibers with zero dispersion near 1550 nm," Hu, E. S., Hsueh, Y. L., Marhic, M. E., Kazovsky, L. G. In *Proc. European Conf. on Optical Communication*, Copenhagen, September 2002; vol. 2, paper 3.2.3.
2. "Ultra-wideband tellurite-based Raman fiber amplifier," Mori, A., Masuda, H., Shikano, K., Oikawa, K., Kato, K., Shimizu, M. *Electron. Lett.*; 2001; vol. 37, pp. 1442–3.
3. "1.5 μm band zero-dispersion shifted tellurite photonic crystal fiber with a nonlinear coefficient γ of 675 $W^{-1}km^{-1}$," Mori, A., Shikano, K., Enbutsu, K., Oikawa, K., Naganuma, K., Kato, M., Aozasa, S. In *Proc. 30th European Conf. on Optical Communication*, Stockholm, September 2004; paper Th3.3.6.
4. "Extruded single-mode, high-nonlinearity, tellurite glass holey fiber," Feng, X., Monro, T. M., Finazzi, V., Moore, R. C., Frampton, K., Petropoulos, P., Richardson, D. J. *Electron. Lett.*; 2005; vol. 41, pp. 835–6.
5. "The fabrication and modelling of non-silica microstructured optical fibers," Hewak, D. W., West, Y. D., Broderick, N. G. R., Monro, T. M., Richardson, D. J. In *Proc. Optical Fiber Communication Conf.*, Anaheim CA, March 2001; vol. 2, pp. TuC4-1–3.

6. "Highly nonlinear and anomalously dispersive lead silicate glass holey fibers," Petropoulos, P., Ebendorff-Heidepriem, H., Finazzi, V., Moore, R. C., Frampton, K., Richardson, D. J., Monro, T. M. *Optics Express*; 2003; vol. 11; pp. 3568–73.
7. "Efficient four-wave-mixing at 1.55 μm in a short-length dispersion shifted lead silicate holey fiber," Asimakis, S., Petropoulos, P., Poletti, F., Leong, J. Y. Y., Ebendorff-Heidepriem, H., Moore, R. C., Frampton, K. E., Feng, X., Loh, W. H., Monro, T. M., Richardson, D. J. In *Proc. European Conf. on Optical Communication*, Cannes, September 2006; vol. 4, paper Th1.3.3, pp. 17–8.
8. "Broadband wavelength conversion over 193-nm by HNL-DSF improving higher-order dispersion performance," Hirano, M., Nakanishi, T., Okuno, T., Onishi, M. In *Proc. 31st European Conf. on Optical Communication*, September 2005, Glasgow.
9. "Dispersion-flattened photonic crystal fibers at 1550 nm," Reeves, W., Knight, J., Russel, P. In *Proc. Optical Fiber Communication Conf.*, Atlanta GA, March 2003; vol. 2, paper FI3, pp. 696–7.
10. "Fully dispersion controlled triangular-core nonlinear photonic crystal fiber," Hansen, K. P., Folkenberg, J. R., Peucheret, C., Bjarklev, A. In *Proc. Optical Fiber Communication Conf.*, Atlanta GA, March 2003; postdeadline paper, vol. 3, pp. PD2-1–3.
11. "Optical parametric amplification in all-silica triangular-core photonic crystal fibers," Poli, F., Adami, F., Foroni, M., Rosa, L., Cucinotta, A., Selleri, S. *Applied Phys. B (Lasers and Optics)*; 2005; vol. B81, pp. 251–5.
12. "375 THz parametric translation of modulated signal from 1550 nm to visible band," Jiang R., Saperstein, R., Alic, N., Nezhad, N., McKinstrie, C., Ford, J., Fainman, Y., Radic, S. In *Proc. Optical Fiber Communication Conf.*, Anaheim CA, March 2006; postdeadline paper, pp. PD16-1–3.
13. "Interband wavelength conversion of 320 Gb/s (32 × 10 Gb/s) WDM signal using a polarization-insensitive fiber four-wave mixer," Watanabe, S., Takeda, S., Chikama, T. In *Proc. 24th European Conf. on Optical Communication*, Madrid, September 1998; vol. 3, pp. 83–7.
14. M. Onishi, private communication.
15. "Temperature control of the gain spectrum of fiber optical parametric amplifiers," Wong, K. K. Y., Marhic, M. E., Kazovsky, L. G. *Optics Express*; 2005; vol. 13, pp. 4666–73.
16. "Experiment of zero dispersion tuning by stretching down-sized HNLF," Takahashi, M., Tadakuma, M., Hiroishi, J., Sugizaki, R., Yagi, T. In *Proc. European Conf. on Optical Communication*, Cannes, September 2006; vol. 4, paper Th1.5.1, pp. 41–2.
17. "Supercontinuum generation in ultraviolet-irradiated fibers," Nicholson, J. W., Westbrook, P. S., Feder, K. S., Yablon, A. D. *Opt. Lett.*; 2004; vol. 29, pp. 2363–5.
18. "Zero-dispersion wavelength mapping in short single-mode optical fibers using parametric amplification," Mussot, A., Lantz, E., Durécu-Legrand, A., Simonneau, C., Bayart, D., Sylvestre, T., Maillotte, H. *IEEE Photon. Technol. Lett.*; 2006; vol. 18, pp. 22–4.
19. "OTDR technique for characterization of fiber optic parametric amplifiers," Olsson, B.-E., Torounidis, T., Karlsson, M., Sunnerud, H., Andrekson, P. In *Proc. Optical Communication Conf. and the National Fiber Optic Engineers Conf.*, Anaheim CA, March 2006; paper OWT3.
20. "Brillouin optical time domain analysis of fiber optic parametric amplifiers," Vedadi, A., Lantz, E., Maillotte, H., Sylvestre, T. In *Proc. European Conf. on Optical Communication*, Cannes, September 2006; paper Th1.3.7.

21. "2.2-W continuous-wave diffraction-limited monolithically integrated master oscillator power amplifier at 854 nm," O'Brien, S., Lang, R., Parke, R., Major, J., Welch, D. F., Mehuys, D. *IEEE Photon. Technol. Lett.*; 1997; vol. 9, pp. 440–2.
22. http://www.m2k-laser.de.
23. http://www.sacher.de.
24. http://www.npphotonics.com.
25. "High power ASE-free tunable laser using a Sagnac ring interferometer within the external cavity," Fulop, L., Souhaite, G., Moulinet, X., Graindorge, P., Lefevre, H. C. In *Proc. Optical Fiber Communication Conf.*, Anaheim CA, March 2001; vol. 2, pp. TuJ6-1–3.
26. "Phase-conjugate pump dithering for high-quality idler generation in a fiber optical parametric amplifier," Wong, K. K. Y., Marhic, M. E., Kazovsky, L. G. *IEEE Photon. Technol. Lett.*; 2003; vol. 15, pp. 33–5.
27. "Nearly 100 nm bandwidth of flat gain with a double-pumped fiber optic parametric amplifier," Marconi, J. D., Chavez Boggio, J. M., Fragnito, H. L. Submitted to OFC 2007, Anaheim CA, March 2007, paper OWB1.
28. "Phononic band-gap guidance of acoustic modes in photonic crystal fibers," Laude, V., Khelif, A., Benchabane, S., Wilm, M., Sylvestre, T., Kibler, B., Mussot, A., Dudley, J. M., Maillotte, H. *Phys. Rev. B*; 2005; vol. 71, pp. 045107-1–6.
29. "Nonlinear optical fibers with increased SBS thresholds," Bickham, S., Kobyakov, A., Li, S. In *Proc. Optical Fiber Communication Conf.*, Anaheim CA, March 2006, paper OTuA3.
30. "Stimulated Brillouin scattering suppression by means of applying strain distribution to fiber with cabling," Yoshizawa, N., Imai, T. *J. Lightwave Technol.*; 1993; vol. 11, pp. 1518–22.
31. "Narrow linewidth wavelength converter with 70 nm signal tuning band by using a strained HNLF," Marconi, J. D., Chavez Boggio, J. M., Fragnito, H. L. In *Proc. Conf. on Optical Amplifiers and Their Applications*, Whistler, Canada, July 2006.
32. "Experimental comparison of a Kerr nonlinearity figure of merit including the stimulated Brillouin scattering threshold for state-of-the-art nonlinear optical fibers," Lee, J. H., Tanemura, T., Kikuchi, K., Nagashima, T., Hasegawa, T., Ohara, S., Sugimoto, N. *Opt. Lett.*; 2005; vol. 30, pp. 1698–700.
33. "Design of highly nonlinear bismuth-oxide holey fibers with zero dispersion and enhanced Brillouin suppression," Poletti, F., Petropoulos, P., Broderick, N. G., Richardson, D. J. In *Proc. European Conf. on Optical Communication*, Cannes, September 2006; paper Tu4.3.2.
34. "Dispersion shifted Bi_2O_3-based photonic crystal fiber," Nagashima, T., Hasegawa, T., Ohara, S., Sugimoto, N. In *Proc. European Conf. on Optical Communication*, Cannes, September 2006; paper We1.3.2.
35. "Al_2O_3–SiO_2 core highly nonlinear dispersion-shifted fiber with Brillouin gain suppression improved by 6.1 dB," Nakanishi, T., Tanaka, M., Hasegawa, T., Hirano, M., Okuno, T., Onoshi, M. In *Proc. European Conf. on Optical Communication*, Cannes, September 2006; postdeadline paper Th4.2.2.
36. "Observation of stimulated Raman amplification in silicon waveguides," Claps, R., Dimitropoulos, D., Raghunathan, V., Han, Y., Jalali, B. *Optics Express*; 2003; vol. 11, pp. 1731–9.
37. "Lossless optical modulation in a silicon waveguide using stimulated Raman scattering," Jones, R., Liu, A., Rong, H., Paniccia, M., Cohen, O., Hak, D. *Optics Express*; 2005; vol. 13, pp. 1716–23.
38. "Tailored anomalous group-velocity dispersion in silicon channel waveguides," Turner, A. C., Manolatou, C., Schmidt, B. S., Lipson, M., Foster, M. A., Sharping, J. E., Gaeta, A. L. *Optics Express*; 2006; vol. 14, pp. 4357–62.

39. "Dispersion tailoring and soliton propagation in silicon waveguides," Yin, L., Lin, Q., Agrawal, G. P. *Opt. Lett.*; 2006; vol. 31, pp. 1295–7.
40. "Broad-band optical parametric gain on a silicon photonic chip," Foster, M. A., Turner, A. C., Sharping, J. E., Schmidt, B. S., Lipson, M., Gaeta, A. L. *Nature*; 2006; vol. 441, pp. 960–3.
41. "Extremely large group-velocity dispersion of line-defect waveguides in photonic crystal slabs," Notomi, M., Yamada, K., Shinya, A., Takahashi, J., Takahashi, C., Yokohama, I. *Phys. Rev. Lett.*; 2001; vol. 87, pp. 253 902/1–4.
42. "Real-space observation of ultraslow light in photonic crystal waveguides," Gersen, H., Karle, T. J., Engelen, R. J. P., Bogaerts, W., Korterik, J. P., van Hulst, N. F., Krauss, T. F., Kuipers, L. *Phys. Rev. Lett.*; 2005; vol. 94, pp. 073 903/1–4.
43. "Enhanced stimulated Raman scattering in slow-light photonic crystal waveguides," McMillan, J. F., Yang, X., Panoiu, N. C., Osgood, R. M., Wong, C. W. *Opt. Lett.*; 2006; vol. 31, pp. 1235–7.
44. "Wide-band tuning of the gain spectra of one-pump fiber optical parametric amplifiers," Marhic, M. E., Wong, K. K. Y., Kazovsky, L. G. *IEEE J. Select. Topics in Quantum Electron.*; 2004; vol. 10, pp. 1133–41.
45. "Continuous-wave ultrahigh-repetition-rate pulse-train generation through modulational instability in a passive fiber cavity," Coen, S., Haelterman, M. *Opt. Lett.*; 2001; vol. 26, pp. 39–41.
46. "Kerr-nonlinearity optical parametric oscillation in an ultrahigh-Q toroid microcavity," Kippenberg, T. J., Spillane, S. M., Vahala, K. J. *Phys. Rev. Lett.*; 2004; vol. 93, pp. 083904/1–4.
47. "Transmission of optical communication signals by distributed parametric amplification," Kalogerakis, G., Marhic, M. E., Wong, K. K. Y., Kazovsky, L. G. *J. Lightwave Technol.*; 2005; vol. 23, pp. 2945–53.

Appendix 1 General theorems for solving typical OPA equations

A1.1 A simple transformation

When dealing with fiber OPAs it is common to encounter ordinary second-order differential equations of the form

$$y'' + ay' + by = 0, \quad (A1.1)$$

where y is an unknown function of the variable z, and a prime means a derivative with respect to x; a and b are known functions of z.

Equation (A1.1) can be put in a more useful form by eliminating the term in y'. This is done by means of the transformation [1]

$$y = Y \exp\left(-\frac{1}{2}\int a\,dz\right). \quad (A1.2)$$

Substitution of Eq. (A1.2) into Eq. (A1.1) leads to

$$Y'' - g^2 Y = 0, \quad (A1.3)$$

where

$$g^2 = -b + \frac{a^2}{4} + \frac{a'}{2} \quad (A1.4)$$

is the invariant of Eq. (A1.1).

The form of Eq. (A1.3) is well known in propagation problems. When g is constant the solution is a linear combination of the exponentials $\exp(gz)$ and $\exp(-gz)$. These correspond to monotonically increasing and decreasing exponentials if g is real, periodic exponentials if g is imaginary, and damped and amplified periodic functions if g is neither wholly real nor wholly imaginary. All these well-known simple cases are encountered in propagation problems where the medium is uniform. The common interpretation of g is then that of a complex propagation constant: its real part corresponds to the gain or attenuation, while its imaginary part corresponds to the phase.

When g is not a constant but varies slowly with z, it is sometimes possible to approximate the solution by regarding g as constant over a short length and writing the solution in terms of exponentials $\exp[\pm \int^z g(\xi)d\xi]$. This type of approximation is very convenient, as it is based upon the well-understood solutions for $g = $ constant and can therefore provide a great deal of physical insight. For this reason, this approach is often used in areas of physics such as waveguide propagation, plasma waves, quantum mechanics, etc.

The best-known formalism is known as the WKB method. We use it in Chapter 13 to obtain approximate solutions for XGM crosstalk.

A1.2 Solution of the OPA equations in the absence of pump depletion

For either one- or two-pump OPAs, the FWM equations for the signal ($k = 3$) and the idler ($k = 4$) are of the form

$$\frac{dA_k}{dz} = ip_k A_k + ir_k \exp\left(i \int_{\xi=0}^{z} q_k d\xi\right) A_l^*, \quad k = 3, 4, \quad l = 7 - k, \quad (A1.5)$$

where A_k represents the complex slowly varying envelope, z is the distance along the fiber, and the asterisk indicates complex conjugation. We assume that p_k and q_k may be functions of z but that r_k is a constant.

Upon letting

$$A_k = B_k \exp\left(i \int_{\xi=0}^{z} p_k d\xi\right) \quad (A1.6)$$

we obtain

$$\frac{dB_k}{dz} = ir_k \exp\left(-i \int_{\xi=0}^{z} \kappa_k d\xi\right) B_l^*, \quad (A1.7)$$

where $\kappa_k = p_k + p_l^* - q_k$.

One way to proceed is to take the derivative of Eq. (A1.7). This leads to

$$\frac{d^2 B_k}{dz^2} = r_k \kappa_k \exp\left(-i \int_{\xi=0}^{z} \kappa_k d\xi\right) B_l^* + ir_k \exp\left(-i \int_{\xi=0}^{z} \kappa_k d\xi\right) \frac{dB_l^*}{dz}. \quad (A1.8)$$

Using Eq. (A1.7) we can eliminate B_l^*, which leads to

$$\frac{d^2 B_k}{dz^2} + i\kappa_k \frac{dB_k}{dz} - r_k r_l^* \exp\left[i \int_{\xi=0}^{z} (q_k - q_l^*) d\xi\right] B_k = 0, \quad (A1.9)$$

which is a second-order differential equation in B_k only and has the form of Eq. (A1.1). Making use of the transformation

$$B_k = C_k \exp\left(-\frac{i}{2} \int_{\xi=0}^{z} \kappa_k d\xi\right), \quad (A1.10)$$

which implies that

$$C_k = A_k \exp\left[-\frac{i}{2} \int_{\xi=0}^{z} (p_k - p_l^* + q_k) d\xi\right]$$

$$= A_k \exp\left[i \int_{\xi=0}^{z} \left(\frac{\kappa_k}{2} - p_k\right) d\xi\right], \quad (A1.11)$$

we obtain

$$\frac{d^2 C_k}{dz^2} - g_k^2(z) C_k = 0, \quad (A1.12)$$

where

$$g_k^2(z) = r_k r_l^* \exp\left[i \int_{\xi=0}^{z} (q_k - q_l^*)d\xi\right] - \frac{\kappa_k^2}{4} + \frac{i\kappa_k'}{2}; \quad (A1.13)$$

the prime indicates a derivative with respect to z.

(An alternative approach can be used. Substituting Eq. (A1.10) into Eq. (A1.7) leads to

$$\frac{dC_k}{dz} = \frac{i\kappa_k}{2} C_k + ir_k \exp\left[-\frac{i}{2}\int_{\xi=0}^{z}(\kappa_k - \kappa_l^*)d\xi\right]C_l^*. \quad (A1.14)$$

Eliminating C_l again leads to Eq. (A1.12).]

To proceed further it is necessary to know $g_k^2(z)$ explicitly, and also the dependences of p_k, q_k, and r_k on z.

Solution for $g_k^2(z) =$ constant

This situation occurs in the case of an OPA with no pump depletion, no loss (hence constant pump power(s)), and constant dispersion properties ($\Delta\beta =$ constant). Then p_k, q_k, r_k, s_k are constants. Often g_k is constant as well. Under these circumstances the general solution of Eq. (A1.12) is of the form

$$C_k(z) = u_k \exp(g_k z) + v_k \exp(-g_k z), \quad (A1.15)$$

where u_k and v_k are constants to be determined from the initial conditions. To do this, let us take the derivative of Eq. (A1.15), which yields

$$\frac{C_k'(z)}{g_k} = u_k \exp(g_k z) - v_k \exp(-g_k z). \quad (A1.16)$$

Adding and substracting Eqs. (A1.15) and (A1.16) and letting $z = 0$ yields

$$u_k = \frac{1}{2}\left[C_k(0) + \frac{1}{g_k}C_k'(0)\right] \quad \text{and} \quad v_k = \frac{1}{2}\left[C_k(0) - \frac{1}{g_k}C_k'(0)\right]. \quad (A1.17)$$

To express everything in terms of the original field envelopes A_k and A_l, we make use of the relations

$$C_k(0) = A_k(0) \quad \text{and} \quad C_k'(0) = \frac{i\kappa_k}{2}A_k(0) + ir_k A_l^*(0). \quad (A1.18)$$

Then $C_k(z)$ can be written as

$$C_k(z) = A_k(0)\cosh(g_k z) + \frac{i}{g_k}\left[\frac{\kappa_k}{2}A_k(0) + r_k A_l^*(0)\right]\sinh(g_k z)$$

$$= \left[\cosh(g_k z) + \frac{i\kappa_k}{2g_k}\sinh(g_k z)\right]A_k(0) + \frac{ir_k}{g_k}\sinh(g_k z) A_l^*(0). \quad (A1.19)$$

The last expression shows that $C_k(z)$ is a linear combination of $A_k(0)$ and $A_l^*(0)$. The coefficient $C_l(z)$ has a similar expression: it is a linear combination of $A_l(0)$ and $A_k^*(0)$. As a result, we can combine the expressions for $C_k(z)$ and $C_l^*(z)$ in the following vector–matrix relationship:

$$\begin{bmatrix} C_k(z) \\ C_l^*(z) \end{bmatrix} = M \begin{bmatrix} A_k(0) \\ A_l^*(0) \end{bmatrix}, \tag{A1.20}$$

where

$$M = \begin{bmatrix} m_{kk} & m_{kl} \\ m_{lk} & m_{ll} \end{bmatrix}$$

$$= \begin{bmatrix} \cosh(g_k z) + \dfrac{i\kappa_k}{2g_k}\sinh(g_k z) & \dfrac{ir_k}{g_k}\sinh(g_k z) \\ \left(\dfrac{ir_l}{g_l}\sinh(g_l z)\right)^* & \left(\cosh(g_l z) + \dfrac{i\kappa_l}{2g_l}\sinh(g_l z)\right)^* \end{bmatrix}. \tag{A1.21}$$

Finally $A_k(z)$ is obtained from

$$A_k(z) = C_k(z)\exp\left[i\left(p_k - \frac{\kappa_k}{2}\right)z\right]. \tag{A1.22}$$

To obtain $A_l(z)$, k and l should be exchanged.

Under certain circumstances $p_k - p_l^*$, q_k, r_k, κ_k and g_k are the same for $k = 3, 4$. We can then drop the subscripts for simplicity. When r and κ are real, $g^2 = r^2 - (\kappa/2)^2$ is real. An OPA has gain if g is real, i.e. if $g^2 > 0$.

Assuming that g^2 is real, M becomes simpler and we have the relations

$$m_{ll} = (m_{kk})^* \quad \text{and} \quad m_{lk} = (m_{kl})^*. \tag{A1.23}$$

The determinant of M is then given by

$$\det(M) = m_{kk}m_{kk}^* - m_{kl}m_{lk}^* = \cosh^2(gz) + \frac{\kappa^2 - (2r)^2}{(2g)^2}\sinh^2(gz)$$

$$= \cosh^2(gz) - \sinh^2(gz) = 1, \tag{A1.24}$$

which shows that M is a unimodular matrix.

Because unitary matrices are such that the magnitude of their determinants is 1, one might think that M is unitary, as stated in [2]. However, the relations between the matrix elements, Eq. (A1.23), are different from those between the elements of unitary matrices; hence in general M is not unitary.

Another way to understand this is that, in optics, unitary matrices represent processes that conserve power, such as propagation through a lossless 2×2 single-mode fiber coupler. Since by definition an OPA has gain, it cannot be viewed as a power-conserving device and thus cannot be described by a unitary matrix.

It is interesting to note that the form of M is generally more complicated than is often assumed when inferring the form of the matrix for a parametric amplifier from basic principles of quantum optics [2]. Yet M does satisfy the most basic principles of quantum optics.

In particular M is *quasi-unitary*, which indicates that the mode operators of both the incident and the amplified light are proper Bose operators [3]. This can be shown by

A1.2 Solution of the OPA equations in the absence of pump depletion

verifying that M has the following properties:

$$|m_{kk}|^2 - |m_{kl}|^2 = 1,$$
$$|m_{ll}|^2 - |m_{lk}|^2 = 1,$$
$$m_{kk}m_{lk}^* - m_{ll}m_{kl}^* = 0. \qquad (A1.25)$$

It can also be shown that

1. The eigenvalues of M are $\exp(gz)$ and $\exp(-gz)$.
2. $MM^* = I$, or $M^{-1} = M^*$.
3. $M = \exp(Nz)$, where

$$N = i \begin{bmatrix} \kappa/2 & r \\ -r & -\kappa/2 \end{bmatrix}. \qquad (A1.26)$$

One way to prove this last property is by solving the OPA propagation equations in a different way. Going back to Eq. (A1.14), it can now be rewritten as

$$\frac{dC_k}{dz} = \frac{i\kappa}{2}C_k + irC_l^*. \qquad (A1.27)$$

Writing this equation for C_l by exchanging k and l and taking the complex conjugate leads to

$$\frac{dC_l^*}{dz} = -\frac{i\kappa}{2}C_l^* - irC_k. \qquad (A1.28)$$

Taken together, Eqs. (A1.27) and (A1.28) can be rewritten as

$$\frac{d}{dz}\begin{bmatrix} C_k \\ C_l^* \end{bmatrix} = N \begin{bmatrix} C_k \\ C_l^* \end{bmatrix}. \qquad (A1.29)$$

Since N is a constant matrix, the solution of this vector differential equation is

$$\begin{bmatrix} C_k(z) \\ C_l^*(z) \end{bmatrix} = \exp(Nz) \begin{bmatrix} C_k(0) \\ C_l^*(0) \end{bmatrix}. \qquad (A1.30)$$

Comparing with Eq. (A1.20), we see that we must have $M = \exp(Nz)$. This proof has the advantage of providing a simple interpretation for N, namely that it is the matrix that governs the basic propagation equation for C_k and C_l, namely Eq. (A1.29).

The matrix N is interesting in its own right, as well as in relation to M. In particular, N has the following properties:

1. $\det(N) = -r^2 + \left(\frac{\kappa}{2}\right)^2 = -g^2.$ \qquad (A1.31)

2. The eigenvalues are $\pm g$.
3. The corresponding eigenvectors are $[ir \ -i\kappa/2 \pm g]^t$ (where the superscript t means the transpose).
4. M has the same eigenvectors as N.
5. $N^2 = g^2 I$.
6. $n_{ll} = (n_{kk})^*$ and $n_{lk} = (n_{kl})^*$ (the same properties as for M).

7. N is purely imaginary, so it can be written as $N = iR$ where R is real but not Hermitian. (This again demonstrates that M is not unitary, since a unitary matrix must be of the form $\exp(iH)$, where H is Hermitian.)

A1.3 Wavelength exchange

In this case, with the frequency assignment described in Section A1.2, Eq. (A1.5) is still valid except that A_l^* is replaced by A_l. As a result, all the derivations leading to Eqs. (A1.12) and (A1.13) are the same as before, except that the asterisk should be omitted everywhere.

In the case where p_k, q_k, r_k, κ_k do not depend on z and are independent of k and l, we can drop the subscripts and obtain

$$g^2(z) = -r^2 \exp\left(2i \int_{\xi=0}^{z} q\,d\xi\right) - \frac{\kappa^2}{4}. \tag{A1.32}$$

Hence the only way in which g can be independent of z is if $q = 0$. In that case we also have $\kappa = -q = 0$. Then $g^2 = -r^2$, and so $g = ir$.

The matrix N is then given by

$$N = ir \begin{bmatrix} 0 & 1 \\ 1 & 0 \end{bmatrix}. \tag{A1.33}$$

Its eigenvalues are $\pm ir$, and the associated eigenvectors are $[1 \pm 1]^t$.

Since ir is imaginary and since the matrix that multiplies it is real and symmetric and hence Hermitian, M is unitary.

The physical reason for this is that, in wavelength exchange, when the signal loses one photon the idler gains one photon and therefore wavelength exchange is a power-conserving process, which must therefore be described by a *unitary* matrix.

The matrix M itself is given by

$$M = \begin{bmatrix} \cosh(gz) & i\sinh(gz) \\ i\sinh(gz) & \cosh(gz) \end{bmatrix} = \begin{bmatrix} \cos(rz) & \sin(rz) \\ -\sin(rz) & \cos(rz) \end{bmatrix}. \tag{A1.34}$$

The last form clearly shows a periodic exchange of power between signal and idler along the fiber. The eigenvalues of M are $\exp(\pm irz)$ and the associated eigenvectors are the same as for N, i.e. $[1 \pm 1]^t$.

The fact that the eigenvectors have equal power for the signal and the idler is consistent with the photon picture. Then the probability of a signal photon being annihilated and a corresponding idler photon being created is equal to that for the reverse process. As a result these two processes cancel out and the initial powers remain unchanged during propagation, which is a basic requirement for an eigenvector.

The form of M in Eq. (A1.33) is the same as that for a lossless beamsplitter [3, 4]. As a result, a wavelength exchanger will have quantum properties similar to those of such a beamsplitter.

A1.4 Case where $r_k = r_k(z)$

Under some circumstances it is possible for r_k to be a function of z. In particular, this occurs when none of the waves remain in the same SOP everywhere along the fiber. This can occur in PMFs when the input waves are not polarized along one of the axes and, more generally, in non-PMFs when the input waves are not polarized along the principal planes of polarization; under these circumstances the SOPs undergo rapid variations and as a result the magnitudes of the four-wave mixing (FWM) terms in the OPA equations are functions of z.

Then Eq. (A1.7) is still valid. We can rewrite it as

$$\frac{dB_k}{dz} = t_k B_l^* \tag{A1.35}$$

where

$$t_k = ir_k \exp\left(-i \int_{\xi=0}^{z} \kappa_k d\xi\right) \tag{A1.36}$$

is a function of z. Taking the derivative of Eq. (A1.34) and using Eq. (A1.35) leads to

$$\frac{d^2 B_k}{dz^2} + i\tilde{\kappa}_k \frac{dB_k}{dz} - t_k t_l^* B_k = 0, \tag{A1.37}$$

where

$$\tilde{\kappa}_k = \frac{i}{t_k}\frac{dt_k}{dz} = \kappa_k + \frac{i}{r_k}\frac{dr_k}{dz}. \tag{A1.38}$$

Letting

$$B_k = C_k \exp\left(-\frac{i}{2}\int_{\xi=0}^{z}\tilde{\kappa}_k d\xi\right) \tag{A1.39}$$

and using Eq. (A1.12) leads to the standard form of Eq. (A1.12) but with

$$g_k^2 = t_k t_l^* - \frac{\tilde{\kappa}_k^2}{4} + \frac{i\tilde{\kappa}_k'}{2}. \tag{A1.40}$$

References

1. *Handbook of Differential Equations*, Zwillinger, D. Academic, Boston; 1992; p. 126.
2. "From quantum optics to quantum communications," Abram, I., Grangier, P. *C. R. Physique, Optical Telecommunications*; 2003; vol. 4, pp. 187–199.
3. "Quantum physics of simple optical instruments," Leonhardt, U. *Rep. Prog. Phys.*; 2003; vol. 66, pp. 1207–49.
4. "Nonclassical fields in a linear directional coupler," Lai, W. K.; Buek, V., Knight, P. L. *Phys. Rev. A*; 1991; vol. 43, pp. 6323–36.

Appendix 2 The WKB approximation

In Chapter 3 we considered OPAs in which the parametric gain coefficient g is a constant along the fiber. This allowed us to obtain closed-form solutions for the gain in several important situations. In practice, however, fibers generally have properties that may cause g to vary as a function of z. Examples of these properties are: fiber loss, which causes the pump power to drop exponentially; non-uniform dispersion, which causes $\Delta\beta$ to fluctuate; birefringence (fixed or randomly varying), which introduces a complex evolution of SOPs. Under these circumstances the constant-g solutions are no longer applicable.

In several areas of physics that involve wave propagation in media with slowly varying properties, one often uses approximate solutions derived by making use of the Wentzel–Kramers–Brillouin (WKB) approximation, also referred to as the phase-integral method [1, 2]. This method was originally introduced in quantum mechanics. In optics, it has been used extensively to study propagation in multimode fibers [3]. It can lead to closed-form solutions if the properties vary in a simple fashion, such as linearly. If the variations are not simple, one may still be able to use the method to obtain some useful expressions involving the average of g along the fiber.

Karlsson first applied the WKB method to fiber OPAs, in the context of modulation instability (MI) [4]. Here we present a slightly different version, starting from the basic OPA equations. In Appendix 1, we showed that these equations can lead to second-order linear differential equations for the fields, where the parametric gain coefficient may be a function of position, $g_k(z)$. We consider the most general case where r_k can be a function of z. We assume for simplicity the common situation where p_k, q_k, r_k, κ_k, t_k, $\tilde{k}_k$ are the same for $k = 3, 4$ and we therefore drop the subscripts of these quantities. We also assume that the resulting g is a slowly varying function of z. Then C_k is governed by

$$\frac{d^2 C_k}{dz^2} - g^2(z) C_k = 0, \qquad k = 3, 4. \tag{A2.1}$$

The first-order WKB approximation yields two approximate independent particular solutions, given by [1]

$$y_\pm(z) = |g(z)|^{-1/2} \exp\left[\pm \int_{\xi=0}^{z} g(\xi) d\xi\right]. \tag{A2.2}$$

The general solution is then obtained as a linear combination of these two solutions, i.e.

$$C_k(z) = a_k y_+(z) + b_k y_-(z), \qquad k = 3, 4, \tag{A2.3}$$

where a_k and b_k are constants to be determined by matching the initial conditions for $C_k(0)$ and $C'_k(0)$, $k = 3, 4$. Equation (A2.3) also implies that

$$C'_k(z) = a_k y'_+(z) + b_k y'_-(z), \quad k = 3, 4. \tag{A2.4}$$

For a particular k, Eqs. (A2.3) and (A2.4) form a system of two equations for a_k and b_k. We can write its solution at $z = 0$ as

$$\begin{bmatrix} a_k \\ b_k \end{bmatrix} = \frac{1}{W(0)} \begin{bmatrix} y'_-(0) & -y_-(0) \\ -y'_+(0) & y_+(0) \end{bmatrix} \begin{bmatrix} C_k(0) \\ C'_k(0) \end{bmatrix}, \tag{A2.5}$$

where $W(z) = y_+(z)y'_-(z) - y_-(z)y'_+(z)$ is the Wronskian of the two WKB solutions at z. Assuming for simplicity that g is real and positive (which is generally true in the desirable cases where the OPA actually has gain), we have at any z

$$y'_\pm = y_\pm \left(\pm g - \frac{g'}{2g} \right) \tag{A2.6}$$

and so

$$W(z) = y_+ y_- \left(-g - \frac{g'}{2g} - g + \frac{g'}{2g} \right) = -2, \quad \text{for all } z. \tag{A2.7}$$

It can be shown that

$$C'_k(0) = \frac{i\tilde{\kappa}(0)}{2} C_k(0) + ir(0) C_l^*(0). \tag{A2.8}$$

We assume for simplicity that there is no idler at the input, i.e. $C_4(0) = 0$. Therefore, we have $C'_4(0) = r(0) C_3^*(0)$. We can now write

$$a_4 = -\frac{1}{2}[y'_-(0) C_4(0) - y_-(0) C'_4(0)] = \frac{ir(0) y_-(0) C_3^*(0)}{2} \tag{A2.9}$$

$$b_4 = -\frac{1}{2}[-y'_+(0) C_4(0) + y_+(0) C'_4(0)] = -\frac{ir(0) y_+(0) C_3^*(0)'}{2}, \tag{A2.10}$$

and

$$C_4(z) = -\frac{ir(0) C_3^*(0)}{2} [y_+(0) y_-(z) - y_-(0) y_+(z)]. \tag{A2.11}$$

Since $y_\pm(0) = |g(0)|^{-1/2}$,

$$C_4(z) = -\frac{ir(0) C_3^*(0)}{2|g(0)|^{1/2}} [y_-(z) - y_+(z)]$$

$$= \frac{ir(0) C_3^*(0)}{|g(0) g(z)|^{1/2}} \sinh\left(\int_{\xi=0}^{z} g(\xi) d\xi \right). \tag{A2.12}$$

Appendix 2 The WKB approximation

To calculate the OPA gain, we need to go back to the actual field envelopes, by means of

$$A_k(z) = C_k(z) \exp\left[i \int_{\xi=0}^{z} \left(p - \frac{\kappa}{2} - \frac{i}{2r}\frac{dr}{dz}\right) d\xi\right]. \tag{A2.13}$$

Equation (A2.12) then leads to

$$A_4(z) = A_3^*(0) \frac{ir(0)}{|g(0)g(z)|^{1/2}} \exp\left[i \int_{\xi=0}^{z}\left(p - \frac{\kappa}{2} - \frac{i}{2r}\frac{dr}{dz}\right)d\xi\right] \sinh\left(\int_{\xi=0}^{z} g(\xi)d\xi\right)$$

$$= iA_3^*(0) \left|\frac{r(0)r(z)}{g(0)g(z)}\right|^{1/2} \exp\left[i \int_{\xi=0}^{z}\left(p - \frac{\kappa}{2}\right)d\xi\right] \sinh\left(\int_{\xi=0}^{z} g(\xi)d\xi\right). \tag{A2.14}$$

Finally, we can calculate the signal-to-idler power conversion efficiency as

$$G_i(z) = \left|\frac{A_4(z)}{A_3^*(0)}\right|^2 = \left|\frac{r(0)r(z)}{g(0)g(z)} \sinh^2\left(\int_{\xi=0}^{z} g(\xi)d\xi\right)\right.$$

$$\left. \times \exp\left[i \int_{\xi=0}^{z}\left(p - p^* + \frac{q - q^*}{2}\right)d\xi\right]\right|. \tag{A2.15}$$

Often the real part of p is $i\alpha/2$ and that of q is $i\alpha$. Under these conditions, Eq. (A2.15) simplifies to

$$G_i(z) = \left|\frac{A_4(z)}{A_3^*(0)}\right|^2$$

$$= \left|\frac{r(0)r(z)}{g(0)g(z)}\right| \sinh^2\left(\int_{\xi=0}^{z} g(\xi)d\xi\right) \left|\exp\left(-2\int_{\xi=0}^{z} \alpha d\xi\right)\right|. \tag{A2.16}$$

The signal gain can then be obtained by the relation

$$G_s(z) = \left|\frac{A_3(z)}{A_3^*(0)}\right|^2 = G_i(z) + \exp\left(-\int_{\xi=0}^{z} \alpha d\xi\right). \tag{A2.17}$$

A key feature of Eq. (A2.16) is that the argument of the sinh function is the integral of the parametric gain coefficient g along the fiber. This is an intuitively satisfying result, as a similar integration is used to calculate exactly the gain in other types of amplifier such as erbuim-doped fiber amplifiers (EDFAs) or Raman amplifiers. The presence of this integral may allow us to draw some important conclusions about the power gain, even if a closed-form solution is not available. For example, if we know that g undergoes rapid spatial fluctuations about an average then it is possible to disregard the fluctuations and calculate the gain from the average, which is a significant simplification.

References

1. *Handbook of Differential Equations*, Zwillinger, D, second edition. Academic, Boston; 1992; chapter 3.
2. *An Introduction to Phase-Integral Methods*, Heading, J. Methuen, London; 1962.
3. *An Introduction to Optical Waveguides*, Adams, M. J. Wiley, Chichester; 1981.
4. "Modulational instability in lossy optical fibers," Karlsson, M. *J. Opt. Soc. Amer. B*; 1995; vol. 12, pp. 2073–7.

Appendix 3   Jacobian elliptic function solutions

In this appendix we provide a detailed solution of the equations coupling four waves having various SOPs. The scalar case of Chapter 3, when all four waves are in the same state of linear polarization, is a particular case of this general solution.

We first treat the non-degenerate case, where the four waves are distinct. This is the case for a two-pump OPA, as well as for wavelength exchange (with different frequency assignments for the pumps).

We then treat separately the degenerate three-wave case where one wave is halfway between the other two. This corresponds to a one-pump fiber OPA.

A3.1 Four-wave case

We showed in Chapter 4 that the coupled equations governing the nonlinear interaction between four distinct waves are given by

$$-i\frac{d\vec{A}_l}{dZ} = [\vec{A}_l, \vec{A}_l^*, \vec{A}_l] + 2\sum_{j \neq l=1}^{4} [\vec{A}_j, \vec{A}_j^*, \vec{A}_l]$$
$$+ 2[\vec{A}_m, \vec{A}_n, \vec{A}_k^*]\, e^{i\Delta\beta_{klmn} Z/\gamma}, \qquad l = 1\text{–}4, \qquad (A3.1)$$

where $\Delta\beta_{klmn} = \beta_m + \beta_n - \beta_k - \beta_l$ and k, l, m, and n satisfy the following conditions:

if $l = 1$ or 2 then $k = 3 - l$, $m = 3$, and $n = 4$;
if $l = 3$ or 4 then $k = 7 - l$, $m = 1$, and $n = 2$.

Dot-multiplying Eq. (A3.1) by $\vec{A}_l^*$ yields

$$-i\frac{d\vec{A}_l}{dZ} \cdot \vec{A}_l^* = [\vec{A}_l, \vec{A}_l^*, \vec{A}_l, \vec{A}_l^*] + 2\sum_{j \neq l=1}^{4} [\vec{A}_j, \vec{A}_j^*, \vec{A}_l, \vec{A}_l^*]$$
$$+ 2[\vec{A}_m, \vec{A}_n, \vec{A}_k^*, \vec{A}_l^*]\, e^{i\Delta\beta_{klmn} Z/\gamma}, \qquad l = 1\text{–}4. \qquad (A3.2)$$

Equation (A3.2) holds for arbitrary SOPs. We have shown that if the waves are initially LP or CP then each one retains its original SOP. Let $\vec{A}_l = A_l \hat{e}_l = \sqrt{P_l}\, e^{i\phi_l}\hat{e}_l$, where A_l is the complex amplitude, $\hat{e}_l$ is the unit Jones vector, P_l is the power, and ϕ_l is the phase of the lth wave. Equation (A3.2) then leads to

$$-i\frac{dA_l}{dZ} A_l^* = a_{ll}P_l^2 + 2P_l \sum_{j \neq l=1}^{4} a_{jl}P_j$$
$$+ 2b\sqrt{P_1 P_2 P_3 P_4}\, e^{i\Delta\beta_{klmn} Z/\gamma + \phi_m + \phi_n - \phi_k - \phi_l}, \qquad l = 1-4, \qquad (A3.3)$$

where $b = [\hat{e}_1^*, \hat{e}_2^*, \hat{e}_3, \hat{e}_4]$ and a_{jl} is defined by Eq. (4.16). It can be shown that b is real and positive for all LP and CP OPAs. Then Eq. (A3.3) yields

$$-i\frac{dA_l}{dZ}A_l^* = a_{ll}P_l^2 + 2P_l \sum_{j\neq l=1}^{4} a_{jl}P_j + 2b\sqrt{P_1 P_2 P_3 P_4}\, e^{i\varepsilon\theta}, \quad l = 1\text{–}4, \quad (A3.4)$$

where: $\theta = \Delta\beta_{1234}Z/\gamma + \phi_3 + \phi_4 - \phi_1 - \phi_2$ and $\varepsilon = 1$ for $l = 1, 2$, $\varepsilon = -1$ for $l = 3, 4$.

The real and imaginary parts of Eq. (A3.4) provide two equations for two real unknowns (which will be θ and the change in one of the powers), which can provide the entire solution to the problem. The imaginary part of Eq. (A3.4) gives

$$-i\frac{dP_l}{dZ} = 2b\sqrt{P_1 P_2 P_3 P_4}(e^{i\varepsilon\theta} - e^{-i\varepsilon\theta}), \quad l = 1\text{–}4, \quad (A3.5)$$

or

$$\frac{dP_l}{dZ} = -4\varepsilon b\sqrt{P_1 P_2 P_3 P_4}\sin\theta, \quad l = 1\text{–}4, \quad (A3.6)$$

i.e.

$$\frac{dP_1}{dZ} = \frac{dP_2}{dZ} = -\frac{dP_3}{dZ} = -\frac{dP_4}{dZ} = -4b\sqrt{P_1 P_2 P_3 P_4}\sin\theta. \quad (A3.7)$$

Equation (A3.7) implies that the powers of the four waves satisfy the Manley–Rowe relations

$$P_1 = P_1(0) - x, \quad P_2 = P_2(0) - x,$$
$$P_3 = P_3(0) + x, \quad P_4 = P_4(0) + x, \quad (A3.8)$$

i.e. that they all go up or down by the same amount x. This is the same conclusion as that reached in subsection 3.2.2 by looking at the interaction in terms of individual photons. We use x as one of the unknowns to solve the equations.

The real part of Eq. (A3.7) yields

$$i\left(\frac{dA_l^*}{dZ}A_l - \frac{dA_l}{dZ}A_l^*\right) = 2a_{ll}P_l^2 + 4P_l \sum_{j\neq l=1}^{4} a_{jl}P_j$$
$$+ 4b\sqrt{P_1 P_2 P_3 P_4}\cos\theta), \quad l = 1\text{–}4. \quad (A3.9)$$

We have

$$\frac{dA_l}{dZ} = \frac{d}{dZ}(\sqrt{P_l}e^{i\theta_l}) = \left(\frac{1}{2\sqrt{P_l}}\frac{dP_l}{dZ} + i\sqrt{P_l}\frac{d\theta_l}{dZ}\right)e^{i\theta_l}, \quad l = 1\text{–}4, \quad (A3.10)$$

$$\frac{dA_l}{dZ}A_l^* = \left(\frac{1}{2}\frac{dP_l}{dZ} + iP_l\frac{d\theta_l}{dZ}\right)e^{i\theta_l}, \quad l = 1\text{–}4, \quad (A3.11)$$

$$i\left(\frac{dA_l^*}{dZ}A_l - \frac{dA_l}{dZ}A_l^*\right) = 2P_l\frac{d\theta_l}{dZ}$$
$$= 2a_{ll}P_l^2 + 4P_l \sum_{j\neq l=1}^{4} a_{jl}P_j + 4b\sqrt{P_1 P_2 P_3 P_4}\cos\theta, \quad l = 1\text{–}4, \quad (A3.12)$$

$$\frac{d\theta_l}{dZ} = a_{ll}P_l + 2\sum_{j\neq l=1}^{4} a_{jl}P_j + \frac{2b}{P_l}\sqrt{P_1 P_2 P_3 P_4}\cos\theta, \quad l = 1\text{–}4. \quad (A3.13)$$

Then
$$\frac{d\theta}{dZ} = \frac{d}{dZ}(\phi_3 + \phi_4 - \phi_1 - \phi_2) + \frac{\Delta\beta}{\gamma}$$

$$= a_{33}P_3 + 2(a_{31}P_1 + a_{32}P_2 + a_{34}P_4) + \frac{2b}{P_3}\sqrt{P_1P_2P_3P_4}\cos\theta$$

$$+ a_{44}P_4 + 2(a_{41}P_1 + a_{42}P_2 + a_{43}P_3) + \frac{2b}{P_4}\sqrt{P_1P_2P_3P_4}\cos\theta$$

$$- a_{11}P_1 - 2(a_{12}P_2 + a_{13}P_3 + a_{14}P_4) - \frac{2b}{P_1}\sqrt{P_1P_2P_3P_4}\cos\theta$$

$$- a_{22}P_2 - 2(a_{21}P_1 + a_{23}P_3 + a_{24}P_4) - \frac{2b}{P_2}\sqrt{P_1P_2P_3P_4}\cos\theta + \frac{\Delta\beta}{\gamma}$$

$$= \frac{\Delta\beta}{\gamma} + u_1 P_1 + u_2 P_2 + u_3 P_3 + u_4 P_4$$

$$+ 2b\sqrt{P_1P_2P_3P_4}\cos\theta\left(\frac{1}{P_3} + \frac{1}{P_4} - \frac{1}{P_1} - \frac{1}{P_2}\right), \quad (A3.14)$$

where
$$\begin{aligned} u_1 &= 2(a_{31} + a_{41} - a_{21}) - a_{11}, \\ u_2 &= 2(a_{32} + a_{42} - a_{12}) - a_{22}, \\ u_3 &= 2(a_{43} - a_{13} - a_{23}) - a_{33}, \\ u_4 &= 2(a_{34} - a_{14} - a_{24}) - a_{44}. \end{aligned} \quad (A3.15)$$

For all the types of OPA being considered here, symmetry considerations imply that: $a_{kl} = a_{lk}$ for all l, k; $a_{ll} = a_{11}$ for $l = 1\text{–}3$; if $k \neq l$ then $a_{kl} = a_{mn}$ where $\{k, l, m, n\}$ is a permutation of $\{1, 2, 3, 4\}$ (for example, $a_{21} = a_{34}$). As a result, $u_1 = u_2 = -u_3 = -u_4 = 2(a_{13} + a_{14} - a_{12}) - a_{11} = u$.

Hence Eq. (A3.14) becomes
$$\frac{d\theta}{dZ} = \frac{\Delta\beta}{\gamma} + u(P_1 + P_2 - P_3 - P_4)$$

$$+ 2b\sqrt{P_1P_2P_3P_4}\cos\theta\left(\frac{1}{P_3} + \frac{1}{P_4} - \frac{1}{P_1} - \frac{1}{P_2}\right). \quad (A3.16)$$

We also have
$$\frac{d\theta}{dZ} = \frac{d\theta}{dx}\frac{dx}{dZ} = \frac{d\theta}{dx}4b\sqrt{P_1P_2P_3P_4}\sin\theta$$

$$= -4b\sqrt{P_1P_2P_3P_4}\frac{d(\cos\theta)}{dx}. \quad (A3.17)$$

Substituting into Eq. (A3.16), and using the fact that $dP_l/dx = -\varepsilon$ leads to
$$\frac{\Delta\beta}{\gamma} + u\left(P_1 + P_2 - P_3 - P_4\right) + 4b\frac{d}{dx}(\sqrt{P_1P_2P_3P_4}\cos\theta) = 0 \quad (A3.18)$$

or
$$\frac{\Delta\beta}{\gamma} + u\Delta P_0 - 4ux + 4b\frac{d}{dx}(\sqrt{P_1P_2P_3P_4}\cos\theta) = 0, \quad (A3.19)$$

where $\Delta P_0 = P_{10} + P_{20} - P_{30} - P_{40}$.

Integrating both sides of Eq. (A3.19) with respect to x, we obtain

$$\left(\frac{\Delta\beta}{\gamma} + u\Delta P_0\right) x - 2\Delta u x^2 + 4b\sqrt{P_1 P_2 P_3 P_4}\cos\theta = K_2, \tag{A3.20}$$

where K_2 is a constant. Since K_2 is independent of x, it can be evaluated at $x = 0$, which yields

$$K_2 = 4b\sqrt{P_{10} P_{20} P_{30} P_{40}}\cos(\phi_{30} + \phi_{40} - \phi_{10} - \phi_{20}), \tag{A3.21}$$

where θ_{l0} is the initial phase of the lth wave. We can now eliminate θ in Eq. (A3.7) by using Eq. (A3.20), to obtain an equation involving only x and Z. We have

$$\frac{dx}{dZ} = 4b\sqrt{P_1 P_2 P_3 P_4}\sin\theta = 4\sqrt{P_1 P_2 P_3 P_4(1 - \cos^2\theta)}$$

$$= 4\left\{(P_{10} - x)(P_{20} - x)(P_{30} + x)(P_{40} + x)\right.$$

$$\left.- \frac{1}{16b^2}\left[K_2 - \left(\frac{\Delta\beta}{\gamma} + u\Delta P_0\right)x + 2ux^2\right]^2\right\}^{1/2}. \tag{A3.22}$$

This can be rewritten as

$$dZ = \frac{dx}{2\sqrt{h(x)}}, \tag{A3.23}$$

where $h(x)$ is a fourth-order polynomial in x:

$$h(x) = 4b^2(P_{10} - x)(P_{20} - x)(P_{30} + x)(P_{40} + x)$$

$$- \frac{1}{4}\left[K_2 - \left(\frac{\Delta\beta}{\gamma} + u\Delta P_0\right)x + 2ux^2\right]^2. \tag{A3.24}$$

A3.2 Solution in terms of Jacobian elliptic functions

The solution $x(Z)$ of Eq. (A3.24) can be expressed in terms of Jacobian elliptic functions [1]. It is obtained in terms of the four roots of $h(x)$, $\eta_1 < \eta_2 < \eta_3 < \eta_4$.

The coefficient of x^4 in $h(x)$ is

$$C_0^2 = 4b^2 - u^2;$$

C_0^2 can be either positive or negative.

$C_0^2 > 0$

If $C_0^2 > 0$ then x oscillates between η_2 and η_3. Specifically,

$$x(z) = \eta_1 + (\eta_2 - \eta_1)\left[1 - \eta\,\text{sn}^2\left(\frac{z + z_0}{z_c}, k\right)\right]^{-1}, \tag{A3.25}$$

where sn denotes a Jacobian elliptic function,

$$z_c^{-1} = \gamma C_0[(\eta_3 - \eta_1)(\eta_4 - \eta_2)]^{1/2}, \qquad \eta = (\eta_3 - \eta_2)/(\eta_3 - \eta_1),$$

$$z_0 = z_c F \left\{ \sin^{-1}\left[\frac{P_{40} - \eta_2}{\eta(P_{40} - \eta_1)}\right], k \right\}, \qquad k^2 = \frac{(\eta_3 - \eta_2)(\eta_4 - \eta_1)}{(\eta_3 - \eta_1)(\eta_4 - \eta_2)},$$

and $F(\phi, k)$ is a standard elliptical integral [1].

An alternative form for the solution can also be used [2]. It is obtained from the preceding one by everywhere exchanging η_2 and η_3 and also η_1 and η_4. It can be shown that these two forms are exactly equivalent. The existence of two forms is useful for checking numerical results: if they do not provide the same results, something is wrong with the numerical calculations.

The special functions encountered in these solutions are available in mathematical packages such as Mathematica [3], and thus in principle it is a simple matter to obtain numerical values of x or related quantities for a given set of parameters. However, numerical problems have been encountered in Mathematica for Jacobian elliptic functions as well as for their inverses, namely elliptic functions such as $F(\varphi, k)$ and $K(k)$ (the difficulties were discovered by comparing the results obtained from the two forms of the solution, as discussed above). Under these circumstances, it is a good idea to check the results against series expansions with well-known coefficients. Particularly useful is the Fourier series expansion for sn, namely

$$\text{sn}(u, k) = \frac{2\pi}{kK} \sum_{m=0}^{\infty} \frac{q^{m+1/2}}{1 - q^{2m+1}} \sin\left[(2m+1)\frac{\pi u}{2K}\right], \qquad (A3.26)$$

where $q = e^{-\pi K'/K}$ is Jacobi's "nome"; K is the real quarter-period of $\text{sn}(u, k)$, given by the complete elliptic integral of the first kind,

$$K = K(k) = F\left(\frac{\pi}{2}, k\right) = \int_{\theta=0}^{\pi/2} (1 - k^2 \sin^2 \theta)^{-1/2} d\theta$$

$$= \frac{\pi}{2} F\left(\frac{1}{2}, \frac{1}{2}, 1, k^2\right) = \sum_{n=0}^{\infty} \frac{[(1/2)_n]^2}{(n!)^2} k^{2n}, \qquad (A3.27)$$

where $(a)_n = a(a+1)\cdots(a+n-1)$ is Pochammer's symbol and $F(a, b, c, d)$ is a hypergeometric function. Also, $K' = K(k')$ where $k' = \sqrt{1-k^2}$.

It is common with fiber OPAs to have no idler at the input. In this case we have $P_{40} = 0$, which leads to $K_2 = 0$. In consequence $\eta_2 = 0$, so that the equation $h(x) = 0$ reduces to a third-order equation. We can then write explicit expressions in terms of the system parameters for the three remaining roots, η_1, η_3, and η_4, and hence for the complete solution itself. This was done in Chapter 3 using Cardano's formula and can be extended to the more general case.

$C_0^2 < 0$

If $C_0^2 < 0$ then the situation is more complicated, as there are now two possibilities. Depending on the sign of η_1, x may oscillate either between η_1 and η_2 or between η_3 and η_4. (If $P_{40} = 0$ then we have in these cases $\eta_1 = 0$ or $\eta_3 = 0$, respectively.)

(i) $\eta_1 \geq 0$ Then the solution can be written as

$$x(z) = \eta_4 + (\eta_1 - \eta_4)\left[1 + \eta\ \text{sn}^2\left(\frac{z+z_0}{z_c}, k\right)\right]^{-1}, \qquad (A3.28)$$

where

$$\eta = \frac{\eta_2 - \eta_1}{(\eta_4 - \eta_2)},$$

$$z_0 = z_c F\left\{\sin^{-1}\left[\frac{P_{40} - \eta_1}{\eta(-P_{40} + \eta_4)}\right], k\right\},$$

$$k^2 = \frac{(\eta_4 - \eta_3)(\eta_2 - \eta_1)}{(\eta_3 - \eta_1)(\eta_4 - \eta_2)}.$$

(ii) $\eta_1 < 0$ Then the solution is

$$x(z) = \eta_2 + (\eta_3 - \eta_2)\left[1 - \eta\ \text{sn}^2\left(\frac{z+z_0}{z_c}, k\right)\right]^{-1}, \qquad (A3.29)$$

where

$$\eta = \frac{\eta_4 - \eta_3}{\eta_4 - \eta_2}$$

$$z_0 = z_c F\left\{\sin^{-1}\left[\frac{P_{40} - \eta_3}{\eta(P_{40} - \eta_2)}\right], k\right\}.$$

Here again alternative forms for the solutions can be used [2]. They are obtained from the preceding forms by exchanging everywhere the subscripts 1 and 2 and the subscripts 3 and 4. It can be shown that these alternative forms are exactly equivalent.

A3.3   Complete pump depletion

Let us now consider the possibility of obtaining complete pump depletion with a two-pump fiber OPA. We restrict the discussion to the case $P_{40} = 0$, which is the most common in practice and leads to simplified expressions (including the fact that $x = 0$ is a root of $h(x) = 0$).

Complete pump depletion implies that, at a certain distance, the entire pump power should be transferred to the signal and idler (in equal amounts, as discussed in subsection 3.2.2). Then necessarily $(P_{10} + P_{20})/2$ must be a root of $h(x) = 0$; hence

$$16b^2 \left(\frac{P_{10} - P_{20}}{2}\right)\left(\frac{P_{20} - P_{10}}{2}\right)\left(P_{30} + \frac{P_{10} + P_{20}}{2}\right)$$

$$= \left(\frac{P_{10} + P_{20}}{2}\right)\left(P_{10} + P_{20} - \frac{\Delta\beta}{\gamma} - u\Delta P_0\right)^2. \qquad (A3.30)$$

Since the left- and right-hand sides have opposite signs, Eq. (A3.30) can be satisfied only if they both vanish. For the left-hand side, this implies that $P_{10} = P_{20}$, which is easy to achieve. For the right-hand side to vanish, we must have $\Delta\beta = u\gamma P_{30}$; whether this can be achieved in practice depends on the fiber dispersion, the signal and pump wavelengths, etc.

Substituting these conditions into $h(x) = 0$, the latter can now be rewritten as

$$h(x) = x(P_{10} - x)^2 \left(4b^2 P_{30} + C_0^2 x\right) = 0, \tag{A3.31}$$

so that the four roots of $h(x) = 0$ are 0, P_{10}, which is a double root, and $-(4b^2/C_0^2)P_{30}$. We see that the last root has a sign opposite to that C_0^2, which depends on the type of OPA used, and that its magnitude is proportional to P_{30}, which can be adjusted. Depending on the position of this root with respect to the others (0 and P_{10}), different situations will arise.

$C_0^2 > 0$

If $C_0^2 > 0$ then the four roots are $\eta_1 = -(4b^2/C_0^2)P_{30}$, and $\eta_2 = 0$, and $\eta_3 = \eta_4 = P_{10}$; x ranges between η_2 and η_3, i.e. P_4 ranges between 0 and P_{10}.

We also find that $k = 1$, which implies that the sn function becomes a tanh function. Then $x(z)$ is not periodic but approaches P_{10} asymptotically for large z. Thus, while complete depletion is in principle possible, mathematically it is achieved only at infinite distance. In practice, however, nearly complete depletion can be obtained after a length of the order of $1/(u\gamma P_{30})$.

$C_0^2 < 0$

If $C_0^2 < 0$ then $-(4b^2/C_0^2)P_{30} > 0$ and therefore $\eta_1 = 0$. The values of the other roots depend on the magnitude of P_{30}, and again we must consider two cases.

(i) $P_{30} < -(C_0^2/4b^2)P_{10}$ Then $\eta_1 = 0$, $\eta_2 = -(4b^2/C_0^2)P_{30}$, and $\eta_3 = \eta_4 = P_{10}$; x oscillates between η_1 and η_2, i.e. P_4 oscillates between 0 and $-(4b^2/C_0^2)P_{30} < P_{10}$. We see that if $P_{30} \ll P_{10}$ then P_4 will always remain quite small and so there will be little pump depletion. We also find that $k = 0$, which implies that sn becomes a sine function; this leads to a simpler expression for the solution, which in this case is periodic.

In this regime, complete pump depletion will be possible only in the limiting case where $P_{30} = -(C_0^2/4b^2)P_{10}$, which requires P_{30} to be of the same order as P_{10} itself. If P_{30} exceeds this value, we enter the case discussed below.

(ii) $P_{30} > -(C_0^2/4b^2)P_{10}$ Then $\eta_1 = 0$, $\eta_2 = \eta_3 = P_{10}$, and $\eta_4 = -(4b^2/C_0^2)P_{30}$; x ranges between η_1 and η_2, i.e. P_4 ranges between 0 and P_{10}. We also find that $k = 1$, so that the solution does not oscillate but approaches its limit asymptotically.

This situation is essentially the same as in the case $C_0^2 > 0$. The main difference is that here a large P_{30} value is necessary for complete pump depletion. This is not a requirement in the case $C_0^2 > 0$, where one has the option of using a smaller P_{30} value (and possibly a longer fiber to achieve the same level of pump depletion).

Let us consider XYXY as an example of an OPA with $C_0^2 < 0$ (it is important because of its polarization insensitivity). In this case $u = -1$ and $C_0^2 = -5/9$, and we find that we need $P_{30} \geq (5/4)P_{10}$ to have the possibility of full pump depletion. Then the conversion efficiency, defined as the ratio of the idler output power and the total input power, is at best $P_{10}/(P_{10} + P_{10} + 5P_{10}/4) = 4/13 \approx 31\%$; this is to be compared with 50% for an OPA with $C_0^2 > 0$, when a very small P_{30} (and a long fiber) is used.

A3.4 Three-wave case

In this case the coupling equations between the three waves take the form

$$-i\frac{d\vec{A}_1}{dZ} = [\vec{A}_1, \vec{A}_1^*, \vec{A}_1] + 2\sum_{j=3}^{4}[\vec{A}_j, \vec{A}_j^*, \vec{A}_1]$$
$$+ 2[\vec{A}_3, \vec{A}_4, \vec{A}_1^*]e^{i\Delta\beta_{1134}z}, \quad (A3.32)$$

$$-i\frac{d\vec{A}_l}{dZ} = [\vec{A}_1, \vec{A}_1^*, \vec{A}_l] + 2\sum_{j=1,k}[\vec{A}_j, \vec{A}_j^*, \vec{A}_l]$$
$$+ [\vec{A}_1, \vec{A}_1, \vec{A}_l^*]e^{-i\Delta\beta_{1134}z}, \quad l = 3, 4, \quad k = 7-l, \quad (A3.33)$$

where $\Delta\beta_{1134} = \beta_3 + \beta_4 - 2\beta_1$.

Dot-multiplying Eq. (A3.32) by $\vec{A}_1^*$ and Eq. (A3.33) by $\vec{A}_l^*$ yields

$$-i\frac{dA_1}{dZ}A_1^* = a_{11}P_1^2 + 2P_1(a_{31}P_3 + a_{41}P_4) + 2bP_1\sqrt{P_3P_4}\,e^{i\theta} \quad (A3.34)$$

$$-i\frac{dA_l}{dZ}A_l^* = a_{ll}P_l^2 + 2P_l(a_{l1}P_1 + a_{lk}P_{7-l}) + bP_1\sqrt{P_3P_4}\,e^{-i\theta},$$
$$l = 3, 4, \quad k = 7-l, \quad (A3.35)$$

where $[\hat{e}_1^*, \hat{e}_1^*, \hat{e}_3, \hat{e}_4] = b$ (which is real and positive) and $\theta = \Delta\beta_{1134}z + \phi_3 + \phi_4 - 2\phi_1$.

The imaginary part of Eq. (A3.34) gives

$$-i\frac{dP_1}{dZ} = 2bP_1\sqrt{P_3P_4}\left[e^{i\theta} - e^{-i\theta}\right] \quad (A3.36)$$

or

$$\frac{dP_1}{dZ} = -4bP_1\sqrt{P_3P_4}\sin\theta. \quad (A3.37)$$

Similarly, we find that

$$\frac{dP_3}{dZ} = \frac{dP_4}{dZ} = -\frac{1}{2}\frac{dP_1}{dZ} = 2bP_1\sqrt{P_3P_4}\sin\theta, \quad l = 1\text{-}4, \quad (A3.38)$$

Equations (A3.37) and (A3.38) imply that the powers of the three waves satisfy the Manley–Rowe relations

$$P_1 = P_1(0) - 2x, \quad P_3 = P_3(0) + x,$$
$$P_4 = P_4(0) + x. \quad (A3.39)$$

These show that the pump goes up or down by twice the common amount x of the signal and idler. Again, this is the same conclusion as that reached in subsection 3.2.2 by looking at the interaction in terms of individual photons.

The real part of Eq. (A3.34) yields

$$i\left(\frac{dA_1^*}{dZ}A_1 - \frac{dA_1}{dZ}A_1^*\right) = 2P_1\frac{d\phi_1}{dz} = 2\left[a_{11}P_1^2 + 2P_1(a_{31}P_3 + a_{41}P_4)\right]$$
$$+ 4bP_1\sqrt{P_3P_4}\cos\theta, \qquad (A3.40)$$

$$i\left(\frac{dA_l^*}{dZ}A_l - \frac{dA_l}{dZ}A_l^*\right) = 2P_l\frac{d\phi_l}{dz} = 2a_{ll}P_l^2 + 4P_l(a_{ll}P_1 + a_{7-l,l}P_{7-l})$$
$$+ 2bP_1\sqrt{P_3P_4}\cos\theta), \quad l = 3, 4. \qquad (A3.41)$$

Hence

$$\frac{d\phi_1}{dz} = a_{11}P_1 + 2(a_{31}P_3 + a_{41}P_4) + 2b\sqrt{P_3P_4}\cos\theta, \qquad (A3.42)$$

$$\frac{d\phi_l}{dz} = a_{ll}P_l + 2(a_{ll}P_l + a_k P_k) + b\frac{P_1}{P_l}\sqrt{P_3P_4}\cos\theta, \qquad (A3.43)$$

where $k = 7 - l$. Then

$$\frac{d\theta}{dZ} = \frac{d}{dZ}(\phi_3 + \phi_4 - 2\phi_1) + \frac{\Delta\beta}{\gamma}$$

$$= a_{33}P_3 + 2(a_{13}P_1 + a_{43}P_4) + b\frac{P_1}{P_3}\sqrt{P_3P_4}\cos\theta$$

$$+ a_{44}P_4 + 2(a_{14}P_1 + a_{34}P_3) + b\frac{P_1}{P_4}\sqrt{P_1P_2P_3P_4}\cos\theta$$

$$- 2\left[a_{11}P_1 + 2(a_{31}P_3 + a_{41}P_4) + 2b\sqrt{P_3P_4}\cos\theta + \frac{\Delta\beta}{\gamma}\right]$$

$$= \frac{\Delta\beta}{\gamma} + u_1 P_1 + u_3 P_3 + u_4 P_4 + bP_1\sqrt{P_3P_4}\cos\theta\left(\frac{1}{P_3} + \frac{1}{P_4} - \frac{4}{P_1}\right),$$
$$(A3.44)$$

where

$$u_1 = 2a_{13} + 2a_{14} - 2a_{11},$$
$$u_3 = a_{33} + 2a_{34} - 4a_{31}, \qquad (A3.45)$$
$$u_4 = a_{44} + 2a_{43} - 4a_{41}.$$

For all types of one-pump OPA being considered here, symmetry considerations imply that: $a_{kl} = a_{lk}$ for all l, k; $a_{ll} = a_{11}$ for $l = 3, 4$; $a_{13} = a_{14}$ and $a_{34} = a_{11}$. As a result, $u_1 = 2(2a_{13} - a_{11})$ and $u_3 = u_4 = 3a_{11} - 4a_{13}$.

We also have

$$\frac{d\theta}{dZ} = \frac{d\theta}{dx}\frac{dx}{dZ} = \frac{d\theta}{dx}2bP_1\sqrt{P_3P_4}\sin\theta = -2bP_1\sqrt{P_3P_4}\frac{d(\cos\theta)}{dx}. \qquad (A3.46)$$

Substituting into Eq. (A3.44) and using the facts that $\dfrac{dP_3}{dx} = \dfrac{dP_4}{dx} = 1$ and $\dfrac{dP_1}{dx} = -2$ leads to

$$\frac{\Delta\beta}{\gamma} + (u_1 P_1 + u_3 P_3 + u_4 P_4) + 2b\frac{d}{dx}(P_1\sqrt{P_3P_4}\cos\theta) = 0 \qquad (A3.47)$$

or

$$\frac{\Delta\beta}{\gamma} + Q_0 + \frac{\Delta u}{2}x + 2b\frac{d}{dx}(P_1\sqrt{P_3 P_4}\cos\theta) = 0, \quad (A3.48)$$

where $Q_0 = u_1 P_{10} + u_3 P_{30} + u_4 P_{40}$, and

$$\Delta u = u_3 + u_4 - 2u_1 = 2(5a_{11} - 8a_{13}).$$

Integrating both sides of Eq. (A3.48) with respect to x, we obtain

$$\left(\frac{\Delta\beta}{\gamma} + Q_0\right)x + \frac{\Delta u}{2}x^2 + 2b P_1\sqrt{P_3 P_4}\cos\theta = K_1, \quad (A3.49)$$

where K_1 is a constant with value

$$K_1 = 2b P_{10}\sqrt{P_{30} P_{40}}\cos(\phi_{30} + \phi_{40} - 2\phi_{10}), \quad (A3.50)$$

where θ_{l0} is the initial phase of the lth wave. We can now eliminate θ in Eq. (A3.37) by using Eq. (A3.49), to obtain an equation relating only x and z. We have

$$\frac{dx}{dZ} = 2b P_1 \sqrt{P_3 P_4} \sin\theta = 2b\sqrt{P_1^2 P_3 P_4 (1 - \cos^2\theta)}$$

$$= 2b \left\{ (P_{10} - x)^2 (P_{30} + x)(P_{40} + x) \right.$$

$$\left. - \frac{1}{4b^2}\left[K_1 - \left(\frac{\Delta\beta}{\gamma} + Q_0\right)x - \frac{\Delta u}{2}x^2\right]^2 \right\}^{1/2}. \quad (A3.51)$$

This can be rewritten as $dZ = dx/[2\sqrt{h(x)}]$, where $h(x)$ is

$$h(x) = b^2 (P_{10} - x)^2 (P_{30} + x)(P_{40} + x)$$

$$- \frac{1}{4}\left[K_1 - \left(\frac{\Delta\beta}{\gamma} + Q_0\right)x - \frac{\Delta u}{2}x^2\right]^2. \quad (A3.52)$$

The solution $x(z)$ of Eq. (A3.52) is obtained exactly as in the four-wave case, from the four roots of $h(x) = 0$. The coefficient of x^4 in $h(x)$ is

$$C_0^2 = 4b^2 - \left(\frac{\Delta u}{4}\right)^2; \quad (A3.53)$$

C_0^2 can be either positive or negative.

A3.5 Complete pump depletion

Let us now consider the possibility of obtaining complete pump depletion with a one-pump fiber OPA, for $P_{40} = 0$. Then necessarily $P_{10}/2$ must be a root of $h(x) = 0$; hence

$$b^2(P_{10} - P_{10})^2\left(P_{30} + \frac{P_{10}}{2}\right) = \left(\frac{P_{10}}{8}\right)\left(-\frac{\Delta\beta}{\gamma} - Q_0 - \frac{\Delta u}{2}\frac{P_{10}}{2}\right)^2 = 0 \quad (A3.54)$$

which leads to

$$\frac{\Delta\beta}{\gamma} = -Q_0 - \frac{\Delta u}{2}\frac{P_{10}}{2} = -\left[a_{11}\frac{P_{10}}{2} + (3a_{11} - 4a_{13})P_{30}\right].$$ (A3.55)

Substituting this condition into $h(x) = 0$, the latter can now be rewritten as

$$h(x) = x\left(\frac{P_{10}}{2} - x\right)^2 (4b^2 P_{30} + C_0^2 x) = 0,$$ (A3.56)

so that the four roots of $h(x) = 0$ are 0, $P_{10}/2$, which is a double root, and $-(4b^2/C_0^2)P_{30}$. We note that these roots are the same as for the two-pump case, except that the double root is now at $P_{10}/2$ instead of P_{10} (these roots are actually the same in both cases if the total pump power is the same, which is likely to be the case in a fair comparison). Because of this similarity, there is no need to discuss again in detail all possible cases: the discussion proceeds as before, with suitable modifications accounting for the value of the double root.

References

1. *Handbook of Mathematical Functions*, Abramowitz, M., Stegun, I., eds. National Bureau of Standards, Washington DC, Applied Mathematics Series; 1964; vol. 55.
2. *Handbook of Elliptic Integrals for Engineers and Scientists*, Byrd, P. F., Friedman, M. D., second edition. Springer-Verlag, Berlin; 1971.
3. http://www.wolfram.com.

Appendix 4 Solution of four coupled equations for the six-wave model

Letting $\xi = i\gamma P_0 z/2$ and introducing the new variables

$$C_1 = B_3 + B_4, \quad C_2 = B_5 + B_6, \quad C_3 = B_3 - B_4, \quad C_4 = B_5 - B_6, \tag{A4.1}$$

Eq. (4.34) leads to

$$\frac{d}{d\xi}\begin{bmatrix} C_1 \\ C_2 \end{bmatrix} = M_1 \begin{bmatrix} C_3 \\ C_4 \end{bmatrix}, \quad \frac{d}{d\xi}\begin{bmatrix} C_3 \\ C_4 \end{bmatrix} = M_2 \begin{bmatrix} C_1 \\ C_2 \end{bmatrix}, \tag{A4.2}$$

where

$$M_1 = \begin{bmatrix} a-b & c-d \\ -(c-d) & -(a-b) \end{bmatrix}, \quad M_2 = \begin{bmatrix} a+b & c+d \\ -(c+d) & -(a+b) \end{bmatrix}. \tag{A4.3}$$

Equation (A4.2) leads to

$$\frac{d^2}{d\xi^2}\begin{bmatrix} C_1 \\ C_2 \end{bmatrix} = M_1 M_2 \begin{bmatrix} C_1 \\ C_2 \end{bmatrix}, \quad \frac{d^2}{d\xi^2}\begin{bmatrix} C_3 \\ C_4 \end{bmatrix} = M_2 M_1 \begin{bmatrix} C_3 \\ C_4 \end{bmatrix}. \tag{A4.4}$$

It can be shown that

$$M_1 M_2 = \begin{bmatrix} u & v \\ v & u \end{bmatrix}, \quad M_2 M_1 = \begin{bmatrix} u & -v \\ -v & u \end{bmatrix}, \tag{A4.5}$$

where $u = a^2 - b^2 - c^2 + d^2$ and $v = 2(ad - bc)$. The matrix elements u and v have the following properties:

$$\begin{aligned} u + v &= (a+d)^2 - (b+c)^2 = \lambda_1^2, \\ u - v &= (a-d)^2 - (b-c)^2 = \lambda_3^2. \end{aligned} \tag{A4.6}$$

Equation (A4.4) implies that

$$\begin{aligned} \frac{d^2 C_1}{d\xi^2} &= uC_1 + vC_2, & \frac{d^2 C_2}{d\xi^2} &= vC_1 + uC_2, \\ \frac{d^2 C_3}{d\xi^2} &= uC_3 - vC_4, & \frac{d^2 C_4}{d\xi^2} &= -vC_3 + uC_4. \end{aligned} \tag{A4.7}$$

The first two equations of (A4.7) lead to

$$\frac{d^2(C_1+C_2)}{d\xi^2} = (u+v)(C_1+C_2) = \lambda_1^2(C_1+C_2), \tag{A4.8}$$

$$\frac{d^2(C_1-C_2)}{d\xi^2} = (u-v)(C_1-C_2) = \lambda_3^2(C_1-C_2). \tag{A4.9}$$

Therefore we must have

$$\begin{aligned} C_1 + C_2 &= A\exp(\lambda_1\xi) + B\exp(-\lambda_1\xi), \\ C_1 - C_2 &= C\exp(\lambda_3\xi) + D\exp(-\lambda_3\xi), \end{aligned} \tag{A4.10}$$

where $A, B, C,$ and D are constants to be determined from the initial conditions. Assuming that only the pumps and the signal are present at the input, we have

$$C_{10} = B_{30} + B_{40} = B_{30}, \quad C_{20} = B_{50} + B_{60} = 0,$$
$$C_{30} = B_{30} - B_{40} = B_{30}, \quad C_{40} = B_{50} - B_{60} = 0.$$

Equation (A4.10) then leads to

$$\begin{aligned} A + B &= C_{10} + C_{20} = B_{30}, \\ C + D &= C_{10} - C_{20} = B_{30}. \end{aligned} \tag{A4.11}$$

Additional conditions are obtained from Eq. (A4.2), namely

$$\left.\frac{dC_1}{d\xi}\right|_{z=0} = (a-b)C_{30} + (c-d)C_{40} = (a-b)B_{30}, \tag{A4.12}$$

$$\left.\frac{dC_2}{d\xi}\right|_{z=0} = -(c-d)C_{30} - (a-b)C_{40} = -(c-d)B_{30}. \tag{A4.13}$$

Adding and subtracting these equations leads to

$$\left.\frac{d(C_1+C_2)}{d\xi}\right|_{z=0} = (a-b-c+d)B_{30} = \lambda_1(A-B), \tag{A4.14}$$

$$\left.\frac{d(C_1-C_2)}{d\xi}\right|_{z=0} = (a-b+c-d)B_{30} = \lambda_3(C-D). \tag{A4.15}$$

Combining Eqs. (A4.11), (A4.14), and (A4.15) yields

$$A = \frac{B_{30}}{2}\left(1 + \frac{a-b-c+d}{\lambda_1}\right), \quad B = \frac{B_{30}}{2}\left(1 - \frac{a-b-c+d}{\lambda_1}\right) \tag{A4.16}$$

$$C = \frac{B_{30}}{2}\left(1 + \frac{a-b+c-d}{\lambda_3}\right), \quad D = \frac{B_{30}}{2}\left(1 - \frac{a-b+c-d}{\lambda_3}\right) \tag{A4.17}$$

The coefficients C_3 and C_4 can be treated in a similar manner. We find that

$$C_3 + C_4 = E \exp(\lambda_3 \xi) + F \exp(-\lambda_3 \xi),$$
$$C_3 - C_4 = G \exp(\lambda_1 \xi) + H \exp(-\lambda_1 \xi),$$
(A4.18)

where

$$E = \frac{B_{30}}{2}\left(1 + \frac{a+b-c-d}{\lambda_3}\right), \quad F = \frac{B_{30}}{2}\left(1 - \frac{a+b-c-d}{\lambda_3}\right) \quad \text{(A4.19)}$$

$$G = \frac{B_{30}}{2}\left(1 + \frac{a+b+c+d}{\lambda_1}\right), \quad H = \frac{B_{30}}{2}\left(1 - \frac{a+b+c+d}{\lambda_1}\right). \quad \text{(A4.20)}$$

From Eqs. (A4.10) and (A4.18) we can obtain the C_k, $k = 1-4$, and from the latter we can obtain the B_k, $k = 3-6$. In particular, we can obtain the signal field B_3 and from it the gain $h_3 = B_3/B_{30}$. We find that

$$h_3 = \frac{1}{4}\left[\left(1 + \frac{a+d}{\lambda_1}\right)\exp(\lambda_1 \xi) + \left(1 - \frac{a+d}{\lambda_1}\right)\exp(-\lambda_1 \xi)\right.$$
$$\left. + \left(1 + \frac{a-d}{\lambda_3}\right)\exp(\lambda_3 \xi) + \left(1 - \frac{a-d}{\lambda_3}\right)\exp(-\lambda_3 \xi)\right]$$
$$= \frac{1}{2}\left[\cosh(\lambda_1 \xi) + \frac{a+d}{\lambda_1}\sinh(\lambda_1 \xi) + \cosh(\lambda_3 \xi) + \frac{a-d}{\lambda_3}\sinh(\lambda_3 \xi)\right]. \quad \text{(A4.21)}$$

This derivation is valid only if the eigenvalues are non-zero. However, we can take the limit of Eq. (A4.21) if $\lambda_1 = 0$ and/or $\lambda_3 = 0$, since $\cosh(x) \approx 1$ and $\sinh(x) \approx x$ for small x. As a result, we obtain the following expressions for the gain in all cases:

$$h_3 = \frac{1}{2}\left[1 + (a+d)\xi + \cosh(\lambda_3 \xi) + \frac{a-d}{\lambda_3}\sinh(\lambda_3 \xi)\right] \quad \text{for } \lambda_1 = 0, \quad \text{(A4.22)}$$

$$h_3 = \frac{1}{2}\left[1 + (a-d)\xi + \cosh(\lambda_1 \xi) + \frac{a+d}{\lambda_1}\sinh(\lambda_1 \xi)\right] \quad \text{for } \lambda_3 = 0, \quad \text{(A4.23)}$$

$$h_3 = 1 + a\xi \quad \text{for } \lambda_1 = \lambda_3 = 0. \quad \text{(A4.24)}$$

Appendix 5 Summary of useful equations

A5.1 Gain spectrum for one- or two-pump OPAs

The following expressions are used extensively. Unless otherwise specified, they apply to either one- or two-pump OPAs, provided that the various quantities are calculated appropriately for the case of interest. It is assumed that all the SOPs are linear and identical, over the entire fiber length. The OPAs are idealized, in the sense that these effects are not included: pump depletion; loss; Raman gain; fixed or random birefringence; ZDW variations; phase-sensitive amplification.

In what follows, the subscripts s and i refer to the signal and idler, respectively.

Signal-to-idler power conversion efficiency

$$G_i = \left|\frac{A_i(L)}{A_s^*(0)}\right|^2 = \left|\frac{\gamma P_0}{g}\sinh(gL)\right|^2. \qquad (A5.1)$$

Signal power gain

$$G_s = \left|\frac{A_s(L)}{A_s(0)}\right|^2 = G_i + 1; \qquad (A5.2)$$

here γ is the fiber nonlinearity coefficient, L is the fiber length, and P_0 is the total pump power. For a two-pump OPA, the pump powers are assumed to be equal.

Parametric gain coefficient g
The parametric gain coefficient g in Eq. (A5.1) is given by

$$g^2 = -\frac{\Delta\beta}{4}(\Delta\beta + 4\gamma P_0) \qquad \text{for one pump,} \qquad (A5.3)$$

$$g^2 = -\frac{1}{4}(\Delta\beta - \gamma P_0)(\Delta\beta + 3\gamma P_0), \quad \text{for two pumps.} \qquad (A5.4)$$

Wavevector mismatch $\Delta\beta$

The wavevector mismatch $\Delta\beta$ is given by

$$\Delta\beta = \beta^{(2)}(\Delta\omega_s)^2 + \frac{\beta^{(4)}}{12}(\Delta\omega_s)^4 \qquad \text{for one pump,} \qquad \text{(A5.5)}$$

$$\Delta\beta = \beta^{(2)}[(\Delta\omega_s)^2 - (\Delta\omega_p)^2] + \frac{\beta^{(4)}}{12}[(\Delta\omega_s)^4 - (\Delta\omega_p)^4] \quad \text{for two pumps,} \quad \text{(A5.6)}$$

where $\Delta\omega_s = \omega_s - \omega_c$ and $\Delta\omega_p = \omega_p - \omega_c$; ω_c is the center frequency. Also, $\beta^{(2)}$ can be expressed as

$$\beta^{(2)} = \beta^{(3)}(\omega_c - \omega_0), \qquad \text{(A5.7)}$$

where ω_0 is the frequency corresponding to the zero-dispersion wavelength (ZDW), for which $\beta^{(2)} = 0$ or equivalently $D = 0$.

Bandwidth

The (half) bandwidth of a well-optimized broadband OPA is of the order of

$$\Delta\omega \propto \left(\frac{\gamma P_0}{\beta^{(2m)}}\right)^{1/2m} \qquad \text{(A5.8)}$$

when $\Delta\beta$ is dominated by the term of order $2m$, $m = 1, 2$.

Maximum gain

For either one- or two-pump OPAs, the maximum value of g is γP_0 and the maximum value of the signal power gain G_s is $G_{s,\max} = \cosh^2(\Phi)$, where $\Phi = \gamma P_0 L$.

For large gain, the maximum signal power gain in decibels is approximately

$$G_{s,\max}^{dB} \approx G_{i,\max}^{dB} \approx 8.6 \times \Phi - 6. \qquad \text{(A5.9)}$$

Gain near the pump of a one-pump OPA

We have

$$G_i = \Phi^2, \qquad G_s = 1 + \Phi^2. \qquad \text{(A5.10)}$$

Double exponential transformation for $\alpha \neq 0$, no dispersion, no pump depletion

In such situations, if the complex vector phasor (SVE) of a field component is $A(z)$ for $\alpha = 0$ then in the presence of loss it is $A'(z)$ and is given by the double exponential transformation

$$A'(z) = e^{-\alpha z/2} A(L_{\text{eff}}) = e^{-\alpha z/2} A\left(\frac{1 - e^{-\alpha z}}{\alpha}\right). \qquad \text{(A5.11)}$$

Complete pump depletion
For one pump,
$$\Delta\beta = \gamma \left(P_{30} - \frac{P_{10}}{2} \right) \quad (A5.12)$$

and $x(z)$ is given by

$$-\frac{16 P_{30}}{7} \left\{ 1 - \left[1 - \frac{7 P_{10}}{7 P_{10} + 32 P_{30}} \tanh^2 \left(\frac{yz}{4} \sqrt{(7 P_{10} + 32 P_{30}) P_{10}} \right) \right]^{-1} \right\}. \quad (A5.13)$$

For two pumps,
$$\Delta\beta = \gamma P_{30} \quad (A5.14)$$

and $x(z)$ is given by

$$-\frac{4 P_{30}}{3} \left\{ 1 - \left[1 - \frac{3 P_{10}}{3 P_{10} + 4 P_{30}} \tanh^2 \left(\gamma z \sqrt{(3 P_{10} + 4 P_{30}) P_{10}} \right) \right]^{-1} \right\}. \quad (A5.15)$$

Index

alternating polarization states, 286
aluminium oxide, 322
amplification, 249–57
amplified spontaneous emission (ASE), 2, 159, 222, 226–8, 268, 303, 325
amplitude-noise reduction, CW OPAs, 269–70
arsenic sulphide, 316
asynchronous pumping, 256–7
attenuation, see fiber loss; pump attenuation

bandwidth, equation for, 359
bidirectional loop mirror, 188
bidirectional singly resonant oscillator (SRO-B), 189, 191–2
birefringence, 4, 15–17, 93–8, 98–107
birefringent fibers, constant linear, 140–3, 148–51
bismuth oxide fibers, 322
bit-level switching, 267
Bragg fibers, 25
Bragg gratings, 14
Brillouin scattering, see stimulated Brillouin scattering

Cardano solution, 45–7
chirp, 177–8
chromatic dispersion, 11, 271
circular polarization (CP), 78, 79, 81, 93
clock recovery, 266
coherent crosstalk, 298–9
communication fibers, design, 21–2
communication systems, categories, 328
cosine-squared modulation, 184–5
coupled equations, 66–7
  pump depletion, 85
  six-wave model, 55–6, 86–91, 355–67
couplers, 213
CP, see circular polarization
cross-gain modulation (XGM), 47, 292–8
  and pump resonators, 325, 327
  and signal-to-noise ratio, 281
cross-phase modulation (XPM), 17, 18
  for arbitrary SOPs, 95
  and bidirectional light, 251

for degenerate OPAs, 71
for one-pump OPAs, 217
in propagation equations, 39, 78, 80–1
and Raman amplifiers, 299
and signal-to-noise ratio, 281, 299–300
in SRO-Bs, 191–2
in two-pump OPAs, 215
crosstalk
  see also cross-gain modulation; cross-phase modulation; four-wave mixing
  coherent crosstalk, 298–9
  one-pump OPAs, 282–91
  pump–signal, 287–90
  two-pump OPAs, 282–4, 291–2
  types of, 281
cubic solution, 45–7
CW fiber OPAs, 233–6, 269–70, 279

degenerate OPAs, 34–5
  noise figure, 205–6
  as pulsed OPAs, 231
  pump depletion equations, 85
  scalar equations, 70–5
demultiplexing, 5, 178–82
differential equations, transform for, 333–4
differential group delay, 261
direct detection, 194
dispersion, 11–13, 69–70
  longitudinal variation, 134–40
  periodic compensation, 138–9
  random variation, 139–40
  tailored, 316–7
  variation, matrix approach, 135–6, 138, 139
dispersion-shifted fibers (DSFs), 14, 21, 26–7, 93, 134, 138, 212, 228–9
dispersion slope, 11–12
distributed-fiber Bragg laser (DFB), 226
distributed parametric amplification (DPA), 5
  advantages, 303–4, 326–9
  ASE output spectra, 306
  bandwidth, 326–7
  BER plots, 307
  comparison with DRA, 307–9

distributed parametric amplification (DPA) (cont.)
    extensions, 309–11
    experiment, 304–9
    polarization dependence, 308
distributed Raman amplification (DRA), 303, 310–1
distributed Raman gain, 326
double Rayleigh scattering (DRS), 303

effective index, 10
ellipse rotation, 93, 99, 160, 161
elliptical polarization, 99
entangled quantum states, 271–2
erbium-doped fiber amplifiers (EDFAs)
    and ASE, 219, 234–5
    bandwidth, 1, 111, 249, 273, 299
    and distributed amplification, 303
    noise, 5
    power, 4–5
    and pulsed pumps, 169
erbium-doped fiber lasers (EDFLs), 182, 223, 319
extruded fibers, 22, 23

Fabry–Pérot cavity, 87, 188, 189, 191
Fabry–Pérot laser, 223
fast light, 264
fiber Bragg grating (FBG), 187, 188, 266
fiber dispersion, and gain spectrum, 111–20
fiber lasers, 3, 319
fiber loss, 10, 49–53, 63–5
fiber optical parametric amplifiers (OPAs)
    bandwidth, 1, 5, 212–3
    characteristics, 1–3, 5, 9
    classification, 78–9
    development of, 3–6
    future developments, 315–29
fiber optical parametric oscillators (OPOs), 187–92
    CW fiber, 240
    high-γ, 315
    pulsed, 232
    pump attenuation in, 191
    stimulated Brillouin scattering, 191
    types of, 21–7
    wavelength conversion, 275
flattening, one-pump OPA gains, 136–7
Fourier method, see split-step Fourier method
Fourier series, 176
fourth-order approximations, OPA gain spectra, 122–3
four-wave case, 344–9
    pump depletion, 85, 349–50, 353–4
four-wave mixing (FWM), 18
    idlers, 282
    one-pump, 282–91

and polarization sensitivity, 250
in propagation equation, 39
pump–pump, 327
pump–signal, 287–90, 325, 327
signal–signal, 284–6, 325, 327
summary, 281–2
two-pump, 291–2
undesired terms, 282–4
vector theory, 78, 80–1, 87–8, 95, 97–8
fusion splices, 27

gain spectrum, summary of equations, 358–60
gallium arsenide, 316
gallium lanthanum sulphide, 316
Gaussian pulses, 174–5, 178–84
Gordon–Haus effect, 268
group velocity dispersion, 322

Hamiltonian for nonlinear interactions, 195–8
Heisenberg evolution equation, 198–200
heterodyne detection, 194
high-γ fibers, 315
highly nonlinear dispersion-shifted fibers (HNL-DSFs), 4, 22
    future development, 316
    nonlinearity coefficient, 25
    polarization-maintaining, 318
    and pulsed fiber OPAs, 228
    zero-dispersion wavelength, 14
highly nonlinear fibers (HNLFs), 4, 22
    coupling, 26–7
    dispersion slope, 11
    and heat dissipation, 134
    length reduction, 324
    refractive index, 195
    and pump gain, 212
    splice loss, 138
    zero-dispersion wavelength, 14
HNLF-based CW fiber OPOs, 240
holey fibres (HFs), 25–6
    and CW fiber OPOs, 240
    future development, 315, 316
    refractive index, 195
    SBS suppression, 321
hollow-core fibers, 195
homodyne detection, 194

idlers, 1, 2, 32
    and demultiplexers, 181, 182
    in DPA systems, 304, 309–10
    frequency modulation, 220
    limitation of, 47
    noise figure, 203
    as output, 270
    phase modulation, 217–8
    in phase-sensitive systems, 328

photon distribution, 33
power, 46, 50
in propagation equation, 82–4
self-phase modulation, 171
SOPs, 79
and undesired terms, 283, 284, 291, 292
and wavelength conversion, 183
integrated optics, developments, 322–3
irradiance, 9
isotropic fibers, 78–93, 147–8

Jacobian elliptic functions, 44, 85, 172, 344–54

Kerr effect, 17, 79, 105, 148, 149

laser pigtails, 213
lead glass, 316
lead silicate, 316
light
  bidirectional, 251
  fast, 264
  propagation speed, 264
  slow, 264, 323
linear polarization, 94–5
linearly polarized lasers, 213
loss, see fiber loss

Mach–Zehner interferometer (MZI), 71–5
Manley–Rowe relations, 33
microstructured fibers, 22, 25–6
mid-span spectral inversion, 261
mode-locked laser, 175
mode field diameter (MFD), 9
mode profile, 9–10
modes of propagation, 9, 143–4

nanowires, 22, 24–5
narrowband tunable gain spectrum, 115–6
noise figure (NF), 2, 5, 194
  degenerate fiber OPAs, 205–6, 231
  non-degenerate fiber OPAs, 200–3, 230, 234–5
  and pump IM, 218–9
  and Raman gain, 206–8
noiseless amplification, 206
non-polarization-maintaining fibers (non-PMFs), 15–17
nonlinear refractive index, 17–18
nonlinear Schrödinger equation (NLSE)
  common SVE derivation, 150–1
  and isotropic fibers, 147–8
  multiple-input-wave case, 152
  no-dispersion-with-loss case, 156
  one-pump OPA gain, 154–5

and Raman gain, 156–7
soliton solutions, 156
and SSFM, 157–62
two-input-wave case, 153–6
nonlinearity coefficient (γ), 3, 4, 18–19, 35–40

one-pump OPAs
  amplitude modulation, 213–5
  bandwidth gain, 235–6
  crosstalk, 282–291
  degenerate, 34–5
  examples, 226–9
  fiber dispersion, effect on gain spectrum, 111–6
  four-wave mixing, 282–91
  frequency assignments, 32
  frequency modulation, 219–20
  gain flattening, 136–7
  maximum gain, 359
  near-pump gain, 359–60
  parameters, 103
  propagation equation, 40, 83, 84
  pump amplitude modulation, 213–4
  and SSFM, 159
optical buffering, see optical delay lines
optical delay lines, 261–4
optical parametric oscillators (OPOs), 187–92
  see also fiber optical parametric oscillators
  $\chi^{(2)}$-based, 188
  doped-fiber, 188
optical pulses, storage, 264–5
optical resonators, 324–6
optical sampling, 169, 269
optical signal-to-noise ratio (OSNR), 218, 271
optical spectrum analyzer (OSA), 227
orthogonal pumps, 328

parametric gain coefficient, 358
periodic pulses, 175–7
periodically poled lithium niobate (PPLN), 2, 320
periodically pulsed pumps, 259–60
phase conjugation, 2, 261
phase-integral method, see Wentzel–Kramers–Brillouin (WKB) approximation
phase-matched gain, 211
phase modulation, 170, 321
phase-sensitive amplification, 4, 5
  degenerate OPAs, 231
  distributed amplification, 309, 311, 326, 328
  non-degenerate OPAs, 253
  two-pump OPAs, 244–5
photonic bandgap fibres, 25
photonic crystal fibres (PCFs), 25

photons, 194
  conservation of energy, 33, 34
  conservation of number, 33, 197–8
  exchange of, 34
  pairs of, 21, 271–2
Poincaré sphere, 93, 98, 101–4, 158
polarization
  see also state of polarization
  alternating, 286
  circular, 78, 79, 81, 93
  elliptical, 99
  linear, 94–5
polarization diversity, 250–2
polarization independence, 5
  distributed, 327–9
  one-pump, 250–2
  two-pump, 252
polarization-insensitive OPAs, 250–2
polarization sensitivity, 250
polarization tracking, 328
polarization-maintaining couplers, 213
polarization-maintaining fibers (PMFs), 4, 15, 17, 148, 317–8
polarization-maintaining fiber (PMF) pigtails, 213
polarization-mode dispersion (PMD), 16, 98
  one-pump models, 164–5
  two-pump models, 165–6
  waveplate model, 163
power, single-mode fibers, 9–10
praesodymium-doped fibre laser (PDFL), 319
propagation constant (β), 10
propagation equations, 35–40
  for isotropic fibers, 79–81
  quantum-mechanical derivation, 195–209
pulse generation, 265
pulse monitoring, 270
pulse walkoff, 171
pulsed pumps OPAs
  for circular polarization, 93
  and CW signal amplification, 253–7
  cosine-squared pulses, 174
  examples, 226–2
  Gaussian pulses, 174–5
  optical signal processing, 266–9
  overview, 169
  pulse trains, 175–7
  raised cosine pulses, 174
  rectangular pulses, 173–4
  sech pulses, 175
  super-Gaussian pulses, 175
  transform-limited pulses, 177–8
pump amplitude modulation, 213–5
pump attenuation
  coupled equations, 84–5
  in distributed parametric amplification, 303–4
  FWM equations, 334–8

and gain spectra, 125–34
  one-pump OPAs, 60–2, 67–9
  in OPOs, 191
  two-pump OPAs, 42–5
pump power
  and gain bandwidth, 110–1, 212–3
  and optical resonators, 324–6
pump resonators, 324–6
pump–pump FWM, 327
pump–signal crosstalk, 287–90
pump–signal FWM, 325, 327
pump-spectrum broadening, 321
pumps
  and Brillouin scattering, 19–20
  frequency modulation, 219–22
  future development, 319–20
  and gain modulation, 221–2
  intensity modulation (IM), 213, 215–9, 221–2
  linewidth, 219
  and OPA gain, 211–2
  phases, 17
  polarization, 17
  propagation equation, 82
  and Raman scattering, 20–1
  relative intensity noise (RIN), 213
  SOPs, 79, 213

quantum communication, 3
quantum electrodynamics, 194
quasi-CW approximation, 146, 169–72, 173
quasi-steady-state approximation, see quasi-CW approximation

Raman amplification (RA), 5, 111, 303, 316
  see also distributed Raman amplification
Raman gain, 65–9, 156–7, 206–8, 223, 252, 268, 326
Raman scattering, 20–1, 65–9
random birefringence, 162–6, 317–8
random longitudinal ZDW variation, 162
Rayleigh scattering, 10, 251, 303
regeneration, 267–8
relative intensity noise (RIN)
  magnification factor, 217
  transfer, 215, 219
residual dispersion, 270–1
Runge–Kutta integration, 105
RZ-DPSK signals, regeneration, 268

Sagnac interferometer, 4, 74–5, 219, 320
saturated gain spectra, 125–34
SBS, see stimulated Brillouin scattering
scaling laws, 40–2
Schrödinger equation, see nonlinear Schrödinger equation (NLSE)

second-order approximations, OPA gain spectra, 121–2
self-phase modulation (SPM), 17, 18
  for arbitary SOPs, 95, 98
  and bidirectional propagation, 251
  in propagation equations, 39
  and pump IM, 215
  of pumps, 138, 170, 171–2
  in SRO-Bs, 191–2
  in vector equations, 78, 80–1
semiconductor Fabry-Pérot laser, 223
semiconductor lasers, 223, 319, 325
semiconductor optical amplifiers (SOAs), 2, 111
shot-noise limit, 272
signal power gain, 358
signal intensity modulation, 215–9
signal SOPs, 79
signal phase modulation, 217–8
signal–signal FWM, 325, 327
signal-to-idler power conversion, 358
signal-to-noise ratio, 271
silica based fibres, 21
silica microspheres, 329
silica photonic crystal fiber, 317
silicon, in optical amplifiers, 322
silicon-on-insulator technology, 323
silicon waveguides, 322, 329
singly resonant oscillator (SRO), 187, 189, 190–2
six-wave model
  coupled equations, 355–7
  gain spectrum, 120
  two-pump equations, 54–60, 85–92
sixth-order approximations, OPA gain spectra, 123
slow light, 264, 323
small-scale gain spectrum, 49, 57, 82–4, 172, 215
solitons, 156, 253, 266, 323
space switches, 262
spectral inversion, 2, 261
split-step Fourier method (SSFM)
  see also Fourier series
  common SVE derivation, 159–62
  and NLSE, 157–62
  and pulsed-pump OPAs, 172–3
  and random longitudinal ZDW variation, 162–6
  software, 166
  SVE-separate-term derivation, 158–9
spun fibers, 16–17, 93
squeezed states, 272–3
SRO-B, 189
state of polarization (SOP)
  and gain independence, 5
  gain spectra, 123–5
  in non-polarization-maintaining fibers, 15–17
  and OPA types, 35
  random variation, 317–8
step-index fibers, 22–3, 322

stimulated Brillouin scattering (SBS), 19–20
  and bidirectional light propagation, 251
  and pulsed one-pump OPAs, 227–8
  and pump depletion, 155
  and pumps, 211, 222
  for SROs, 191
  suppression of, 233, 320–2, 325
  two-input-wave case, 155
  in vector theory, 78–9
stimulated Raman scattering (SRS), 20–1, 78–9, 322–3
storage of optical pulses, 264–5
stress variations, 321–2
slowly varying envelope (SVE), 80, 150

tapered fibers, 22, 23–4
tellurite, 299, 316
temperature distribution, 138–9
thermally expanded cores (TECs), 26
third-order nonlinearity, 17–19
three-wave case, 351–3
  pump depletion, 85
three-wave mixing, 18
time-division demultiplexing (TDM), 5, 178–82, 266–7
transfer matrix, 200, 336–8
transform, for differential equations, 333–4
tunable diffraction grating (TDG), 188
two-photon absorption (TPA), 21
two-pump OPAs
  fiber dispersion and gain spectrum, 117–20
  four-wave mixing, 282–4, 291–2
  frequency assignments, 32
  frequency modulation, 220
  maximum gain, 359
  parameters, 91–2, 104
  propagation equations, 35–40, 83, 84
  pump amplitude modulation, 214–5
  scaling equations, 42–60
  types of, 91–2
two-pump crosstalk, 282–4, 291–2

ultraviolet light, and ZDW, 318
unidirectional rings, 187, 188
unidirectional singly resonant oscillator (SRO-U), 189, 190–1
unitary matrices, 336–7

wafer-scale fabrication, 322
wave-division multiplexing (WDM) system, 146
wavelength conversion, 1–2, 3, 5
  any-to-any, 258
  continuous-wave (CW), 236–9
  efficiency, 258
  fixed-pump, 257–8
  by FWM, 250

wavelength conversion (*cont.*)
  by OPOs, 275
  by pulsed OPAs, 231–2, 259–60
  Gaussian pulses, 182–4
  high-power, 125, 133–4, 273–4
  multiband, 260
  and multiple ZDWs, 317
  phase-matching condition, 258–9
  and pump depletion, 128–9
  tunable-pump, 258–9
wavelength exchange, 34, 53–4, 70, 203–5, 240–4, 338
wavelength range, increasing, 317

wavelength-division multiplexing, in DPA, 304
waveplate model, 163
wavevector mismatch ($\Delta\beta$), 359
Wentzel–Kramers–Brillouin (WKB) approximation, 340–2
Whittaker functions, 51, 52

ytterbium-doped fiber lasers (YDFLs), 273

zero-dispersion wavelength (ZDW), 1, 11
  longitudinal variation, 14, 318–9
  multiple, 327
  temperature dependence, 14–15

Printed in the United States
By Bookmasters